Food Traceability and Authenticity
Analytical Techniques

Books Published in *Food Biology* series

Food Biology Series

Food Traceability and Authenticity
Analytical Techniques

Editors

Didier Montet

Food Safety Team Leader
UMR Qualisud, CIRAD
Montpellier, France

and

Ramesh C. Ray

Principal Scientist (Microbiology)
ICAR - Central Tuber Crops Research Institute
Bhubaneswar, Odisha, India

CRC Press is an imprint of the
Taylor & Francis Group, an **informa** business

A SCIENCE PUBLISHERS BOOK

CRC Press
Taylor & Francis Group
6000 Broken Sound Parkway NW, Suite 300
Boca Raton, FL 33487-2742

First issued in paperback 2021

© 2018 by Taylor & Francis Group, LLC
CRC Press is an imprint of Taylor & Francis Group, an Informa business

No claim to original U.S. Government works

Version Date: 20170912

ISBN-13: 978-0-367-78166-8 (pbk)
ISBN-13: 978-1-4987-8842-7 (hbk)

Library of Congress Cataloging-in-Publication Data

Names: Didier, Montet, editor. | Ray, Ramesh C., editor.
Title: Food traceability and authenticity analytical techniques / editors,
    Montet Didier, food safety team leader, UMR Qualisud, CIRAD, Montpellier,
    France, and Ramesh C. Ray, principal scientist (microbiology),
    ICAR--Central Tuber Crops Research Institute, Bhubaneswar, Odisha, India.
Description: Boca Raton, FL : CRC Press, 2017. | Series: Food biology series
    | "A science publishers book." | Includes bibliographical references and
    index.
Identifiers: LCCN 2017033063 | ISBN 9781498788427 (hardback : alk. paper)
Subjects: LCSH: Food industry and trade--Quality control. | Food industry and
    trade--Safety measures. | Food adulteration and inspection. | Food supply.
Classification: LCC TP372.5 .F68 2017 | DDC 664--dc23
LC record available at https://lccn.loc.gov/2017033063

Visit the Taylor & Francis Web site at
http://www.taylorandfrancis.com

and the CRC Press Web site at
http://www.crcpress.com

Preface to the Series

Food is the essential source of nutrients (such as carbohydrates, proteins, fats, vitamins, and minerals) for all living organisms to sustain life. A large part of daily human efforts is concentrated on food production, processing, packaging and marketing, product development, preservation, storage, and ensuring food safety and quality. It is obvious therefore, our food supply chain can contain microorganisms that interact with the food, thereby interfering in the ecology of food substrates. The microbe-food interaction can be mostly beneficial (as in the case of many fermented foods such as cheese, butter, sausage, etc.) or in some cases, it is detrimental (spoilage of food, mycotoxin, etc.). The *Food Biology* series aims at bringing all these aspects of microbe-food interactions in form of topical volumes, covering food microbiology, food mycology, biochemistry, microbial ecology, food biotechnology and bio-processing, new food product developments with microbial interventions, food nutrification with nutraceuticals, food authenticity, food origin traceability, and food science and technology. Special emphasis is laid on new molecular techniques relevant to food biology research or to monitoring and assessing food safety and quality, multiple hurdle food preservation techniques, as well as new interventions in biotechnological applications in food processing and development.

The series is broadly broken up into food fermentation, food safety and hygiene, food authenticity and traceability, microbial interventions in food bio-processing and food additive development, sensory science, molecular diagnostic methods in detecting food borne pathogens and food policy, etc. Leading international authorities with background in academia, research, industry and government have been drawn into the series either as authors or as editors. The series will be a useful reference resource base in food microbiology, biochemistry, biotechnology, food science and technology for researchers, teachers, students and food science and technology practitioners.

Ramesh C. Ray
Series Editor

Preface

The word 'Traceability' is formed etymologically from the verb 'trace' which is derived from the Latin: *tractiare* (to drag) and *tractus*, past participle of *trahere* (to pull). It refers to track or trail, to follow or study in detail or step by step (Merriam Webster Dictionary 2003).

Historically, traceability was used to get a better flow of the matters inside enterprises with the objectives to save time and money. Traceability became a legal tool in 2005 in Europe that is applied to all food stuffs that are tradable with the objective to assure their food safety (Article 18, UE regulation 178/2002) and giving food safety responsibility to all business operators (Article 19).

Actually, the word 'traceability' is employed to describe a system that permits to document the history of a product along its entire production chain from primary raw materials to the final consumable products (MacDaniel and Sheridan 2001). It has to be noticed that the sentence "ability to trace" is used in ISO standard, "ability to follow" in Codex standard, while "ability to trace and follow" is used by the EU. In USA, they use "the creation and maintenance of records".

Food traceability also became a growing consumers' concern worldwide. Traceability is undertaken primarily at the administrative level, and the use of advanced analytical tools is emerging. Currently, there is no analytical method available that permits the efficient determination of foodstuff origin or that allows to trace food during international trade. Nevertheless, the determination of geographical origin is a demand of the traceability system for the import and export of foodstuffs (UE regulation 178/2002).

The book highlights all aspects of food traceability through many angles: the history of traceability, legislations and rules, the actual traceability techniques, and the potential analytical techniques for food traceability; these topics are hotspots of contemporary food science research. Analytical techniques for food traceability include molecular methods (e.g., DGGE, SSCP), next generation sequencers (NGS), bio-captors, chromatographic techniques that are used for discrimination of organic food, fish, oils and meats. The chromatographic techniques help in the use of volatile compounds analysis. Ambient mass spectrometry is used for studying mycotoxins and alkaloids in foodstuffs and their management, food and feed authentication in olive and other plant oils, and wine. Vibrational methods (e.g., NMR and NIRS) are used to trace food by global spectrum. States of art of the actual and future techniques including metabolomic techniques and probes are discussed.

Didier Montet
Ramesh C. Ray

Contents

History of Food Traceability

Didier Montet[1],* and Gargi Dey[2]

1. Introduction

In the past the food industry has had its fair share of scandals, accidents, and incidents. It must be pointed out that reported food scares were not always associated with microorganisms; many of them were connected to new technology, environmental pollution or changes in co-product management. For example the food colorant (tartrazine and amaranth) incident reported in mid-1980 in UK; mercury poisoning in oranges reported in 1979; mercury poisoning in fish reported in 1970; radioactivity in lamb reported in 1986; glass, pin and caustic soda found in baby food product reported in UK in 1989 which resulted in the recall of 100 million jars off the shelves and repackaging of another 60 million. These incidents are very much in the memories of the general public (Gregory 2000). In recent times there have been several more notable incidences that have been reported: for instance the pork scare reported in Dec 2008 where pork products sold and exported from Northern Ireland were recalled from supermarkets across Europe since they contained dioxins that were 80 to 200 times more than acceptable limit; similarly the egg scare reported in Dec 2008, where seven shipping containers of chicken eggs produced by China's Dalian Hanwei Enterprise Group that reached Hong Kong were recalled, while two others that were

---

[1] Cirad, UMR 95 Qualisud TA B-95/16 73, rue Jean-François Breton 34398 Montpellier Cedex 5, France.
[2] School of Biotechnology, KIIT University, Bhubaneswar, Pin 751024, India.
  E-mail: drgargi.dey@gmail.com
* Corresponding author: didier.montet@cirad.fr

already on the island were sealed off. It was reported that the chickens that laid the eggs were given feeds that contained melamine; the coffee and tea scare where seven instant coffee and milk tea products from China believed to contain melamine were recalled in the US. Food giant Unilever also recalled four batches of its Lipton Milk Tea sold in Hong Kong and Macau after the company found traces of melamine in the products; one of the biggest food scandal, which the World Health Organization had to deal with as it affected countries worldwide was related to milk. In July 2008, it was reported that 53,000 children got sick from consuming the infant formulas and milk products, and at least four infants died due to kidney stone and other kidney problems. These products that were exported by China were found to contain melamine; in August 2007, the California Department of Public Health found aldicarb sulfoxide in fresh ginger imported from China. Aldicarb sulfoxide is a pesticide that can cause headache, nausea, and blurred vision; in July 2007 the report of Chinese-made products White Rabbit Creamy Candy, Milk Candy, Bairong Grape Biscuits and Yong Kang Food Grape Biscuits were ordered recalled in Philippines since they were believed to contain formaldehyde; similarly in March 2007, contaminated products caused renal and kidney failure among pets that reportedly led to the death of 8,500 animals. These cat and dog foods that contained Chinese-made wheat gluten were ordered recalled in the US, since they were found contaminated with melamine used for making plastic and leather; earlier in 2005 there was report of contamination of noodles with formaldehyde, a cancer-causing substance used in embalming and for making glue, plastic, and disinfectant. The Indonesian government found that 60 percent of food establishments in Jakarta had been serving noodles that contained formaldehyde. The same food scare happened in Vietnam in 2007 when it was reported that formaldehyde was used to lengthen the shelf life of noodle products, including Pho, Vietnam's national dish (GMA news on line April 27, 2009).

Interestingly, a thorough survey of literature indicates that food scares and adulteration have been reported ever since middle ages, Table 1 summarizes the reported cases of food scares and food adulteration incidences since middle-ages to 2016.

As a result of globalization in trade, food travels far and wide from the producer to reach the consumer. Consequentially the number of food crises reported has also increased, for instance the Bovine spongiform Encephalopathy (BSE) or mad cow disease, Dioxin in food and feed, Foot and Mouth disease (FMD), outbreaks of food borne illness from *Salmonella*, *Campylobacter*, *E. coli* O157:H7 (Montet 2009). Furthermore, the food industry is rapidly becoming a customer driven industry with rising customer demands for safe food. Under these circumstances, it has become mandatory to establish a robust traceability system to minimize the production and

Table 1. Incidences of Food Scares and Food Adulteration.

Time Period	Type of incidence	Locations	References
Middle Ages	Numerous incidents of human poisoning due to the consumption of rye bread made from grain infected with ergot fungi.	Europe	https://en.wikipedia.org/wiki/List_of_food_contamination_incidents
1857	Adulteration of bread with alum in, causing rickets.	London	Snow 2003
1857	Poisoning of bread with arsenic targeting the colonial community.	Hong Kong	Lowe 2015
1900	Beer contaminated with arsenic. Traced to sugar manufactured with sulphuric acid that was naturally contaminated with arsenic from Spanish pyrites. An epidemic of 6070 cases in London, including 70 deaths.	London	Reynolds and Ernest 1901
1910–1945	Cadmium from mining waste contaminated rice irrigation water. Illness known as Itai-itai disease affected more than 20% of women aged over 50 years.	Japan	Kasuya et al. 1992
1920	80 people suffered poisoning from eating bread contaminated with naturally occurring pyrolizidine alkaloids.	South Africa	Kakar et al. 2010
1900–1947	Severe and widespread neurological disorders due to bleaching of bread flour with the agene process for bleaching of flour with nitrogen chloride, a process no longer in use. The denatured protein in the treated flour is toxic and causes a condition of hysteria in dogs eating biscuits made from the flour.		Shaw and Bains 1998
1950	Mercury poisoning in fish contaminated by industrial discharge causing the Minamata disease. By 2010 more than 14,000 victims had received financial compensation.	Japan	George 2002
1955	Arsenic in milk powder. Disodium phosphate additive was inadvertently contaminated with sodium arsenate. The incident was known as the "Moringa dried milk poisoning". By 2002 there were an estimated 13,400 cases and over 100 deaths attributed to consumption of the milk powder.	Japan	Dakeishi et al. 2006

Table 1 contd. …

...Table 1 contd.

Table 1. Incidences of Food Scares and Food Adulteration.

Time Period	Type of incidence	Locations	References
1957	Chicken feed and thence chickens were contaminated with dioxins from polychlorinated treated cow hides. 300,000 chickens were killed or destroyed to avoid consumption.	USA	Firestone 1973
1972	Mercury poisoning killed 100 to 400 as seeds treated with mercury as a fungicide that are meant for planting are used as food.	Iraq	Bakir et al. 1973
1973	Widespread poisoning of populace following cattle contamination from feed contamination with flame retardant.	Michigan	Dunckel 1975, Fries and Kimbrough 1985
1974–1976	Widespread poisoning (an estimated 7800 people affected with hepatic veno-occlusive disease (liver damage) and about 1600 deaths) was attributed to wheat contaminated with weed seeds known as charmac (*Heliotropium popovii*. H Riedl) that contains pyrrolizidine alkaloids.	Afghanistan	Kakar et al. 2010
1985	Adulteration of wines with diethylene glycol.	Austria	Schanche 1986
1986	Adulteration of wines with ethylene glycol killed more than 18 people.	Italy	Schanche 1986
1987	Beech-Nut Nutrition Corporation paid $2.2 million, then the largest fine issued, for violating the Federal Food, Drug, and Cosmetic Act by selling artificially flavored sugar water as apple juice. John F. Lavery, the company's vice president for operations was convicted in criminal court and sentenced to a year and a day in jail; Niels L. Hoyvald, the president of the company was also convicted, served six months of community service. Each of them also paid a $100,000 fine.	USA	Traub 1988
1989	Milk contamination with dioxin.	Belgium	Bernard et al. 2002
1998	Adulteration of edible mustard oil with Argemone mexicana seed oil caused epidemic dropsy in thousands of people. Epidemic dropsy is a clinical state resulting from consumption of	India	Sharma et al. 1999

Table 1 contd. ...

…Table 1 contd.

Table 1. Incidences of Food Scares and Food Adulteration.

Time Period	Type of incidence	Locations	References
	edible oils adulterated with Argemone mexicana seed oil that contains the toxic alkaloids sanguinarine and dihydrosanguinarine. The epidemic is the largest so far, in which over 60 persons lost their lives and more than 3000 victims were hospitalized.		
1998	Meat and milk were found with elevated dioxin concentrations. The dioxin was traced to citrus pulp from Brazil that had been neutralized with dioxin-contaminated lime. 92,000 tons of citrus pulp was discarded.	Germany and the Netherlands	Malisch 2000
1999	Animal feed contaminated with dioxins and polychlorinated biphenyls affected more than 2500 poultry and pig farms. This incident led to the formation of the Belgium Federal Food Safety Agency. The loss to the Belgium economy was estimated at €1500–€2000 M.	Belgium	Covaci et al. 2008
1999–2000	There were an estimated 400 cases of liver damage and over 100 deaths due to pyrrolizidine poisoning. The food source was not identified.	Afghanistan	Kakar et al. 2010
2001	Olive pomace oil was contaminated with polycyclic aromatic hydrocarbons. Contaminated product was recalled.	Spanish	http://www.food.gov.uk/multimedia/faq/olivepomoilqa
2002	Nitrofurans were detected in 5 (of 45) samples of chicken imported from Thailand and Brazil. The product was withdrawn and destroyed.	Northern Ireland	http://www.reading.ac.uk/foodlaw/news/uk-02035.htm
2002	Nitrofurans were detected in 16 (of 77) samples of prawns and shrimps imported from SE Asia. Affected batches were withdrawn and destroyed.	UK	http://www.food.gov.uk/multimedia/faq/51434/
2002	The banned antibiotic, chloramphenicol, was found in honey from China.	UK and Canada	http://www.hc-sc.gc.ca/dhp-mps/vet/faq/faq_chloramphenicol_honey-miel-eng.php#a4

Table 1 contd. …

...Table 1 contd.

Table 1. Incidences of Food Scares and Food Adulteration.

Time Period	Type of incidence	Locations	References
2003	Dioxins were found in animal feed that was contaminated with bakery waste that had been dried by firing with waste wood.		Thomson et al. 2012
2003	The banned veterinary antibiotic nitrofurans were found in chicken from Portugal. Poultry from 43 farms was destroyed. Nitrofurans are banned from food because of concerns including a possible increased risk of cancer in humans through long-term consumption.	Poultry	http://www. reading.ac.uk/ foodlaw/news/uk-03018.htm
2004	Detection of chloramphenicol in honey. Chloramphenicol is banned for use in food-producing animals, including honey bees, in Canada as well as in a number of other countries. The Canadian Food Inspection Agency (CFIA) informed Health Canada that five lots of honey labelled as "Product of Canada" were distributed in British Columbia and were found to contain residues of the banned drug chloramphenicol. A voluntary food recall occurred.	Canada	http://www.hc-sc. gc.ca/dhp-mps/ vet/faq/faq_ chloramphenicol_ honey-miel-eng.php
2004	Soy milk manufactured with added kelp contained toxic levels of iodine. Consumption of this product was linked to five cases of thyrotoxicosos. The manufacturer ceased production and re-formulated the product line.	New Zealand	O'Connell et al. 2005
2004	Corn flour and corn flour-containing products were contaminated with lead, thought to have occurred as a result of bulk shipping of corn (maize) contaminated by previous cargo in the same storage. Affected product was distributed in New Zealand, Fiji and Australia. Four products were recalled.	New Zealand	http://www. foodsafety.govt.nz/ elibrary/industry/ Source_Lead-Nzfsa_ Confident.htm
2004	Aflatoxin-contaminated maize resulted in 317 cases of hepatic failure and 125 deaths.	Kenya	Azziz-Baumgartner et al. 2005
2004	EHEC O104:H4 contamination of hamburgers as a possible cause.	South Korea	Kim et al. 2011

Table 1 contd. ...

...Table 1 contd.

Table 1. Incidences of Food Scares and Food Adulteration.

Time Period	Type of incidence	Locations	References
2005	Worcester sauce was found to contain the banned food coloring, Sudan I dye, that was traced to imported adulterated chili powder. 576 food products were recalled.	UK	FSA (2005)
2006	Pork, containing clenbuterol when pigs were illegally fed the banned chemical to enhance fat burning and muscle growth, affected over 300 persons.	China	Manilla Bulletin 2012
2008	Baby milk scandal: 300,000 were babies affected, 51,900 hospitalizations and 6 infant deaths. Lost revenue compensation ~$30 M, bankruptcy, trade restrictions imposed (by 68 countries, 60 or more arrests, two executions, one life sentence, and loss of consumer confidence Melamine from the contaminated protein worked into the food chain a year later.	China	Gossner et al. 2009
2008	Wheat flour contaminated with naturally-occurring pyrrolidizine alkaloids is thought to be the cause of 38 cases of hepatic veno-occlusive disease including 4 deaths.	Afghanistan	Kakar et al. 2010
2008	Irish pork and pork products exported to 23 countries was traced and much was recalled when animal feed was contaminated with dioxins in the feed drying process. The cost of cattle and pig culling exceeded €4 M, compensation for lost revenue was estimated to be €200 M.	Ireland	BBC News 2010
2009	Pork containing the banned chemical clenbuterol when pigs were illegally fed it to enhance fat burning and muscle growth. 70 persons were hospitalized in Guangzhou with stomach pains and diarrhea after eating contaminated pig organs.	China	Manilla Bulletin 2012
2009	Hoola Pops contaminated with lead.	Mexico	MSNBC 2009

Table 1 contd. ...

...Table 1 contd.

Table 1. Incidences of Food Scares and Food Adulteration.

Time Period	Type of incidence	Locations	References
2009	Bonsoy-brand Soymilk enriched with 'Kombu' seaweed resulted in high levels of iodine, and 48 cases of thyroid problems. The product was voluntarily recalled and a settlement of 25 million AUS$ later reached with the victims.	Australia	Crawford 2010
2010	Edible Snakes were contaminated with clenbuterol when fed frogs treated with clenbuterol. 13 people were hospitalized after eating contaminated snake. There were 113 prosecutions in 2011 relating to clenbuterol, with sentences ranging from three years imprisonment to death.	China	Manilla Bulletin 2012
2011	Poor-quality illegal alcohol resulted in an estimated 126 deaths. The alcohol may have contained ammonium nitrate and/or methanol.	West Bengal, India	BBC News 2011
2011	*E. coli* O104:H4 outbreak was caused by EHEC O104:H4 contaminated fenugreek seeds imported from Egypt in 2009 and 2010, from which sprouts were grown.	Germany	Harrington et al. 2011
2011	Vinegar contaminated with ethylene glycol when stored in tanks that previously contained antifreeze, led to 11 deaths and an estimated 120 cases of illness.	China	The China Times 2011
2011	Meat, eggs and egg products contaminated from animal feed containing fat contaminated with dioxins. 4,700 German farms affected. 8,000 hens and hundreds of pigs were culled. Imports from Germany to China were banned.	Germany	Harrington et al. 2011
2013	200 farms were suspected of selling eggs as "organic" but not adhering to the conditions required for the label.	Germany	https://en.wikipedia.org/wiki/List_of_food_contamination_incidents

Table 1 contd. ...

...Table 1 contd.

Table 1. Incidences of Food Scares and Food Adulteration.

Time Period	Type of incidence	Locations	References
2013	A batch of 1800 almond cakes with butter cream and butterscotch from the Swedish supplier, Almondy, on its way to the IKEA store in Shanghai were found by Chinese authorities to have a too high amount of coliforms and were subsequently destroyed.	IKEA store in Shanghai	https://en.wikipedia.org/wiki/List_of_food_contamination_incidents
2013	A vegetable seller realized that the lettuce he had been selling throughout the day contained rat poison. The poison appeared as small blue kernels.	Germany	https://en.wikipedia.org/wiki/List_of_food_contamination_incidents
2013	Halal Lamb Burgers contained samples of Pork DNA, affected 19 schools.	Leicester, UK	https://en.wikipedia.org/wiki/List_of_food_contamination_incidents
2013	Bihar school meal poisoning incident.	India	https://en.wikipedia.org/wiki/List_of_food_contamination_incidents
2015	Caraga candy poisonings.	Philippines	https://en.wikipedia.org/wiki/List_of_food_contamination_incidents
2015	United States *E. coli* outbreak.		https://en.wikipedia.org/wiki/List_of_food_contamination_incidents
2016	Mars Chocolates contamination incident, in which plastic found in candy bars lead to a recall affecting 55 countries.	55 countries	https://en.wikipedia.org/wiki/List_of_food_contamination_incidents

distribution of unsafe or bad quality food. In fact, food traceability has become a global concern which has been addressed at a multilateral level by different countries throughout the world, for instance, Food Modernization and Safety Act (FSMA, 2011) in USA, the US Bioterrorism Act (2001, H.R. 3448), Full Beef traceability system in Korea (GAIN Report N° KS1033. 2010), National Agriculture and Food Traceability system (Center for food in Canada 2012) in Canada, "Internet of Things" and establishment of a future cloud computing centre in Shanghai's Jinshan District, China (Anon et al. 2011), Regulation no. 178/2002 and EU rapid Alert system for Food and Feed (RSFF), tracking and tracing software Grapenet for export of table grapes from India to EU (Buiji et al. 2012), Merger of GS1 Australia and Efficient Consumer Response Australia (ECRA), establishment of Recall and withdraw of food products by the Australian Food & Grocery Council (Australia GS1 2010).

2. Principles of Traceability

At the outset it is important to define what "Traceability" means since it can act as either a constraint or an opportunity for the consumers, professionals, or for the authorities.

For consumers, traceability satisfies:

- their safety needs in terms of health and well-being, and
- their expectations in terms of information.

In terms of their buying motivation, traceability can answer their questions in terms of:

- Financial reasons (rate/quality/value),
- Expiration date,
- Product composition,
- Geographical origin,
- Medical grounds,
- Ideological reasons (health, body image, nutritional contribution, religion),
- Allergic risks, hazards,
- Environmental reasons, ethical concerns.

For professionals within the food chain, traceability gives them:

- A better risk management,
- A crisis management tool,
- A way to better define their limits of liability,
- A tool to allow the establishment of a relationship of trust with consumers.

For authorities traceability constitutes:

- A means of risk prevention,
- A means of localization and expertise in case of food crises.

The main objective for companies is to ensure continuity in the supplier's traceability chain to the final consumer and *vice versa* by optimizing and streamlining the physical flow of goods and information.

In order to meet the challenge various organizations and legislations have put together the framework for traceability. The ISO 8402 (1994) quality standards defined traceability as "the ability to trace history, application and location of an entity by means of recorded identification". In ISO 9000

(2005) standards, traceability is defined as "the ability to trace the history, application or location of that which is under consideration". It further specifies that traceability may refer to the origin of materials, parts, the processing history and the distribution and location of the product after the delivery (Peres et al. 2007). The European Union (EU) regulation 178/2002 (EU 2002) narrows the definition to the food industry by defining traceability as "the ability to trace and follow a food, feed, food producing animal or substance intended to be, or expected to be incorporated into a food or feed through all stages of production, processing and distribution. The definition given by the *Codex Alimentarius* Commission (CAC 2005) is "the ability to follow the movement of a food through specified stages of production, processing and distribution". The Food Safety Agency (FSA 2002) specifies three basic characteristics for traceability system: (1) identification of units/batches of all ingredients and products (2) information on when and where they are moved and transformed (3) a system to link these data. Furthermore, FSA also specifies three types of traceable units, viz., batch; trade unit; logistic unit. A batch is defined as a quantity going through the same process. A trade unit is a unit which is sent from one company to the next company in a supply chain (e.g., box, bottle) and a logistic unit is a type of trade unit that designates the grouping that a business creates before transportation or storage (e.g., pallet, container) (Karlsen et al. 2010).

Depending on the flow of direction, traceability can be back traceability or suppliers' traceability, internal traceability or process traceability and forward traceability or client traceability (Perez-Aloe et al. 2007). Currently traceability has been broadly divided into tracking and tracing.

2.1 Tracking and tracing

Downstream traceability (tracking) allows a company to trace a food material from the beginning of its life (raw material) to the final product. This technique is essentially used to manage the flux of merchandise and to optimize the full process of production. Tracing allows the determination of a food product both in space and in time. It is used especially in case of withdrawals or recalls of products whenever they are suspected to potentially have a detrimental effect on the health of the consumers.

Upstream traceability (tracing) allows a company to trace the history of a food product through production, back to the origin of its ingredients and packaging. From a safety point of view, ascending traceability allows organizations to detect issues that could render the products unsafe for consumption. Food companies should therefore be able to provide all information necessary about the life of the product (origin of the food, ingredients, processing operations, inputs used in production, phytosanitary

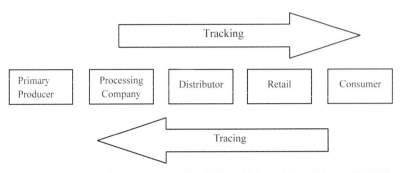

Figure 1. Tracking and Tracing Along Food Chain (Adapted from Schwägele 2005).

treatments, etc.). Tracing could be used also to check the accuracy of declared characteristics (organic, specific origin, appellations, fair trade).

2.2 A case study of tracking: contamination of chopped steaks in Leclerc supermarket, France in 2006

The French press (Le Figaro, Métro, 20 minutes, Dépêche du Midi, Sud-Ouest, Libération, and La nouvelle République) during the weekend of Easter in October 2006, reported about "the biggest epidemic of hemolytic uremic syndrome (HUS) associated with food poisoning ever recorded in France". Then it summarized the incident in brief, citing "an epidemic due to frozen chopped steak" (Montet 2009).

The newspaper also explained the crisis management by the supermarket chain Leclerc in a second article. The free daily Metro described efforts made by the distributor to return 13,000 packets of suspect meat, especially from consumers who may have eaten it, saying that "half of them were consumed almost immediately after purchase even though these products were frozen". The regional daily La Dépêche du Midi said that "Two pediatricians, one from Pau hospital and the other one from Bordeaux hospital, gave the first alert on Tuesday. Consumers have been warned, commented the newspaper 'Sud-Ouest', which noted that "the situation appears stabilized in France" but the alarm button was switch on also in Spain and Portugal, where some suspicious products were exported. In the west of France, La Nouvelle République was concerned more about the economic health of Soviba, the company that produced the contaminated meat, and noted that the plant of Lion d'Angers, which employs 700 people and produces around 160,000 tons of meat per year, had not experienced such a publicity since the catastrophic episode of mad cow disease in the mid-1990s. Edouard Leclerc, the General Director, said in his blog that he was completely distraught because 13 children were hospitalized.

This bacterial accident occurred in the company Soviba, a big supplier of brand Leclerc. Soviba supplies most of the large distribution and catering. It underwent industrial audits whose results were satisfactory. The usual microbial tests did not reveal the presence of the germ. However, samples kept by the company for security, revealed the contamination. In a case like this, it was difficult to explain the origin of the contamination. The infectious agent is often present in the digestive tract of cattle, but in a slaughterhouse, offal should never be in contact with the muscles from which the chopped steaks are made. There was a failure within the company or contamination at a stage of the process either within the organization or external (although the possibility of an external source of contamination remains quite unlikely). The disruption of the cold chain though could have been involved.

The management of the crisis through traceability: Leclerc decided on Friday evening (21 October 2006) to withdraw the products already on sale. The investigations were not very easy because cases were found in three French regions. The health services questioned the parents. Schools were closed because of holidays, and it was thus difficult to establish a causal relationship. The investigation turned to common food consumption. Finally, some certainty emerged on October 27th: all children had consumed at least one of the products purchased in a Leclerc store. On Sunday afternoon, the monitoring results from the supplier left no doubt about the role of frozen ground beef as a propagator (Montet 2009).

On Thursday 27th at 6 pm, Leclerc and DGAL (French Direction Générale de l'Alimentation) released information about the number of poisoning detected from 5th to 26th October. Leclerc immediately organized a crisis management team: quality manager, product buyer, head of the market (with suppliers), veterinary surgeons, specialists of industry logistics, internal and external communication services, all in direct relationship with government (health and agriculture) and Edouard Leclerc himself by phone who was travelling in Italy.

Friday 28th at about 6 pm, the Institute for Public Health Surveillance (INVS) confirmed the presumption on the packaged steak not yet consumed by a sick customer. At 7 pm, a decision was taken to withdraw all batches of the product from sale in the south-west and an email was sent to all sellers under the precautionary principle. At 5 pm, the 14 cases were confirmed and three lot numbers were identified by INVS. At 6 pm, the DGAL organized a telephone conference on the crisis.

On Sunday 30th a release was sent to the media and to Agence France Presse. The recall of products was displayed in all stores. 7500 people were expected to buy these steaks on this Sunday. On Monday 31st, all media reported the problem and 750 employees analyzed all cashier receipts to identify the buyers of steak and called them by phone.

On Monday 31st, three new cases were identified. Leclerc's employees continued to draw customers of which 85% had credit cards, therefore, were on a file with their telephone numbers. For other clients additional bank information (CC, checks) was required. Three to five people per store were mobilized. The crisis lasted four days and had an average cost of some hundred thousand euros for the Leclerc Company, but the risk was very severe that the brand was discredited by consumers.

It is encouraging to note that in recent years there has been enormous progress in tracking technique from "Food to farm" chain through Automatic Identification & Data capture (AIDC). Electronic tracking devices requires relatively clean environment which may not be always possible on farms, however, Radio Frequency IDentification (RFID) may be a better option since it operates on radio signals rather than line of sight for identification. The European Article Numbering Association Codes (EAN-UCC 2002) is the accepted system for an electronic tracking device. The EAN-UCC systems consists of three components: (a) Identification number (b) Data carrier-bar codes and Radio frequency tags used to represent these numbers (c) Electronic messages that integrates the physical flow of goods with electronic flow of information for, e.g., the Harmonized Electronic Data Interchange, HEDI used in meat supply chain. The tracking technology has also seen several advancements in data handling systems like Wireless Sensor network (WSN) and tracking system like Global Positioning System (GPS).

Interestingly, different systems have been proposed for different food groups, for instance, for fresh, non-processed food products a framework based on Physical markup language (PML) has been proposed which may be adopted for web enabled business applications. Similarly for wine logistics chain a system was established to trace the wine bottles from producer cellar to a shop by application of Flexible Tag data logger (FTD) which collects data on light, humidity and temperature. The data can be later accessed by smart phone or Personal Digital Assistant (PDA). Such PDA-based portable record keeping has also been proposed in aquaculture farms to trace the stages of live fish supply chain.

2.3 A case study of European perspective-an example of crisis: Dioxin in Belgian chocolate

To relate the relation crisis/media/consumers/doubt, the following example is given. This is an interesting anecdote which happened to the author (Montet 2009) in 1998 while working in the Asian Institute of Technology in Bangkok, an international research center. The author met a Belgian chocolate maker who complained that he could no longer sell the Belgian chocolates in Thailand. The reason stated was a broadcast on local television announcing that traces of dioxin were found in Belgian milk in Belgium. Incidentally, Thai consumers had made a quick association between dioxin in Belgian milk and Belgian chocolate! That is a very normal reaction for a consumer but, surprisingly, Thai consumers had clearly forgotten that the Belgian chocolates were made in Thailand, with milk from Australia or New Zealand or even during the last few years with Thai milk. Thus, there was no chance of getting (Belgium) dioxin in it.

On perception of consumers, Menozzi et al. (2015) compared the perception towards food traceability among French and Italian consumers and observed that it is product-specific and country specified. They found that consumers from both countries relied on traceable food items and considered them safer. The intention to buy traceable food was also strong among both the groups of consumers. The study also showed that French consumers habitually looked for information on origin and production process of the food product, since they were more concerned with quality attributes. Italian consumers, on the other hand, were more concerned with food safety. Furthermore, some of the past studies have emphasized the need for improving consumer's trust on the food safety policies (Mazzochi et al. 2008, Stefani et al. 2008). It is also interesting to note that the consumers' intention to buy traceable or safe food is strongly influenced by their family's doctors', and nutritionists' opinions. In fact doctors and nutritionists have a dominant role in disseminating information about traceable food and in shaping consumer attitude towards it.

3. Influence of Mad Cow Disease on The Traceability World

It is generally accepted that the term "Traceability" became important in 1996 during or after the mad cow crisis in Europe. It was effectively used in Europe as the starting point of a technical, political, judicial agitation that instigated the writing of the European Food law 178/2002 that was applied in the European Union on the 1st of January, 2005. It was a global shock because this law included the imported and exported food stuffs. The word Traceability became the title of the article 18 of the EU regulation 178/2002.

Mad-cow disease is the bovine spongiform encephalopathy (BSE). It is a fatal neurodegenerative disease in cattle that causes a spongy degeneration of the brain and spinal cord. One of the main characteristics of BSE is that it has a long incubation period from 30 months to 8 years and it usually affects adult cattle at four to five years. We think that this crisis started in the United Kingdom as this was the country which was most affected with more than 179,000 infected cattle and 4.4 million were slaughtered during the eradication program (epidemic in 1986–1998). At this time, nobody knew the exact origin of the disease which was later described as a prion infection. It created a lack of confidence in the scientific world and mainly in the political world. It was characterized by the collapse in beef consumption in the 1990s when consumers were informed about the transmission of BSE in humans through the ingestion of cow meat. This problem was of such severity that in the General Food Law created to give to Europe a food safety system it was specified that "The Authority should be an independent scientific source of advice, information and risk communication in order to improve consumer confidence."

Coming back to the EU regulation 178/2002, the article 18 which title is "Traceability" says literally:

1. The traceability of food, feed, food-producing animals, and any other substance intended to be, or expected to be, incorporated into a food or feed shall be established at all stages of production, processing and distribution.
2. Food and feed business operators shall be able to identify any person from whom they have been supplied with a food, a feed, a food-producing animal, or any substance intended to be, or expected to be, incorporated into a food or feed. To this end, such operators shall have in place systems and procedures which allow for this information to be made available to the competent authorities on demand.
3. Food and feed business operators shall have in place systems and procedures to identify the other businesses to which their products have been supplied. This information shall be made available to the competent authorities on demand.
4. Food or feed which is placed on the market or is likely to be placed on the market in the Community shall be adequately labeled or identified to facilitate its traceability, through relevant documentation or information in accordance with the relevant requirements of more specific provisions.

This article does not give any information on the means that can be used to trace food and the operators. In fact, food operators could use all of the technical tools that permit following food during its processing. It could be a book note, a bar code or some electronic or informatics devices.

4. Historical Aspects of Traceability

Considering historical aspects (Table 2), we could locate a first document in *Historia Anglicana* in 1275 written by Thomas of Walsingham enabling the tracing back of the origin of an epizootic in Europe. He concluded that a ewe had been infected by both sheep pox and mange. England was infected by these diseases for 28 years. This was the first example that we find in literature. Another disease called rinderpest touched Western Europe (1711).

Table 2. History of Food Traceability and Political Reactions.

Date	Events	For consumers	For industry	For politics	Reactions
1275	Sheep pox and mange crisis.	Cattle diseases and death, fear.	High loss	Fear, reactions	Tracing back
October 7th, 1952	Original patent of bar code Created by two American students, Norman Joseph Woodland and Bernard Silver.	All food are now trace in developed countries.	Better knowledge on food flux.	Methods to automate the registration of manufacturers.	Worldwide diffusion and application.
1965	Web link, (hypertext system to automatically switch to a document) invented by Ted Nelson.	Application on mobile phone.	Low cost advertising	Useful for everybody. No reaction.	Worldwide application on mobile phone.
1970	George Laurer added numbers under the vertical bars to identify the product.	All food are now traced in developed countries. Addition of origin information.	This development replaced the concentric lines too easily unreadable.	Importance of food origin in political discussion.	Worldwide diffusion and application.
January 23, 1973	First RFID U.S. Patent by Mario W. Cardullo who claimed to have received the first patent for an active RFID tag with rewritable memory.	Real-time information is available.	Information is readable at long distance. Better Supply chain management. Real-time information is available. Administration and planning are improved.	Limitation of problems and food crisis.	RFID ensures that the right goods are available in the right place. The supply chain is considerably more precise and improves the efficiency and reliability of the entire chain.

Table 2 contd. ...

...Table 2 contd.

Date	Events	For consumers	For industry	For politics	Reactions
June 26, 1974	First food with a bar code (chewing gum) in the city of Troy (Ohio).	Same as 1970			Analyses the content of safety messages to provide UE Members for risk analysis.
1979	Creation of RASFF	Safer food in European Union.	Safety information on imported food.	Better control at the European border.	
1994	ISO standard 8402:1994	Information on origin are now available.	Definition of Traceability for all goods.		Definition of Traceability by ISO.
1996	Mad cow crisis	Fear, doubts	Discovering of a new hazard (prion).	Fear and important reactions.	Discussion on a regulation on traceability and discussion of the European regulation on food safety.
1997	*Codex Alimentarius*	Traceability arrive in all the countries of the world.	Product Tracing as a Tool within a Food Inspection and Certification System.	International reaction on traceability.	Definition of traceability for the food sector.
January 1st, 2005	EU/EC regulation 178/2002	Food have to be safe for all consumers of EU including exported and imported food.	More responsibility to producers.	Responsibility is given to producers. Less governmental controls.	Unique regulation for food and feed in Europe. Impact on international regulations.
18 December 2006	Regulation EC No. 1907/2006. REACH 2006	Fear, doubts on chemicals. Research of an organic life and more safety.	Registration, Evaluation, Authorization and Restriction of Chemicals (REACH). Labeling of dangerous substances.	More security and less responsibility.	Establishing a European Chemicals Agency.

2007	ISO 22005:2007 establishes the principles and specifies the basic requirements for the design and implementation of a food chain traceability system.	Safer life and better knowledge of food origin.	This system is designed to be flexible enough to allow organizations in the food chain to achieve identified objectives. It is a technical tool to assist an organization to meet its stated objectives, and it can be used to determine the history or location of a product or its components.	General principles and basic requirements for system design and implementation.	Traceability in the feed and food chain.
April 23rd, 2015	US patent on DNA marking of previously undistinguished items for traceability.	Possibility to family to trace their children, animals….	Could be useful for the certification of living organisms (that have DNA). DNA could be introduced in any products as marker of identification.	New deontology problems.	Provides a method of marking an item with naturally-derived or synthetic non-natural polymeric marker molecules, such as a DNA or peptide marker.

By tracing back, an ox was identified as being responsible in 1711 for the introduction of the epizootic into Italy that killed more than 200 million of cattle during this period in Western Europe.

It is evident that without any analytical techniques, health/safety officials at this time, conducted epidemiological surveys very successfully, to determine the origin of epizootics.

In 1994 a definition of Traceability by ISO (ISO 8402, 1994) was published. At this time on the internet, it was nearly impossible to find a site concerning Traceability. The ISO definition was *"Traceability is the aptitude to find the history, the utilization or the localization of an entity (activity, process, product…) by the mean of registered identification"*.

This general definition that was not created for foodstuffs but for all products was very succinct and is still applicable nowadays. If we apply this definition to food stuffs, 'History' could be the process used to produce the food; 'Utilization' could be the description of the recipe by the producer to the consumers; 'Localization' could be the origin of ingredients and where the final food was produced.

In 1997, *Codex Alimentarius* (FAO/WHO 1997) defined traceability for the food sector as *"the ability to follow the movement of a food through specified stage(s) of production, processing and distribution"*. In 2000, ISO 9000 and 22005 gave a slightly different definition of traceability and proposed that traceability should be defined as *"The ability to trace the history, application or location of that which is under consideration"*. The EU General Food Law (EU 178/2002) defined in article 18 traceability as *"The ability to trace and follow a food, feed, food producing animal or substance intended to be, or expected to be incorporated into a food or feed, through all stages of production, processing and distribution"*.

Some details were given in 2004 by the EU (Guidance on the implementation of articles 11, 12, 16, 17, 18, 19, 20 of EC No. 178/2002).

Some important features of these articles are the following:

Article 18 requires food business operators:

- to be able to identify from whom and to whom a product has been supplied;
- to have systems and procedures in place that allow for this information to be made available to the competent authorities upon their request.

The requirement relies on the "one step back"-"one step forward" approach which implies for food business operators that:

- They shall have in place a system enabling them to identify the immediate supplier(s) and immediate customer(s) of their products.
- A link "supplier-product" shall be established (which products supplied from which suppliers).

- A link "customer-product" shall be established (which products supplied to which customers). Nevertheless, food business operators do not have to identify the immediate customers when they are final consumers.

In addition to its impact on product flow in factories, Traceability is also a good system that was used and is still used to manage food crises. A crisis is a complex situation tending towards a perturbation of people and organizations. It usually creates a brutal breaking of the usual working functions in the social life, producing important financial and food loss. It also increases distrust towards administration and spread out of irrational fear.

Food crisis ➡️ Medialization ➡️ Doubts

Crisis characteristics

In recent times, the most important crisis related to the food sector were mad cow disease, dioxin in chickens, *Listeria* in delicatessen, avian flu, and more recently the fraud in the beef meat market.

5. To Avoid the Crisis: Towards a Preventive Policy

A food crisis can have economic repercussions on all stakeholders of a specific chain. Due to that, the EU establishes the common principles and responsibilities (Article 61, EU 178/2002) of all actors of the food chain and establishes the principle of management of food crises. In summary, food operators have to process the following steps: (1) Risk analysis, (2) Corrections and project management, (3) Means and tools and (4) Tests and simulation.

In reality, companies have to create a Crisis group cell with well determined members and they have to identify an official company spokesman who will relate to the media the importance of the crisis and also how the crisis will be solved by the company. The important things to remember are: The choice of the moment for communication? How to write a press release? What means of communication to be used? Companies have also to organize the technical follow-up of the crisis by proposing all corrective actions (Products withdrawal, Control of the remaining batches, Destruction of contaminated products, Corrective measures).

A consultancy of FAO/OMS experts was organized in Rome in February, 1998 on the application of communication on food safety. They described the objectives of the risk communication and its role in the risk management, the role and the responsibilities of the actors in risk communication and something very important, they gave the elements of an efficient risk communication.

At International level (essentially in Europe), one of the main interesting tools used to trace back an important food problem is the Rapid Alert System for Food and Feed (RASFF, article 35, EU 178/2002). In order to perform the monitoring the food safety and nutritional risks, the European Authority shall be the recipient of messages circulating in the early warning system. It analyzes the content of these messages to provide to the Commission and the Member States, all information necessary for risk analysis.

As an example (Table 3), if you use the key words "pesticides residues in vegetables" on the first four months (January–April) of 2016, you will find 53 notifications of serious hazards. If you use the key words "mycotoxins in nuts and seeds" on the first four months (January–April) of 2016, you will find 130 notifications of serious hazards.

As a part of preventive measures, upcoming techniques like immune-sensors can make a significant contribution on food chain traceability. The antigen-antibody recognition systems have been established for a number of target compounds like *Yersinia pestis*, alphatoxin, ricin, brevetoxin, okadic acid and atrazine (Vaughan et al. 2003, Schipper et al. 1998). In past immune-sensors have been used to type foot and mouth disease virus (Gajendragad et al. 2001).

To avoid crisis and as a part of the next generation food traceability system, the Critical Tracking Event (CTE) approach is gaining popularity. The CTE system focuses more on the events that manipulated the product in the supply chain than the food product itself. The IFT definition of CTE is "any occurrence involving an item at a specific location and time associated with collection and storage of data useful for associating the item (or related items) to the specific occurrences at a later time and is determined to be necessary for identifying the actual path of the item through the supply chain". The CTE requires minimum handling data for example, a code to identify the item, a code to identify the particular event (e.g., "received at XYZ Ltd. At door #3) and a date and time stamp. Other CTEs like mixing, repacking …, require additional information to link the inbound and outbound product codes. When an outbreak occurs, investigators can use the CTE traceability system to ask questions like "who has seen the item code 123?" CTE servers can also alert a food company regarding a formal request made by an appropriate authority, to which the company might reply in affirmative (the item was seen at these locations on these dates and times) or negative (the item was never seen by our CTEs). Additional codes of CTEs regarding mixing, repackaging … will further specify the link between the product and ingredients. This gives the investigators a clear picture of the supply chain for the item with respect to its location, time

Table 3. Extract from RASFF using the following Search criteria: **Notified from** 02/01/2016, **Notified till** 21/04/2016, **Product category** fruits and vegetables, **Hazard category** pesticide residues, Search result: 53 notifications.

Classification	Date of case	Reference	Country	Type	Risk decision	Product category/subject
Border rejection	21/04/16	2016.AOX	Denmark	food	undecided	fruits and vegetables: dimethoate (0.2 mg/kg—ppm) and unauthorised substance dinotefuran (0.023 mg/kg—ppm) in eggplant (*Solanum melongena* L.) from Thailand
Border rejection	20/04/16	2016.AOU	Italy	food	undecided	fruits and vegetables: unauthorised substance propargite (0.04 mg/kg—ppm) in fresh green beans from Egypt
Information for attention	18/04/16	2016.0467	Germany	food	serious	fruits and vegetables: prohibited substance dieldrin (0.17 mg/kg—ppm) in organic carrots from Israel
Border rejection	18/04/16	2016.AOF	Italy	food	serious	fruits and vegetables: omethoate (1.62 mg/kg—ppm) and dimethoate (9.53 mg/kg—ppm) in fresh cowpea (*Vigna unguiculata*) from the Dominican Republic
Alert	08/04/16	2016.0421	Netherlands	food	serious	fruits and vegetables: tebuconazole (0.64 mg/kg—ppm) and unauthorised substance carbendazim (0.71 mg/kg—ppm) in mango from Peru
Alert	29/03/16	2016.0374	Belgium	food	serious	fruits and vegetables: imazalil (11 mg/kg—ppm) in grapefruit from Turkey, via the Netherlands
Border rejection	22/02/16	2016.AIF	United Kingdom	food	serious	fruits and vegetables: unauthorised substance carbofuran (0.01 mg/kg—ppm) in aubergines from Thailand

and dates. With adequate commitment from operators to collect and store minimal set of data that may be retrieved later, the CTE promises to be a rapid and effective approach to traceability (IUFoST, Science Bulletin 2012).

6. Traceability and Genetically Modified Organisms (GMOs)

It is impossible to finish a chapter on the history of traceability without mentioning the genetically modified organisms (GMOs). We think this is the topic where the greatest amount and most controversial information are found, some originating from politicians who are thinking of electoral gain. The most difficult aspect in the GMO field is to identify the important information and disinformation for the public.

The various anti-GMO groups have been successful over the last ten years in making the public believe that GMO is synonymous with hazard. This relationship GMO = hazard is, in fact, based on some incidents (Monarch butterfly; allergic risk: Brazil nuts, Starlink corn; possible toxicity: Putzstaï case; agronomic spread: Percy Schmeiser case, rats in France) profiled to the extreme by the media hungry for scoops and fears for their readers.

It is clear that consumers do not see the interests of these new genetically modified plants (GMP), because they are primarily intended to improve farming practices (decreased inputs: herbicides and pesticides that are expensive and toxic).

GMOs are either plants or animals, or microorganisms which received a fragment of DNA from quite often (for those which currently exist) a bacterium which has an interesting property to the scientist's eyes. The majority of GMOs on the market so far have had received a portion of a gene from a bacteria called *Bacillus thuringiensis* (whose acronym is Bt) that codes for the production of a toxin, active against some insect pests. Some GMOs also contain a fragment of a gene that allows them to tolerate certain herbicides. The aim was to reduce the use of insecticides and herbicides that are toxic molecules for human.

An important fact is that organic farming covered by the regulation (EC No. 834/2007 from Council of 28 June 2007 and application regulation EC No. 889/2008) refused this technology for its consumers. Some countries like France for example refused also to allow GMO in agriculture but allowed them in food through importation. For that, at the EU level, the states have been working for a very long time on positive lists of plants allowed for consumption. This list is regularly updated when someone wants to introduce to the market a new plant selected by conventional or other means or imported from third countries. However, in the case of GMOs, they must meet very stringent guidelines when they are intended for human and/or animal consumption.

In fact, Europe applies the precautionary principle to food that was first clearly defined on January 1st, 2005 (European Regulation 178/2002). Contrary to the widespread idea among consumers, GMO are the plants most closely monitored and controlled over the world because they are always analyzed by experts before being placed on the food market. Moreover, when a GMO is proposed to the European market, it is necessary that the 25 committees of experts from each EU country give a favorable opinion!

All these actions are traced all over European Union.

7. The Precautionary Principle

The precautionary principle became mandatory in Europe on the first of January 2005 (EU 178/2002). This principle influences the traceability of food. In fact, governments or industries are now obliged to implement it when they encounter or suspect a problem. For simplicity, the precautionary principle must be applied to a problem identified either by companies or by consumer associations. When the problem is stated, it must be considered by public services through its expert committees. They have to fix a period to study the problem which is, of course, based on the estimated risk for consumers. This can vary from a few hours, as in the recent crisis of salmon containing dioxins, to some years as for GMOs in Europe.

7.1 Example of the precautionary principle: The new European regulation on chemicals (REACH)

The acronym REACH means Registration, Evaluation, Authorization and Restriction of Chemicals). Regulation (EC) No. 1907/2006 and Directive 2006/121/EC amending Directive 67/548/EEC were published in the Official Journal on 30 December 2006. REACH came into force on 1st June 2007. Companies that produce or import more than one ton of a chemical substance per year will be required to register in a central database managed by the new European Chemicals Agency. This agency will provide tools and guidance while Member States will provide a support service to the concerned companies.

The new regulation aims to improve protection of human health and the environment while preserving the competitiveness of the chemical industry in the EU and their ability to innovate. Under REACH, industry also assumes greater responsibility in managing the risks associated with chemicals and the communication of information on the safety of the substance.

REACH will require registration, over a period of 11 years, of some 30,000 chemicals. As part of this registration process, manufacturers and importers are required to generate data for all chemicals produced or imported into the EU in quantities above one ton per year. The registrants must also identify appropriate risk management and share them with users. In addition, REACH will further evaluate substances where there are concerns and provide a system for authorizing the use of substances of very high concern. This applies to substances that cause cancer, infertility, genetic mutations or birth defects and those that are persistent and accumulate in the environment.

The system will require companies to adopt progressive, safe alternatives when they exist. All requests for authorization must include an analysis of alternatives and a substitution plan when a suitable alternative exists. REACH also ensures that animal testing is limited to an absolute minimum and that the use of alternative methods is encouraged.

8. Conclusion

The term Traceability has appeared recently in international regulations but traceability has been used by industry for a long time to optimize their process and finance. The mad cow crisis, which highlighted the well-known crisis of 1996, showed the deficiencies in monitoring food and feed in the EU. Traceability is a tool that enables industrial flow management effectively in two ways (tracing and tracking).

Food has to be followed from production to consumption to remove it as quickly as possible from the market in case of incidents that can be detrimental for the health of the consumers. Trade is becoming more intense and spread throughout the world. Consumers and governments are increasingly demanding and are sensitive to the quality and origin of food products they purchase or control. The mad cow crisis, dioxin, rejection of GMOs, avian influenza had developed a sense of mistrust of consumers in respect to the quality and food safety. To ensure fair trade and the development and promotion of food label (appellation of origin, protected geographical indication), it becomes essential to know and ensure the nature and geographic origin of food products.

Keywords: History, traceability, tracking, tracing, origin

References

Agriculture and Agri-Food Canada; Government of Canada - http://www.ats-sea.agr.gc.ca/trac/sys-eng.htm.

Anon. 2011. Getting technical over food traceability, China Daily. http://www.china.org.cn/business/2011-07/11/content_22965087.htm.

Australia GSI. 2010. Australia serving up safety and traceability. http://www.ferret.com.au/c/GS1-Australia.

Azziz-Baumgartner, E., Lindblade, K., Gieseker, K., Rogers, H.S., Kieszak, S., Njapau, H., Schleicher, R., McCoy, L.F., Misore, A., Decock, K., Rubin, C. and Slutsker, L. 2005. The aflatoxin investigative group: case-control study of an acute aflatoxicosis outbreak, Kenya, 2004. Environmental Health Perspectives 113(12): 1779–83.

Bakir, F., Damluji, S.F., Amin-Zaki, L., Murtadha, M., Khalidi, A., Al-Rawi, N.Y., Tikriti, S., Dahahir, H.I., Clarkson, T.W., Smith, J.C. and Doherty, R.A. 1973. Methylmercury poisoning in Iraq. Science 181(4096): 230–241.

BBC News. 2010. Minister heard feed toxic on TV. Available at http://newsvote.bbc.co.uk/mpapps/pagetools/print/news.bbc.co.uk/2/hi/uk_news/northern_ireland/8567267.stm?ad=1.

BBC News. 2011. Who, what, why: Why are Indians dying from alcohol poisoning? Available at http://www.bbc.co.uk/news/magazine-16197280?print=true.

Bernard, A., Broeckaert, F., De Poorter, G., De Cock, A., Hermans, C., Saegerman, C. and Houins, G. 2002. The Belgian PCB/dioxin incident: analysis of the food chain contamination and health risk evaluation. Environmental Research 88(1): 1–18.

Bujji, M. 2012. Tracing the Footsteps, FnB News (India). http://www.fnbnews.com/article/detnews.asp?articleid=26342&sectionid=32.

CAC. 2005. Codex procedural manual (15th ed.) Retrieved from. In: ftp://ftp.fao.org/codex/Publications/ProcManuals/Manual_15e. pdf (Accessed 12.07.12).

Canadian Traceability. National Agriculture and Food Traceability System (NAFTS 2013).

Center for food in Canada. 2012. Forging Stronger Links: Traceability and the Canadian Food Supply Chain.

Codex Alimentarius (FAO/WHO, 1997). Principles for Traceability/Product Tracing as a Tool within a Food Inspection and Certification System. Reference CAC/GL 60-2006.

Congress.gov, http://thomas.loc.gov/cgi-bin/query/z?c107:H.R.3448.ENR:Agri-Food Trade Services for Exporters.

Council Regulation (EC) No. 834/2007 on organic production and labelling of organic products and repealing Regulation (EEC) No. 2092/91.

Covaci, A., Voorspoels, S., Schepens, P., Jorens, P., Blust, R. and Neels, H. 2008. The Belgian PCB/dioxin crisis-8 years later. Environmental Toxicology and Pharmacology 25(2): 164–70.

Crawford, B.A., Cowell, C.T., Emder, P.J., Learoyd, D.L., Chua, E.L., Sinn, J. and Jack, M.M. 2010. Iodine toxicity from soy milk and seaweed ingestion is associated with serious thyroid dysfunction. The Medical Journal of Australia 193(7): 413–415.

Dakeishi, M., Murata, K. and Grandjean, P. 2006. Environmental Health: A Global Access Science Source 5(1): 31.

Directive 2006/121/CE of the European Parliament and of the Council of 18 December 2006 amending Directive 67/548/EEC on the approximation of laws, regulations and administrative provisions relating to the classification, packaging and labeling of dangerous substances to adapt it to Regulation EC No. 1907/2006 concerning the Registration, Evaluation and Authorization of Chemicals and restriction of chemicals (REACH) and establishing a European chemicals agency.

Dunckel, A.E. 1975. An updating on the polybrominated biphenyl disaster in Michigan. Journal of the American Veterinary Medical Association 167(9): 838–41. PMID 1184446.

EAN-UCC. 2002. European Article Numbering Association. EAN International and the Uniform Code Council. Available from http://www.ean.ucc.org.

EU Regulation (EC) No. 178/2002 of the European parliament and of the council. 2002.

Firestone, D. 1973. Etiology of chick edema disease. Environmental Health Perspectives 5: 59–66.

Fries, G.F. and Kimbrough, R.D. 1985. The Pbb episode in Michigan: an overall appraisal. Critical Reviews in Toxicology 16(2): 105–56.

FSA. 2002. Traceability in the food chain a preliminary study. UK: Food Standard Agency. Retrieved from www.food.gov.uk/multimedia/pdfs/traceabilityinthefoodchain.pdf (Accessed 15.07.12).

FSA. 2005. Action taken to remove illegal dye found in wide range of foods on sale in UK. Available at http://www.food.gov.uk/news/newsarchive/2005/feb/worcester.

FSMA. 2011. FDA U.S. Department of Health and Human Services, Food Safety Modernization Act (FSMA), https://www.fda.gov/food/guidanceregulation/fsma/.

GAIN Report N° KS1033. 2010. Traceability Requirements for Imported Beef.

Gajendragad, R., Kamath, K.N.Y., Anil, P.Y., Prabhudas, K. and Natarajan, C. 2001. Development and standardization of a piezo electric immunobiosensor for foot and mouth disease virus typing. Veterinary Microbiology 78: 319–330.

George, T.S. 2002. Minamata: Pollution and the Struggle for Democracy in Postwar Japan. Cambridge, Massachusetts: Harvard University Asia Center. ISBN 0-674-00364-0.

GMA Affairs Poison on my plate: List of food scares around the world - http://www.gmanetwork.com/news/story/158734/news/nation/poison-on-my-plate-list-of-food-scares-around-the-world#sthash.VwArGxG3.dpuf). http://www.gs1.org/traceability.

GMA news on line April 27. 2009. Available at: http://www.gmanetwork.com/news/story/158734/news/nation/poison-on-my-plate-list-of-food-scares-around-the-world#sthash.VwArGxG3.dpuf.

Gossner, C.M.E., Schlundt, J., Ben Embarek, P., Hird, S., Lo-Fo-Wong, D., Beltran, J.J.O., Teoh, K.N. and Tritscher, A. 2009. The melamine incident: implications for international food and feed safety. Environmental Health Perspectives 117(12): 1803–1808.

Government of Canada, Health Canada, Health Products and Food Branch, Veterinary Drugs Directorate - http://www.hc-sc.gc.ca/dhp-mps/vet/faq/faq_chloramphenicol_honey-miel-eng.php#a4.

Government of Canada, Health Canada, Health Products and Food Branch, Veterinary Drugs Directorate - http://www.hc-sc.gc.ca/dhp-mps/vet/faq/faq_chloramphenicol_honey-miel-eng.php http://www.reading.ac.uk/foodlaw/news/uk-02035.htm, http://www.reading.ac.uk/foodlaw/news/uk-03018.htm.

Gregory, N.G. 2000. Consumer concerns about food. Outlook on Agriculture 29(4): 251–257.

GS1 Traceability. Supporting Visibility, Quality and Safety in the Supply Chain.

Guidance on the implementation of articles 11, 12, 16, 17, 18, 19 and 20 of Regulation (EC) No. 178/2002 on general food law conclusions of the standing committee on the food chain and animal health. 20 December 2004.

Harrington, R. 2011. Dioxin-contaminated liquid egg distributed in UK, Contamination worse than feared in German dioxin scandal. Available at http://www.foodnavigator.com/content/view/print/351701.

ISO 22005:2007. Traceability in the feed and food chain. General principles and basic requirements for system design and implementation.

ISO 9000, Quality Management. 2000. http://www.iso.org/iso/fr/iso_9000

ISO 9000. 2005. Retrieved from http://www.pqm-online.com/assets/files/standards/iso_9000-2005.pdf (Accessed 02.08.12).

ISO standard 8402:1994. International Organization for Standardization (1994), http://www.iso.org. Definition of Traceability by ISO.

IUFoST Scientific Information Bulletin (SIB) March 2012, pp 1–14.

Joint FAO/WHO Expert Consultation. The application of risk communication to food standards and food safety, Rome, 2–6 February 1998.

Kakar, F., Akbarian, Z., Leslie, T., Mustafa, M.L., Watson, J., Van Egmond, H.P., Omar, M.F. and Mofleh, J. 2010. An outbreak of hepatic veno-occlusive disease in Western Afghanistan associated with exposure to wheat flour contaminated with pyrrolizidine alkaloids. Journal of Toxicology 1–7.

Karlsen, K.M., Olsen, P. and Donnnelly, K.A. 2010. Implementing traceability: practical challenges at a mineral water bottling plant. British Food Journal 112(2): 187–197.

Kasuya, M., Teranishi, M., Aoshima, K., Katoh, T., Horiguchi, H., Morikawa, Y., Nishijo, M. and Iwata, K. 1992. Water pollution by cadmium and the onset of Itai-itai disease. Water Science and Technology 26: 149–56.

Kim, J., Oh, K., Jeon, S., Cho, S., Lee, D., Hong, S., Cho, S., Park, M., Jeon, D. and Kim S. 2011. *Escherichia coli* O104:H4 from 2011 European Outbreak and Strain from South Korea. Emerging Infectious Diseases 17: 9, 1755–1756.

List of food contamination incidents, https://en.wikipedia.org/wiki/List_of_food_contamination_incidents.

Lowe, K. 2015. Caution! The bread is poisoned: The Hong Kong mass poisoning of January 1857. The Journal of Imperial and Commonwealth History 43(2): 189–209.

Malisch, R. 2000. Increase of the PCDD/F-contamination of milk, butter and meat samples by use of contaminated citrus pulp. Chemosphere 40(9-11): 1041–53.

Manila Bulletin. 2012. Skinny pigs, poison pork: China battles farm drugs. Available at http://www.mb.com.ph/articles/300535/skinny-pigs-poison-pork-china-battles-farm-drugs.

Mazzocchi, M., Lobb, A., Traill, W.B. and Cavicchi, A. 2008. Food scares and trust: a European study. Journal of Agricultural Economics 59(1): 2–24.

Menozzi, D., Halawany-Darson, R., Mora, C. and Georges, G. 2015. Motives towards traceable food choice: a comparison between French and Italian consumers. Food Control 49: 40–48.

Ministry for Primary Industries, http://www.foodsafety.govt.nz/elibrary/industry/Source_Lead-Nzfsa_Confident.htm.

Montet Didier. 2009. Editeur. Les vraies-fausses informations de l'alimentaire. 162 pages, Edilivre.com, ISBN: 9782812102219.

MSNBC: Food firm recalls lead-contaminated lollipops. MSNBC (Calexico, California). May 2, 2009. http://www.msnbc.msn.com/id/30530838/.

O'Connell, R., Parkin, L., Manning, P., Bell, D., Herbison, P. and Holmes, J. 2005. A cluster of thyrotoxicosis associated with consumption of a soy milk product. Australian and New Zealand Journal of Public Health 29(6): 511–512.

Peres, B., Barlet, N., Loiseau, G. and Montet, D. 2007. Review of the current methods of analytical traceability allowing determination of the origin of foodstuffs. Food Control 18: 228–235.

Perez-Aloe, R., Valverde, J.M., Lara, A., Carrillo, J.M., Roa, I. and Gonzales, J. 2007. Application of RFID tags for overall traceability of products in cheese industries. *In*: Proceedings of the 1st annual RFID Eurasia, Isatanbul, Turkey.

RASFF Portal, the Rapid Alert System for Food and Feed. European Commission: https://webgate.ec.europa.eu/rasff-window/portal/.

Regulation EC No. 178/2002 of the European Parliament and of the Council of 28 January 2002 laying down the general principles and requirements of food law, establishing the European Food Safety Authority and laying down procedures in matters of food safety.

Regulation EC No. 1907/2006. REACH of 18 December 2006 concerning the Registration, Evaluation, Authorization and Restriction of Chemicals (REACH) and establishing a European Chemicals Agency.

Regulation EC No 889/2008 of 5 September 2008 laying down detailed rules for the implementation of Council Regulation EC No 834/2007 on organic production and labelling of organic products with regard to organic production, labelling and control.

Reynolds, S. and Ernest, A. 1901. An Account of the Epidemic Outbreak of Arsenical Poisoning Occurring in Beer-Drinkers in the North of England and the Midland Counties in 1900. The Lancet 157(4038): 409–452.

Schanche, D.A. 1986. Death Toll Stands at 18 Italy Jolted by Poisoning, Halts All Exports of Wine. Los Angeles Times, April 9, 1986.

Schipper, E.F., Rauchalles, S., Kooyman, R.P.H., Hock, B. and Greve, J. 1998. The waveguide mach-zender interferometer as atrazine sensor. Analytical Chemistry 70: 1192–1197.

Schwägele, F. 2005. Traceability from a European perspective. Meat Science 71: 164–173.

Sharma, B.D., Malhotra, S., Bhatia, V. and Rathee, M. 1999. Classic diseases revisited: Epidemic dropsy in India. Postgraduate Medical Journal 75(889): 657–661.

Shaw, C.A. and Bains, J.S. 1998. Did consumption of flour bleached by the agene process contribute to the incidence of neurological disease? Medical Hypotheses 51(6): 477–81.

Snow, J. 2003. On the adulteration of bread as a cause of rickets. International Journal of Epidemiology 32: 336–337.

Stefani, G., Cavicchi, A., Romano, D. and Lobb, A.E. 2008. Determinants of intention to purchase chicken in Italy: the role of consumer risk perception and trust in different information sources. Agribusiness 24: 523–537.

The China Times. 2011. Eleven people died from poison in vinegar. Available at http://www.thechinatimes.com/online/2011/08/1121.html.

The website of the Food Standards Agency, http://www.food.gov.uk/multimedia/faq/51434/

The website of the Food Standards Agency, http://www.food.gov.uk/multimedia/faq/olivepomoilqa.

Thomson, B., Poms, R. and Rose, M. 2012. Incidents and impacts of unwanted chemicals in food and feeds. Quality Assurance and Safety of Crops & Foods 4(2): 77–92.

Traub, J. 1988. Into the Mouths of Babes. The New York Times, July 24, 1988.

US Bioterrorism Act. 2001. H.R. 3448. Consulted on 1/15/2015: http://thomas.loc.gov/cgi-bin/query/z?c107:H.R.3448.ENR.

Vaughan, R.D., Geary, E., Pravda, M. and Guilbault, G.G. 2003. Piezoelectric immunosensors for environmental monitoring. International Journal of Environmental and Analytical Chemistry 83: 555–571.

2

Actuality on Food Traceability

Pierre Chartouny

1. Introduction

"The sciences of life are not limited to the sciences of Humans"

Food traceability has become a major global issue. Globalization, which has been progressing for decades, has transformed the perception of the various sectors involved, from agricultural production at the top of the cycle to recycling of wastes to the very end, including all intermediate processes of production (processing, distribution and consumption). Globalization has both boosted and complicated the traceability process. The different crises suspected, perceived or experienced generated a collective awareness. For the same crisis, the behavior or reactions vary from one continent to another, questions of means or priorities.

The different vertical integrations (processing, distribution and consumption), have shown that the borders of yesterday or those in preparation today in Europe or beyond the Atlantic can no longer protect consumers from one continent to another. There are no frontiers that can stop migratory birds from crossing continents, no matter how advance our societies are or will become. These migrations render useless the sometimes radical actions taken, such as the plans for the eradication of avian influenza affecting ducks or geese raised in the southwest of France.

Whatever is the nature of the stakeholder in the production cycle, from the agricultural sector to the food industries, production, processing, distribution, consumption and recycling—in all, the common characteristic

Company ITAQUE, 8 rue de l'abbé Hazard, 92000 Nanterre, France.
  E-mail: pchartouny@gmail.com

is that all the players in the food chain are and will remain consumers. We are all responsible and concerned. No science can resist the essential nature of human societies, namely greed. The lure of profit and unconsciousness are often the two pillars of food crises. Traceability alone will not be the solution.

All the current topics remind us that:

News coverage of today was the news of yesterday, and history repeats itself.

- The new form of globalization is advancing more than the procedures set in place to react. That is why we are becoming more aware of the limitations of late and very partial responses on the regional, national or even global levels.
- The assertion of food traceability as an essential issue in our societies has made it possible for us to detect crises, which were unnoticeable yesterday because of lack of indicators or tracers. There are more crises but there are still no sufficient solutions.
- Crisis analyzes over the past 20 years are no longer national but global. The causes of crises are often undetected. See the crisis of the "Spanish cucumber" in 2011. This epidemic of gastroenteritis and hemolytic uremic syndrome in Europe, mistakenly called the "cucumber crisis" because of the initial suspicion of cucumber batches from southern Spain, was an epidemic due to *Escherichia coli* O104:H4 that killed more than fifty people mainly in Germany.

  Reactions based on fears, economic and societal issues undermine our ability to analyze such kind of food crisis and find the right reasoning and the right solutions. An intra-European incoherent debate took place, which was replaced a few months later by an another explanation of this cucumber crisis (2011) in Egypt and North Africa, without really understanding what actually happened or making an attempt to trace back the root of the problem.

- We always remain in the reaction rather than in the prepared and thought before action. We remain in the management of a crisis situation rather than in the anticipation.
- All adequate responses with new measures are often implemented late and quickly surpassed by global trade openings.
- There is a clear lack of anticipation from national authorities in the country, at regional (e.g., Europe) or international (WHO) levels. The means of these actors are very limited not only financially but also by the evolution of globalization, the historical opacity of some global actors, the commercial and financial stakes.
- We are "crying wolf" most of the time rather than working on efficient solutions.

- The complexity of all the sectors shows the need to arrive at a solution applicable and usable in the field of food traceability.
- The analysis of developments of traceability in the tobacco sector gives us an example of misuse of traceability, even when it exists and is implanted.
- The French experience in food labelling shows a willingness to do better; but it also demonstrates the obvious limitations of such an experiment.

In this chapter, we will give a panorama of the news evolving around the food traceability field that has been rich in events recently. We will discuss two real recent examples:

- The first one concerns the tobacco sector in Europe. The tobacco sector, by its nature, could be a good example for a feasible study on the matter of traceability especially that it concerns a large spectrum of economical actors.
- The second will be the experience of food labelling in France.

One day, a UN Conference on Food Traceability called TOP21, will have to be launched, such as COP 21 (www.gouvernement.fr/action/la-cop-21), the United Nations Climate Change Conference (www.unfcc.int).

2. Outline for a Partial Panorama: A Global Problem with Local or Sectorial Solutions

The diagram (Table 1) shows that it is technically possible to make major advances in the world of food traceability. Several solutions exist; some existed for more than 20 years. Others have emerged in the past three years.

But it remains clear that even if the nature of the product permits its analysis (tobacco for example); there is not, to our knowledge, a complete and integral traceability solution. Even if new technologies have enabled new tools, new perspectives and major advances, these necessary tools remain insufficient without the adherence of national and international public authorities, major and global actors like global companies.

For Man, a passport is and will remain the talisman that opens the borders, or closes them. For services or intellectual values, globalization has been a vector of fantastic propagation and enrichment during the last thirty years. The borders are fading.

For the products, the border factor remains always decisive. The organization of work with regard to any agricultural or food product makes logistics essential. The border remains up-to-date. All the underlying ways of product trade between two economies have evolved rapidly as have the services as well. On the other hand, the components of an agricultural product begins always at a local level as the land in which it was produced,

Table 1. Panorama of Food Traceability.

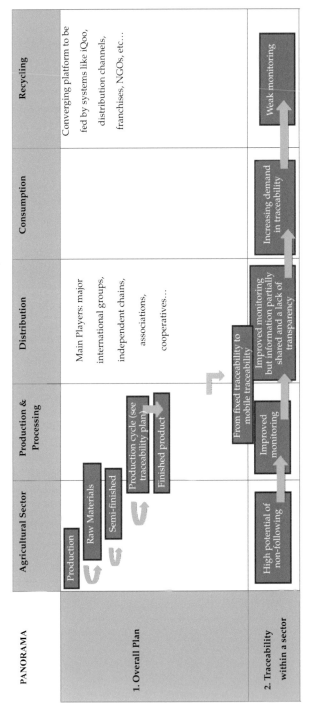

PANORAMA	Agricultural Sector	Production & Processing	Distribution	Consumption	Recycling
1. Overall Plan	Production / Raw Materials / Semi-finished / Production cycle (see traceability plan) / Finished product		Main Players: major international groups, independent chains, associations, cooperatives…		Converging platform to be fed by systems like iQoo, distribution channels, franchises, NGOs, etc…
2. Traceability within a sector	High potential of non-following	Improved monitoring / From fixed traceability to mobile traceability	Improved monitoring but information partially shared and a lack of transparency	Increasing demand in traceability	Weak monitoring

	High potential of breaking load	Low to high chances of breaking load	High potential of breaking load / Deed of assignment and free of charge transfer (to NGOs, associations, etc...)	Frequent breaking loads	Very weak monitoring
3. Traceability between differents sectors					
4. Standard Solutions	Qualiboo (20 years of experience)	iQoo, Codentify and other solutions	iQoo/Codentify		iQoo
5. Examples	Monitoring of the cooperatives, farms and farms operators	Factories and agri-food industry: slaughterhouses, fish slicing	Retail, tobacco shop, distributor, Tesco, Wallmart, etc...	Consumer protection associations	All stakeholders are involved
6. Public Entities	Ministry of Agriculture/ European Union	Ministry of Industry and Environment, E.U.	Ministry of Commerce and Industry, E.U.	Ministry of Health, E.U.	Ministry of Industry and Environment, E.U.

passed through production, storage, processing, shipment, reception, repackaging, and the manufacturing process. The original product could become a sub-form of a new product that integrates the distribution sector. It could be sold or consumed locally or exported or even resold to the country of origin as an ingredient of the final product.

A process of traceability is essential to accompany the life cycle of an ingredient from its conception to its consumption and beyond (a problem could emerge from the consumed product). It is obvious that in an increasingly globalized world, it is impossible to carry out this monitoring process and to trace it. Multiple factors prevent this traceability. Multiple and multidimensional criteria are insurmountable obstacles, often compounded by artificial factors created to prevent this traceability from going upstream. It should be noted that a real traceability system must be in both directions, from downstream to upstream, but also from upstream to downstream. Otherwise, the traceability is induced in error and/or not complete. Major advances have been made in technology these recent years. The good surprise is that there are many resolutions of difficulties considered insurmountable yesterday. These advances allow us to trace a safer path, but also are richer in discoveries and in finding difficulties that were unrecognized yesterday.

Till date, the following four bases have been connected:

A. The base of the temperature
B. The basis of labelling
C. The basis of procedures and Sanitary Master Plans
D. The basis of Traceability *"stricto sensu"*

These elements are essential because food traceability cannot be summarized simply by finding the origins of the product. It is necessary to publish a passport that will accompany it throughout its process of transformation until its consumption. This passport will contain different information such as the temperature conditions of its transportation and, one must analyze the process of its evolution to show and verify that it has undergone a transformation in the rules of the art by the professionals of the sector, therefore not only in terms of its production and processing but also its distribution and consumption.

So, food must be traced from the farmers to the consumers in Europe or in the world. Customer relationship management (CRM) tools become necessary to track the life of a product once it has been distributed. These tools help with contact management, sales management, workflow processes, and productivity. Recipe sites should also include traceability. But it is known that a system, no matter how powerful it is, is never enough. We need the support of the players and actors involved in all the processes (producer, packaging industry, distributer and even the official government authorities in charge of regulating and controlling). Like the story of the drop of water that begins its journey when it starts from a rain drop on the side of a mountain, which flows in streams, carrying passing identities on its way. Thus it flows into the sea and mixes and leaves again in the sky with evaporation to join the clouds and return to the earth and strike the mountains and leave again. Like the salmon from its birth, journey then demise. The panorama (Table 1) shows the complexity of integral and complete food traceability process. This traceability, when it is desired, is very complicated to put in place. It has evolved well over the last 30 years. To date, we can find absolutely acceptable levels within a sector of activity, agricultural, production or processing, etc.

Consumer pressures, organized and unorganized, and national, regional or international governments have shifted attitudes and legislations. The traceability has even become a factor of promotion and sales, a search to want to consume products of which we know the origins, for example, French meat, organic bread with organic wheat, organic nuts, etc. But the evolution remains limited to consumers with the economic profile of socio-professional groups (High Rated Social Professional Category), with a decisional dilemma between 'I consume traced products, therefore often they are more expensive, but I do not want to pay more.' Hence the rising trend of labels, many of them were created in recent years (Bio label, 100% Vegan label, the eco-responsible label, the ethical production, and recycling related labels). It is obvious that we are unable to trace an agricultural product and then food from A to Z. As much as it is feasible to do it within the same sector, things get tough when this product is moving from one sector to another, or even continue its journey in the five main sectors taken in this panorama (Table 1). Within the same chain, several solutions exist, with varying degrees of completeness.

3. Example of Food Industries: Food Traceability or Collective Catering Solution

One of the most solid and perennial solutions is Qualiboo (Proposed by company ITAQUE www.itaque.fr). This solution was created following the mad cow crisis in Europe, some 20 years ago, with the participation, in the beginning, of Purpan University of Toulouse, France. Since then, it has not stopped evolving, with an incremental contribution of added value obtained from the progress of research and development, the evolution of technological platforms, the contribution of "A multitude of customers" from different sectors such as meat, slaughterhouses to cutting, processing, fish, bakery, vinegar, biscuits, and all sectors of the Agro-Food Industries, from production to processing industries. It is also present in some central kitchens, public hospitals and schools, etc.

A global food company, the Swiss multinational "Nestlé", through one of its investment funds, has provided financial resources and a global vision at the initial stages of the project. Till date, this food traceability solution appears to be the most complete. It goes beyond the regulatory framework imposed and allows for almost total control and follow-up within a food factory, laboratory or sub-sector. It differs from solutions developed within large national or a global group in so far as it does not serve a specific context but provides standardized operational solutions quickly. Currently it operates in more than 400 production locations in France, Switzerland and Belgium.

The diagram (Fig. 1) explains the overall standard process that begins with receipt of goods, raw materials, and ends with shipments. Everything is traced in both directions, from upstream to downstream, but also from downstream to upstream. Everything is controlled, within the unit itself, through the cycle of production and transformation. This scheme of food traceability minimizes the risk. This solution is reinforced by the fact that it is adopted by professionals in different industries and especially in different packaging companies which gives it a renowned status in the agro-food industries.

Its installation strengthens and consolidates the HACCP (Hazard Analysis Critical Control Point) processes, followed by a serious and normal traceability. Without invalidating the risks inherent in the food sector, it contributes effectively to dramatically reducing these risks and, above all, to having a very rapid reactivity if a crisis occurs.

3.1 The limits of such a solution

Qualiboo alone, or any other system or solution by itself, cannot solve all the problems and blockages of food traceability. It is necessary beforehand, to be aware of the different sectors concerned; it is necessary to accept the methods and ways of work required sometimes to install this solution. It is necessary to go through a complete reform of the working methods in place at the origin of the food preparation, generating food risks. The whole process of production or transformation is involved. We have to accept change. This is not always obvious. We must also accept to respect new constraints in terms of production and processing. The consequences are sometimes an increase in the cost of production, and therefore sometimes difficult to shoulder in times of economic crises. Here we touch on an essential point related to the costs of production and comparison between an agri-food factory in Europe or the United States and a factory in China or India.

The food traceability factor will have to be integrated, in the same way as the ecological factor, when comparing costs in order to compare between

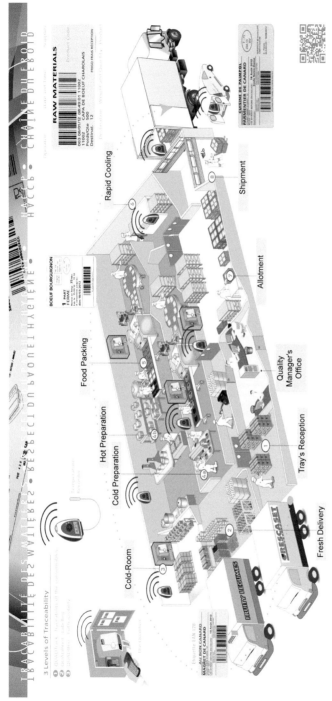

Figure 1. Example of a Full Complete Industrial Solution of Food Traceability (in a factory, central kitchen, packaging, food manufacturing …)
Qualiboo Solution—by ITAQUE

comparable or even identical natures in order to improve these comparisons. It is essential for making objective decisions. Otherwise talking about fair competition and balance between trades is impossible, at worst it will generate an economic imbalance and consumer lures.

This solution is ideal within a sector, but is *de facto* limited when looking for a cross-channel solution because by nature, it has been designed for so-called "heavy" professional sectors. It will not be able to transgress the artificial blockages fixed by the actors who want to be involved as little as possible and avoid a real transparency *vis-à-vis* the other actors in the chain, or *vis-à-vis* consumers. A very good example is the trade of garlic, and thus all garlic-making, albeit simplistic, but revealing what human factor is capable of doing. A garlic produced in China is transformed by smugglers importers, into Romanian or Hungarian garlic, then is introduced into the French market. There it is further transformed into French garlic, introduced in the processed food sector. Then in the composition of French dishes, it is labeled, certified or labellized, sold and consumed, in France or abroad, etc. The above example shows us how a raw material that has nothing to do with France becomes at the end of the process labeled as French made. A processing plant equipped with Qualiboo will integrate this garlic as a French product in the reception of raw materials, will permit to trace it most effectively in the world from the beginning to its release, but will not be able to trace its origins which are duly stamped in France. Thus, the whole process of traceability is well followed within a production or processing plant, but it is infallibility suffers greatly from practices between the various sectors at the time of passage from one sector to another. No system can solve this "the human interventionism" especially when the intervention is bad intentioned.

4. An Example of a Mobile Food Traceability Solution in the Distribution Sector

The creation of other solutions that secure the varied sectors are helping to extend the traceability of the food product. Solutions like Qualiboo (Itaque) for production and processing are supported by iQoo (Itaque, Fig. 2) or SureCheck in the USA that are part of Walmart that were created especially for the distribution sector.

The strength of IQoo when compared to others is that this solution is compatible with Qualiboo and therefore the gap between the two sectors—Production/Transformation and distribution—can now be addressed and solved provided that a real political and ethical will exists to impose this necessity. It is clear that the major global and national distribution chains also have manufacturing plants (production and processing). They have had them for decades now but their productions only partially cover the

Figure 2. Description of iQoo solution by ITAQUE.

perimeter of their products and the almost independence enjoyed by these major players and their freedom to maneuver, throws a lot of suspicions and added to the multiple scandals in Europe and in the world.

Traceability systems were developed internally within these groups, in order to comply with the legislation, but they are not very transparent except in case of control or the partial transparency for the control would be sufficient. In fact, as western and developed societies have evolved towards greater transparency in fiscal traceability, transparency on food traceability systems will become necessary to secure consumers. The contribution of IQoo is essential in that it ensures the convergence of several previously untreated or separately addressed needs. Thus, this solution integrates the management of traceability, monitoring of the cold chain, management of labels and monitoring and compliance with the sanitary master plan. Moreover it is easily configurable and, given its different uses and adaptability, highly appreciated in different sectors of distribution, retail and future consumption. It is a true "All in One" tool, that is currently working and it will be improved to be made more complete. The design of such a tool took place over a long period. Several intermediate steps with failures in the process of Research & Developments have marked the way to the first prototype of IQoo. The power of such a system is also perceptible in the ongoing project of a Database which ensures the real-time management of a multitude of food products, belonging to large groups of distributions and production making it possible to:

- A better tractability and follow up for certain food products, with expiry dates almost reaching the maturity, in order to be sent to Non Government Organizations working in the food sector such as the Food Bank, Restos du cœur (French charity organization), etc. without any risk for the end consumer and thus to enable the needy populations to eat better.
- Better fight against waste and therefore make zero waste plans more realistic.
- Better and more reliable information given to the recycling sector,
- Better track of information according to its origin, nature and destination (foodstuff, packaging, etc.).

In its category, IQoo stands out from the competition in a significant way.

Beside the above, other new products based on measurement and approach by spectrometer, allied to Smartphone and a certain genius of inventors is very promising. One of the latest inventions is the realization carried out by a Massachusetts Institute of Technology (MIT) team allowing a Smartphone connected to a device with an electroluminescent diode projector to determine from the skin of an apple the level of maturity of the fruit.

It is possible to envisage, in the long term, various and multiple applications allowing for research openings, to increase the perimeter of traceability, which can never be one-dimensional but multidimensional, in order to provide necessary answers that cannot be obtained today. A convergence of these tools, belonging to different sectors of Research & Development (R & D), will be essential in the long term. It will allow the desired and hoped-for meeting of several technologies and several know-how belonging to various sectors of R & D.

Innovative experiments conducted in Europe over the last few years, such as the one related to the concept of food DNA, or the experience of the Neo-Food project, which despite their relative failures, have opened up new horizons in this field. In short, a multitude of projects by different public or private organizations represent important advances to allow a more complete and almost total food traceability. It will always lack an authority for arbitration between the different technological and economic actors as well as the arbitration across borders. This arbitrary authority will have the heavy task of carrying out this necessary convergence.

We will find (Fig. 1) the simplified general overview, followed by the production traceability scheme, addressed by Qualiboo and will end with the follow-up at the distribution level, addressed by IQoo. We do not pretend to exhaustiveness but to explanatory examples.

5. Traceability in Tobacco or Traceability That Makes the Tobacco?

A fox in the henhouse or "How tobacco manufacturers impose their Traceability system and ... promote smuggling"

Multiple players have a role in this chain: Manufacturers of tobacco, National States, WHO, Anti-tobacco organizations and anti-tobacco crusaders, Tobacconists, Economic actors such as INEXTO, IMPALA, Philip MORRIS, etc.

Facts: The leader of the tobacco manufacturers Philip Morris International, created in 2007 a code of traceability affixed on its packages of cigarettes: CODENTIFY, an international patent to protect it. From 2007 to 2010, the code was promoted to the various actors, in particular the European Union. Negotiations were held with the European Anti-Fraud Office.

- In 2010, other tobacco manufacturers were invited to share the code. In 2011, the Digital Coding & Tracking Association in Zurich, Switzerland was created. It included British American Tobacco, Imperial Brands, Japan Tobacco International and Philip Morris International. IDENTIFY becomes the "standard" of anti-fraud manufacturers.
- In June 2016, the CODENTIFY code was transferred to INEXTO, a Swiss subsidiary of IMPALA, a French investment company. Amount of the transfer: one Swiss franc. On the surface, there were no longer any links between CODENTIFY and tobacco manufacturers.

In reality, the head office of INEXTO was located in Lausanne (Switzerland), in a building where a manager of Philip Morris resides. The CEO was a former Philip Morris employee, several executives of Philip Morris, of whom inventors of the code form were in the Board of Directors. How they could explain the transfer of ownership for only one symbolic franc Swiss by INEXTO?

Conclusion: Tobacco manufacturers have anticipated the evolution of the fight against tobacco. They have integrated the importance of traceability into their strategy, have ruined it by creating their own code and system:

they invited their competitors and brothers to share, organized their "true false" cession, diverted it from the normal objective displayed initially and oriented for their profits. Thus, they became the real puppeteers of contraband. Food traceability is too serious and strategic to entrust it solely to private and particular interests.

6. From the French Experience in Food Labelling or "From Labeling To Tag Labels"?

The free movement of people and their products has always been a major issue in the history of mankind and our times.

A Recent French Experience: Food labeling in the retail sector. In this section, four different food labelling systems will be presented and compared:

- A. Labeling of mass distribution,
- B. Labeling of agri-food products,
- C. Labeling using "Tri colored lights" system used in the United Kingdom,
- D. Labeling using the five colors developed by the researchers of Paris XIII/Inserm (Nutri-score).

The multiplicity of systems shows once again:

- The situation changes according to the profile and context of each actor,
- The non-integration of actors depends on the economic activity for the same product,
- The divergence of interests, caused by the will to defend the sectoral interests, personal, national, pressure groups, etc.

Aims and Objectives: Combat the increase in cardiovascular diseases, diabetes and obesity that affects more than 18% of the French population. The struggle involves a concerted effort by several actors: the ministry of health, agro-food industries, mass distribution, research organizations and associations involved in food safety and consumer information.

Methodology: The operation consisted of carrying out a full-scale experiment of a nutritional labeling on products sold in the supermarkets, in about forty stores, in three different commercial signs.

Actors: The Ministry of Health, GD: Directorate General of Health, INSERM/Paris XIII University, EREN: Nutrition Epidemiology Research Team (University of Paris XIII) Avicenne Hospital, OPENFOODFACTS association; BDD collaborative free and open food products from around the world FFAS: French Fund for Food and Health, financed by the agro-food industries, Big distributers and supermarkets, including: Casino, Carrefour, Auchan Simply Market, etc., FOODWATCH, The Scientific Committee set up by the Ministry of Health to monitor, the in-store evaluation of nutritional labeling systems (13 members initially), including nutrition specialists, food science specialists.

Facts:

- The products concerned were catered products, ready-made meals, breads, pastries and industrial pastries,
- The testing process has been criticized throughout the experiment,
- The lack of trust between the various stakeholders with opposing interests has cast suspicion resignation in a cascade of six out of 13 members in six months of existence, with reasons including the non-scientific nature of this study,
- Members of this committee had proven links with the Food and Agriculture Industries and in particular with the FFAS (French fund for food and health).

Conclusion: Despite the failure, this experience brings a benefit to all actors and consumers.

It is necessary that a national, regional or global public actor imposes the procedures, and the rules of the game. It is advisable to consult with the essential private actors. But it is not normal to ask for a share of participation from these private actors. Actions carried out for years, with public funding, work of researchers, provide a compulsory and salutary passage. The INSERM (National Institute of Health and Medical Research) Data Base seems like a very interesting embryo of work for the continuation of the operations.

7. Accessibility and Democratization

Major technological advances over the last ten years have allowed development of new tools, major advances and new perspectives. A phenomenon of global democratization aimed at given access to technological tools has radically transformed the approach and the accompaniment of food traceability. Open operating systems, the democratization of "smart phones" genuine convergence of several computer technologies, internet and telephone communication have put powerful tools of traceability at the disposal of the economic actors. We will note that on the traceability table above, the boundaries between different sectors concerning the life of a product are a little more porous today, allowing a better follow-up of the life of a product through the complex movement of this product.

8. Conclusions

Integral and total traceability will not be possible in our world filled with a multitude of natural or artificial obstacles. In the best case, it will be only partial. Faced with the structural blockades listed in our chapter, linking private and particular interests, closure of intra-state, sectoral administrative boundaries, business corporatism, more global private structures as well as national states or regional entities, food traceability will always be tossed by the winds of societal evolutions, often dismissed by private interests, but never succeeding properly. In line with the major evolutions in our societies over the last few years, traceability will move towards greater transparency and integrity and thus a concept of "Trans-traceability" (Transparency-Traceability) will have to be invented. Tracing and drawing would be a perceptible future in the next decade and be likely a "technical" response to the current insurmountable complexity.

A double revolution in progress

Increased convergence and revolutionary advances in computing, bio-surgery, nanotechnology, cybernetics, artificial intelligence, medicine, the explosion of connected objects, and so many others will shake up scientific approaches and the philosophy of tracers. NBIC, namely nanotechnologies, biotechnologies, new generation computing—from Big Data to connected objects, cognitivism (Robotics: artificial intelligence; robotic intelligence) are underway and are already revolutionizing our societies and lives. With the development of "GAFA" (Google, Apple, Facebook and Amazon) and twitter, Microsoft and many others will facilitate, thanks to their world presences, the fabrics of the Big data's accessibility in real time, that could be used to improve Traceability. A new form of traceability, the "Trans-traceability" will probably emerge. It will revolutionize our different approaches and transgress the present physical and perceptual boundaries. There will remain the moral will of integrity that no system can fill, and which will be the prerogative of our human will. What human beings have not been able or are willing to achieve in the last fifty years on food traceability, a "trans-human" society will probably emerge, but will it?

Keywords: Food traceability, avian flu, scandals, food allergy, food crises, food intolerances, cigarettes

References

We have read and analyzed many articles and information over the years. We have not taken sentences or paragraphs but have synthesized and analyzed a lot of information that has come to support our explanation.

Anshuman, J.D., Akshat, W., Ishan, K. and Raskar, R. 2016. Ultra-portable, wireless smartphone spectrometer for rapid, non-destructive testing of fruit ripeness. Nature Research. www.nature.com.

Commonwealth Scientific and Industrial Research Organisation, Australian Government http://www.csiro.au/

Dr. Vera Luisa da Costa e Silva, Responsible for tobacco control at WHO.

Food DNA Project, Internal project in the R & D department of UBIQUE Group.

Food Safety Authority of Ireland, www.fsai.ie

Foodsafety Magazine, www.foodsafetymagazine.com

Foodsafety News, www.foodsafetynews.com

Foodonline, www.foodonline.com

Food News International, www.http://foodnewsinternational.com/

Global Food Traceability Center, www.ift.org

Groupe UBIQUE, 8 rue de l'Abbé Hazard, 92 000 Nanterre, France. www.ubique.fr

IQoo, www.iqoo.fr

QUALIBOO, www.qualiboo.eu

Mme Emmanuelle Béguinot, Director of the National Committee Against Tobacco (CNCT, France).

Mme Anne-Laure Barrett, Investigative journalist at the Journal du Dimanche, notably on the actions of the cigarettiers.

Mme Pascale Santi, journalist at the newspaper Le Monde, notably on public health issues.

Nestlé Group, www.nestle.com

Néo-Food Project France-Europe, Internal project in the R & D department of UBIQUE Group.

3

Traceability in French and European Law

Henri Temple

1. Introduction

The food rules are as old as the first exchanges of foodstuffs between fishermen, pastors, farmers, and the artisans of early inhabitances. In France, the Act on Frauds established on the 1st August 1905, has long served as a model for other European legislation (French regulation on frauds and falsifications 1905). The origin of the product appeared as one of the key components for which the lack of trustworthiness was the cause of criminal penalties. Therefore, traceability was practiced in foods with a registered designated origin.

Then the European Community regulations were published commencing with the General Food law in 2002 (EC 2002) following on from the directives for nutrition labelling for foodstuffs (EC 1990). These Directives or Regulations are now commonplace for food specialists, and have brought even more detail to the "legal corpus", common to the European countries, which is being built incrementally. Bad habits were taken at that time (since 2005) as specialists were faced with the risk of "submersion" in this deluge of increasingly complex, numerous and

14 Bd du Jeu de Paume 34 000 Montpellier, France.
  E-mail: icabe.temple@free.fr

voluminous texts that are often changing. Nonetheless, the modern notion of traceability appeared in these European texts. However, technicians, scientists and businessmen understood that the Standards whether those provided by the International Organization for Standardization (ISO) or Norme française (NF) types were on the same level of coercion as the technical rules in the European regulations and that the special rules were sufficient and predominant on the general legal principles. Since this recent period of food scandals, business leaders have been brought to justice on the basis of their criminal responsibility. Some of them had limited to no insurance cover or whereby the activities conducted voided their policy with the consequence of bankruptcy for the company.

The general food law provided as European Regulation 178/2002 of 28 January 2002 (EC 2002) and the numerous accompanying texts as "hygiene package" (Table 1) provides the key requirements for implementation by all the dimensions of the agro-food business, incorporating by examples the necessary multidisciplinary collaborations of scientists, salesmen, logistics, computer specialists and specialized lawyers. The latter have an irreplaceable role in assessing the nature, the scope and the meaning of the substantive rules, the connections between them, their sanctions and the risks and judicial techniques. Nonetheless, lawyers may lack

Table 1. Key regulations in Europe Adapted from Bernd et al. 2013.

Year	Regulation
2002	Regulation 178/2002 (GFL)
2003	Regulations 1829/2003 and 1830/2003: GMO package
2004	Regulations 852–854/2004 Hygiene package
	Regulation 882/2004 Official feed and food controls
	Regulation 1935/2004 Food contact materials
2005	Regulation 2073/2005 Microbiological Criteria for Foodstuff
2005	Allergen labelling requirements included in Directive 2000/13
2006	Commission Regulation (EC) No 1881/2006: MRLs for certain contaminants
2006	Regulation 1924/2006 Nutrition and health claims
2007	White Paper A Strategy for Europe on Nutrition, Overweight and Obesity related health issues
2008	Regulations 1331–1334/2008: Food Improvement Agents Package (FIAP); additives, flavourings and enzymes
2011	Regulation 1169/2011 Food information to consumers
2013	Commission Regulation (EU) No 1337/2013 has been adopted setting out the modalities requiring the indication of the place of rearing and the place of slaughter for prepacked fresh, chilled and frozen meat of swine, sheep, goats and poultry

appreciation of the technical application of these texts, or the conformity of the company to perform these practices relative to the legal requirements. These requirements within the European general food law (Regulation 178/2002) resulted from the combination of the national laws of each EU country. Thus, for France, the national requirement is the French Code on Consumers Rights.

Modern rules of traceability are mandatory, legally sanctioned under criminal law, and above all, breaches of these rules can lead to the bankruptcy and closure of the company. However, to understand fully the mandatory rules of traceability these must be described within the legal context of the food law.

2. The Legal Context of Traceability

It is essential to know the legal context of traceability to enable the officially authorized implementation. However, the diversity, and even the dispersion, of the texts of food law were due to the disparity of sources, more precisely of the political authorities, who initiated the texts.

2.1 At the international level

The authorities and global sources at the international level are the World Health Organization (WHO), Food and Agriculture Organization (FAO), the World Trade Organization (WTO) and ISO for the provision of standards. Its European counterpart, CEN (European Committee for Standardization), is intervening to standardize the food sector (EN WS/86).

Local authorities and European sources have a large number of Directives and Regulations. It should be made clear that the directives do not have a direct internal effect and that it is for each EU member State to determine, under the control of the Commission and possibly the Court of Justice, how and what type of legislative instrument could be used to effect the national transposition. As for the regulations, apparently their application in the Member States is easier since transposition is not necessary, they are compulsory *ipso jure*, directly in the 27 national legal systems. In the case of the United Kingdom which is in the processing of leaving the EU (BREXIT), it will retain most of the Community legal acquis, including the texts on traceability.

However, in each European country the administrative and judicial application of traceability is complex when a national criminal enforcement of a Community regulation is pursued.

2.2 At the national level

National texts are used for the application, monitoring and sanctioning of traceability. Thus, in France, compulsory texts were first known: the highest figure is the constitution, then the **Acts,** voted by Parliament, and which alone has sufficient force to institute criminal offenses. The Decrees on the application of laws have much less authority than laws, from which they derive their legitimacy, and which they have the function of supplementing with a view to their concrete application. Criminal proceedings based on the infringement of a Decree (or Orders) could fail if the person brought to justice availed himself of the objection of illegality, admissible in defense. There are also ministerial Decrees (Orders) which are not directly linked to the implementation of a law (for example, Decrees that temporarily suspend the marketing of a foodstuff on national territory). Finally, there are ministerial, departmental, or municipal decrees and orders, emanating from the ministers, Prefects, Mayors, who can take measures prohibiting the sale of food.

2.2.1 The "great codes"

The above texts are often grouped by subject in codes, thus making the practice easier. Since the food law is by nature multidisciplinary, the use of several codes and uncodified laws are frequently used. Therefore, for the purpose of this article the most useful codes will be described commencing with the Consumer Code. Food law is an integral part of consumer law. It is a sub-part of product law, itself a component of consumer law. This will explain why, in many cases, the use of general rules applicable to products, and therefore also to food, will be used as for example on the traceability of meat within Article R.214-14 of the Consumer Code.

Though we must not forget other codes which contain useful provisions including traceability, such as: the Civil Code (e.g., contracts, guarantee of hidden defects, civil liability), the Rural Code (countryside) and sea fishing, the Criminal Code, the Code of Criminal Procedure, the Customs Code and so on.

All the preceding texts are either coercive, because they concern public health and fair trading, or incentives to proposing signs of quality (appellations of origin, labels, etc.) which the producers could adopt or not. However, the approach once adopted is binding on both producers and third parties, and is likely to lead to specific penalties for breaches.

2.2.2 *The voluntary standards*

Besides these texts discussed, there are also standards elaborated and supervised by the public authorities that are not applied following a voluntary option of the producer, but that are elaborated in a quasi-private approach involving producers. In France, it is the role of the Association Française de Normalisation (AFNOR) to lead and coordinate the standards development process and to promote the application of those standards. A growing number of AFNOR standards are dedicated to traceability. It is increasingly common, however, that these national standards duplicate ISO international standards. The ISO 9000 family of standards has been developed to assist organizations, of all types and sizes, to implement and operate effective quality management systems. Though if we examine the definition of "traceability" given within the ISO 9000/2005 standard, it is observed that the term is described within the framework of quality management as the ability to trace the history, application or location of that which is under consideration (ISO 9000/2005). Therefore, within the framework of this standard, traceability must not be separated from the technical operations related to it such as computer monitoring, attestations, checks and self-checks (or self-monitoring), calibrations and verifications. Although this standard defining traceability is the one preferred it lacks sufficient precision.

Thus, another standard defining the international vocabulary of the basic and general terms of metrology (VIM) defines: metrological traceability as the *"properties of measurement according to which this result can be connected to a reference through an uninterrupted and documented chain of calibrations each of which contributes to the uncertainty of measurement"* (NF ISO/IEC GUIDE 99: 2011 2.41: 6.10).

It is necessary to specify, from the perspective of techniques and management, these two convergent but distinct aspects of traceability. Technical traceability is always ensured by a connection using an uninterrupted chain from the previous place to the next place. Documentary traceability is ensured by compliance with the requirements of quality insurance standards such as those of the NF EN ISO/IEC 17025 standard. The information provided by the standardization bodies, can be misleading and misrepresentative and thereby extremely detrimental to the supply chain when *"Technical traceability is only to be ensured when the company cannot demonstrate technically that the absence of traceability has no influence on the result of the measurements or their associated uncertainty (see NF EN ISO/IEC 17025, §5.4.6). A preliminary assessment of the need and the causes of uncertainty is the best way to demonstrate the influence of a non-connection". Traceability: Ability to retrieve the history, implementation or location of what is being reviewed.*

3. Complexity and central principles (Temple 2013)

The complicatedness, stacking and overabundance of sources and texts inevitably add additional complexity to the understanding of traceability especially because it is necessary to combine these different rules for implementation. Importantly, dynamic decisions will be required in order to avoid a technical drift that would lead to unrecognized or illegal practices such as in the case of horsemeat additions within beef lasagna. This will require an in depth understanding of the hierarchy of legal sources, the right choice of the applicable rule as well as its conditions and methods of application. Indeed, the options taken by professionals in terms of production, processing, packaging, labeling, traceability, in summary, all the management of the food supply chain, may be challenged in the future through the obligations of the French or foreign administration and/or the judicial bodies, to the resentment of the supplier, the customer, the consumer and the professional partners (e.g., insurers-banks).

It is not unnecessary to recall the imprudence of certain non-lawyers who would venture into these tasks, committing their personal services in the event of misconduct in the advice given. Let us reiterate that multidisciplinary cooperation and transparency is now necessary.

3.1 Unifying central principles

Fortunately for lawyers, there are key obligations required of food and feed business operators for understanding a "guide-plan" for intellectually classifying and prioritizing this mass of rules towards traceability (Table 2). Among these key areas we shall merely recall the seven general

Table 2. Seven key obligations of food business operators in Europe from EU regulation 178/2002.

Obligation	Action
Safety	Operators shall not place on the market unsafe food or feed
Responsibility	Operators are responsible for the safety of the food and feed which they produce, transport, store or sell
Traceability	Operators shall be able to rapidly identify any supplier or consignee
Transparency	Operators shall immediately inform the competent authorities if they have a reason to believe that their food or feed is not safe
Emergency	Operators shall immediately withdraw food or feed from the market if they have a reason to believe that it is unsafe
Prevention	Operators shall identify and regularly review the critical points in their processes and ensure that controls are applied at these points
Co-operation	Operators shall co-operate with the competent authorities in actions taken to reduce risks

obligations contained in the Consumer Code, the Civil Code, and Regulation 178/2002, which weigh on producers, *lato sensu*, of food. These include the general obligations for the provision of information and transparency which incorporates traceability in part; the general safety obligations in particular reference to food safety and its determinants such as traceability for prevention and emergency withdrawal-recall; the general obligation of conformity and co-operation (to the texts) and its consequences such as self-monitoring, technological and legal supervision with respect to traceability and the general obligation of responsibility with reference to product liability and its insurance consequences.

The same principle is fixed in many legal systems and affects, as well, food and non-food: business operators are requested to place safe products on the market. It is the responsibility of the business operator (art. 14, EU R.178/2002 for food; art.3.1, EU D.95/2001 for non-food) (Marchalant 2013).

3.1.1 *Risk analysis, assessment, management* (Marchalant 2013)

To be deemed safe a product must undergo a risk analysis process by the operator. Modern legislations and food safety authorities all invoke this approach throughout the food supply chain (ex: art 6, EU Regulation 178/2002). The risk assessment is generally performed through the application of a Hazard Analysis Critical Control Point (HACCP) evaluation as a standard principle. This allows the risks to be identified and evaluated at different stages of the supply chain whereby suitable risk management procedures can be implemented. Though in developing countries these principles are often ignored by food operators.

3.1.2 *Precautionary principle* (Godard 2013)

Modern legislation establishes the principle of precaution (ex: EU, art. 7 Reg. 178/2002), and the WTO mentions this as problematic in relation to trading practices (SPS agreement, art.7.7). In the case of a scientific uncertainty about the safety of a product, a state can suspend the commercialization, for a limited time, of a good (food and non-food). Exporting countries can demand justifications of these measures (Art.7.8), and, eventually complain to the WTO for a trial.

3.1.3 *Traceability* (Viruega 2009)

Although completely ignored in some developing countries traceability became an international requirement at the end of the 20th Century following the outbreaks of "mad cow" disease whereby infected meat entered the supply chain. The requirement was for any food business operator to "be able to identify any person from whom they have been supplied" as well as "the other businesses to which their products have been supplied" (R. UE 178/2002, Art.18). Thus traceability had to be "established at all stages of production, processing and distribution". A similar rule concerns the nonfood producers and retailers on the general safety of products (EU Directive 2001).

3.1.4 Self monitoring

Self-monitoring is the new and non-reversible approach of the firm's management and of the relationships between operators and administrations. It means a larger involvement of a business in its own control. Non-monitored products will be refused by more and more countries or buyers as at a higher risk to be non-conforming.

3.1.5 Conformity (Mouillet 2013)

Food operators whether at production, processing or distribution are submitted to a legal obligation to "ensure the food satisfies the requirements of the law" (art.17, R EU 178/2002). The same obligation to traceability requirements are to be complied by non-food operators (EU Directive 3rd dec. 2011, n° 2001/95).

3.1.6 Vigilance and responsibility, withdrawal, recall (Verger 2013)

If a food, or a non-food, is not complying, or is suspected of non-compliance, the operator "shall initiate procedures to withdraw the [product] in question from the market…and if necessary, recall from consumers products already supplied…" (Article 19, EU regulation 178/2002). The expectation is that there are preliminary vigilance, traceability, and ready to use procedures for recall.

3.1.7 Information (Fourgoux-Jeannin 2013)

Transparency in the provision of information for consumers and customers are legal requirements. The rules on advertisements and labeling are important as well to insure a fair competition between firms. In the case of suspicions or real concerns for unsafe or non-conforming products, the business operators must inform the competent authorities.

4. Traceability Legal Rules and Sanctions (Temple 2008)

4.1 Legal rules

The legal requirements are well defined in Regulation 178/2002 of 28 January 2002. This text should be the rigorous starting point for any act of management in the agro-enterprise food industry and any intervention by its agents. Regulation 178-2002 establishes the general principles of food law, in addition to explicitly establishing the obligation of traceability, is also very specific as to the scope and conditions of its implementation. It is a compulsory text and its non-fulfillment should be severely punished.

It should be emphasized that the regulation lays down the safety requirement at an extreme level in its assessment, since it encompasses "precaution", simple scientific uncertainty within article 7 and risk analysis to the likely effect in the long term not only on the consumer but also on their offspring taking into account probable cumulative effects or sensitivities of a specific category of consumer (UE Regulation 178/2002, Articles 14 (4), (a), (b) and (c)).

As for the requirement for food traceability, it is also detailed in its implementation whereby the actors in the sector must be able to identify any person who has supplied them with a food product (Article 18, 3). The article 18 on traceability is presented (Fig. 1). Traceability is well defined as the obligation, for all operators from producers and distributors, to be

1. *The traceability of food, feed, food-producing animals, and any other substance intended to be, or expected to be, incorporated into a food or feed shall be established at all stages of production, processing and distribution.*

2. *Food and feed business operators shall be able to identify any person from whom they have been supplied with a food, a feed, a food-producing animal, or any substance intended to be, or expected to be, incorporated into a food or feed.*

3. *To this end, such operators shall have in place systems and procedures which allow for this information to be made available to the competent authorities on demand.*

4. *Food and feed business operators shall have in place systems and procedures to identify the other businesses to which their products have been supplied. This information shall be made available to the competent authorities on demand.*

5. *Food or feed which is placed on the market or is likely to be placed on the market in the Community shall be adequately labelled or identified to facilitate its traceability, through relevant documentation or information in accordance with the relevant requirements of more specific provisions.*

6. *Provisions for the purpose of applying the requirements of this Article in respect of specific sectors may be adopted in accordance with the procedure laid down in Article 58 (2)*

Figure 1. Article 18 Traceability.

able to identify any person from whom they have been supplied with a food, a feed, a food-producing animal, or any substance intended to be, or expected to be, incorporated into a food or feed.

This requires having an established traceability management system in place as systems and procedures to identify the other businesses to which their products have been supplied. This information should be made available to the competent authorities on demand. Therefore, the mapping of a business's supply chain is essential. Such a system should not be limited to computerized tracing processes or even fine-tuned analyzers. It is clear then that the drafting of consecutive sales contracts for food products will solicit the creative talent of corporate lawyers.

The traceability requirement for a business, in addition to supplier companies, extends even to the final consumer of food. Indeed, Article 19 of the Regulation 178/2002 requires the trader to recall the products already supplied to the consumers whereby they shall immediately initiate procedures to withdraw the food from the market if deemed non-conforming or unsafe. This in itself extends the pathway of the food supply chain (Article 19, 2) to trace foods to the final consumers (Article 19, 3) from the primary producer (Viruega et al. 2005, 2007). It has thus been well understood that the logistical and computational legal structure of such an obligation between and within each point in the supply chain is a technically formidable exercise (Viruega 2004).

It is also necessary to mention, as comparison to their highly technical nature, French regulatory requirements specific to the traceability of certain products. Similar requirements apply to the traceability of beef (Rural Code Article L232-1-1), which is based largely on the materials and processes used to identify livestock. AFNOR standards complement and enforce obligations. Similar traceability regulations are applied for genetically modified organisms (GMOs) whereby the traceability measures enforced for these organisms enables enhanced biological monitoring of the territory. In this case, protection is focused more on the environment than on consumption for food safety (Law n ° 99-574 of 9 July 1999). Finally, there are global applications utilizing traceability in many sectors including cosmetics, medicines, chemicals and aeronautical engineering.

4.2 Penalties for inadequate traceability

The penalties for lack of traceability and defective traceability vary considerably depending on the damage or the risk involved. The professional whether primary producers, processors or distributors must carry out the operations allowing complete traceability such as self-monitoring, and take the necessary measures with to withdraw or recall if the product is dangerous or non-compliant.

4.2.1 The legal basis for penalties

The foundations for the penalties that may be imposed are the result of European texts, in particular Regulation 178/2002, Article 17 which lays down the responsibilities for feed and food business operators (Fig. 2).

It can be deemed that insufficient traceability may be at least in part the cause of detrimental health effects for the consumer. In France, this will entail the risk of serious penal sanctions, the textual basis of which may vary (from 222-19 to R 622-1 of the Penal Code), with a maximum of two years' imprisonment and complementary penalties (221-8 et seq.), some of which strike the legal representatives of the organizations involved (221-7, 222-21). It is stated that by causing to another person total incapacity to work for more than three months due to a breach of an obligation of security or prudence imposed by law or regulation is punishable by two years of imprisonment. If this breach is manifestly deliberate, these aggravating circumstances justify a three-year prison sentence (article L 222-19 of the Criminal Code).

Moreover, the offenses of deception through falsification that are still in force could lead to even heavier penalties. Similarly, for a business to deceive its customer or consumer on the origin, substantive qualities, suitability for use or quality control checks performed is punishable by

---

1. *Food and feed business operators at all stages of production, processing and distribution within the businesses under their control shall ensure that foods or feeds satisfy the requirements of food law which are relevant to their activities and shall verify that such requirements are met.*

2. *Member States shall enforce food law, and monitor and verify that the relevant requirements of food law are fulfilled by food and feed business operators at all stages of production, processing and distribution.*

3. *For that purpose, they shall maintain a system of official controls and other activities as appropriate to the circumstances, including public communication on food and feed safety and risk, food and feed safety surveillance and other monitoring activities covering all stages of production, processing and distribution.*

4. *Member States shall also lay down the rules on measures and penalties applicable to infringements of food and feed law. The measures and penalties provided for shall be effective, proportionate and dissuasive.*

Figure 2. Article 17 Responsibilities.

two years' imprisonment (article L213-1 of the Consumers Rights Code). Obviously, the sale of products supposedly traced when this is not the case would constitute a deception. However, if this deception should enable the use of the goods dangerous to human or animal health the penalty will be doubled to four years in prison.

Legal entities such as commercial companies may be closed or prohibited from conducting economic or commercial activities. Likewise the managers or directors of such companies may be prohibited from these activities. Though if the damage to health has not been realized, the offense of endangering the life of others consists solely of the placing on the market of a product which transgresses a particular obligation of safety or for which there is no reason why traceability should not be included (Article 223-1 of the Criminal Code). In additional to legal actions, non-traceability can also cause civil actions to companies in the sector. The French Consumers Code states that the products must comply with the requirements in force concerning human health, fair trading and consumer protection at the time of first placing on the market (Article L212, 1). There is no doubt that traceability is part of the prescriptions in force. The consumer, moreover, in the same way as the professional buyer (importer, wholesaler, distributor), could propose the nullity of the contract by the illegality of the object (articles 1134 et seq. of the Civil Code).

4.2.2 Penalties and legal actions

When people says right, by essence and always, conducts to potential sanction. This potentiality (which weighs at all times on the choices and practices of the food sector) finds expression in legal actions.

4.2.3 Sanctions

The sanctions of the food law, including traceability deficiencies, are of three legal types not forgetting the economic repercussions. There are criminal and administrative sanctions and civil damages.

4.2.3.1 Criminal sanctions

These are the most serious because they can infringe on the freedom of the entrepreneur or his accomplices or colleagues. The prison sentence as described may be up to four years in the case of aggravating circumstances is provided for in numerous texts not only applicable in the case of activities dangerous to public health but also for fraudulent activities. Now, contrary to the first sentiment on this subject, it is rather frequent that the sentence imposed is pronounced as a suspended sentence not imprisonment. There are many examples. A fine may be used to supplement or complement imprisonment, but one must not lose sight of the so-called complementary or ancillary penalties such as prohibition of running a business, publication in the press losing brand reputation, or closure.

4.2.3.2 Administrative sanctions

These are more serious for the company than for its leader and include the provisional or final closure of the establishment; provisional and targeted ban on the marketing of a product through withdrawal or recall; customs seizure of goods whether for export or import; refusal or withdrawal of an authorization or administrative authorization and the definitive ban on a food component

4.2.3.3 Civil damages

The purpose of civil damages is to stop or compensate for damage to the affected person or persons. In particular, the harm caused to a consumer but also to business partners such as processors or distributors. Thus Articles 1386-1 et seq. of the Civil Code establish, in accordance with the European Directive of 25 July 1985, an automatic system of no-fault liability, favorable to victims for this reason. The new system (law of 19 May 1998) and derogation from the common law of civil liability rests on two simple ideas that firstly different legal rules should not be applied to the victims of a product according to whether it is a contracting party or not and secondly that the victim no longer has to prove the fault of the producer but only the defect of the product, the damage suffered and the link between the defect and the alleged damage.

However, the system established is, in its implementation, too complex to be explained here in detail (Calais-Auloy and Temple 2010). In particular, questions of prescription and development risks have given rise to doctrinal doubts and contested judicial decisions (Temple, note under ECJ 2 Dec. 2009, Lamy Dr. des Affaires 2010 p.40). This complexity is compounded by the fact that it may be necessary to combine this responsibility for faulty products with the latent defects systems of the civil code and consumer code (Calais-Auloy and Temple 2010) and for problems in the case of international sale and for the determination of the applicable law applying The Hague Convention of 2 October 1973.

However, civil sanctions can also be aimed at compensating economic operators when they are the victims of a defect, non-security or non-conformity of food goods imported or purchased on the domestic market. Parties to a food sales contract may invoke non-compliance, hidden defects, civil or criminal fraud as by deception either in proceedings before the criminal courts or before the civil or commercial courts. These actions may lead to either obtaining the nullity of the contract with reimbursement of the price and costs or the resolution for non-performance of this contract. The claimant may be awarded damages from the judge covering the whole of the damage. If the non-compliance to the traceability was intentional one can even invoke the civil fraud or the penal fraud. There are a few cases in the (Supreme) Court of Cassation rulings which have ruled on penal sanctions. However, a judgment of 2015 had on this occasion a case concerning minced meat unsuitable for consumers through *E. coli* O157, H7 contamination as a major food safety hazard. The High Court relies on Article 19 of Regulation 178/2002 to require traders to withdraw unsuitable lots from the market. It noted that the traceability of the lots had been poorly carried out, as the accused themselves admitted, and that the operator had waited to inform customers of the withdrawal, while the guide to withdrawal prepared by the Ministry gave clear instructions. The Court of Cassation thus confirmed the criminal convictions: € 30,000 in fines and, above all, the inclusion in the mainstream newspapers of a far-reaching commercial sanction (Cass. Crim., 27 Oct. 2015, n°. 4475).

5. Impact assessment of traceability for the commercial management of companies

There is now a requirement to assess through the interactions of multiple disciplines the impact of traceability in the commercial management of companies in their relations with the upstream and downstream links of the sector. Though there are many questions posed in how this should be performed. Though it is recognized that there should be a new culture adopted within the organization of food companies.

5.1 New culture and organization of food companies

It is absolutely essential to insist, at first glance, on the radical change in the risk management culture that commenced with the general food law at the turn of the millennium. Indeed, companies must now self-monitor the products, their safety and their compliance. The role of the administration, if it still applies the established traditional role is, above all, to verify the existence and relevance of private self-control systems in and in relation to the company (argument L212-1 Code of Consummation; R. 178-2002, passim, v 17, 18, 19).

5.2 Practical considerations for traceability

There are at this time many questions being posed for how in practice the self-control of traceability can be assured during the previous supply steps and on the following economic pathway of the product thereby throughout the food supply chain. As described herein the inclusion of a contract linking each link of the chain to the previous link, or to the next link should be a clause (term) of traceability. Of course, it is not a question of removing business secrets and forcing partners to deliver their address books. However, there is a necessity for business operators to commit themselves to being able, according to internal management protocols whether through the use of barcodes, batches, chips, RFID tags to demonstrate validated procedures by specialized auditors and patented in traceability, to trace the product.

This commitment of traceability includes, as infinitely as the mirrors in the mirrors, the obligation to acquire, then to transmit only products traced *ab origine*, throughout their journey, including their inputs. Scientific methods of verification such as analyzes, genetics, geology, chemistry, electron microscopy, spectroscopy techniques including nuclear magnetic resonance spectroscopy (NMR) are now available to detect frauds or approximations on the exact origins of the product. There is the recognition that downstream traceability, or tracking, within a supply chain is more complex and random. The producer will perform sufficient surveys on the evolution of his products on the shelves of the distributor. However, at the end of the food chain, the distributors or retailers can find the consumers who paid by check or by card but it will be more difficult to locate the passing tourist who settled in cash.

Two other sensitive issues regarding imported products and internal traceability within the company should be discussed with regards to the implementation of traceability. There needs to be consideration of imported products that are also subject to the obligation of traceability for the business. Difficulties in enforcement can arise as French or EU Community laws do not apply outside their domain directly to foreign operators. It is through the result of the contractual clauses that the importer will impose EU requirements on its supplier to prove that it has in place, in its country, traceability and self-checking protocols identical to the requirements laid down in the countries of implementation on the market. In many foreign countries, exporters have already complied. The other sensitive issue is internal traceability within the company and whereby there is still the obligation of traceability (28th recital, EC 178/2002) even if the company repackages or transforms. Businesses must put in place an efficient system and manage their policy of identifying lots of goods.

5.3 Economic impacts

The economic impacts may be more severe than the legal sanctions themselves, but of which they are the consequently induced. The effects can go as far as the closure of the company or its redemption by a competitor who profits from its indiscretions and loss of creditability. Lawsuits can come from victims, consumers and consumer associations, but also from the supervisory authorities, and even from competitors or partners, upstream or downstream of the sector. These lawsuits can have commercial and industrial repercussions in France or internationally and paralyze the company, whereby others then take up its activity. All sanctions referred to herein are either the result of judicial proceedings or lead to their initiation. Legal proceedings that result in sanctions are far too technical to be covered in full as there are several levels of jurisdiction, civil, criminal, and administrative, that may involve issues of traceability. A great procedural difficulty is due to the rigidity of the administrative order with the judicial order. For example, if a problem is of two kinds at the same time, because two public services (e.g., veterinary officers or IFREMER analysts) are involved in the occurrence of damage due to a food, it is not possible to manage the dispute in its factual coherence with a unique judge (for example, Court of Appeal Toulouse 22 Feb. 2000, JCP 2000, II, 10429).

5.4 Methodology of international conflicts in traceability

Increasingly widespread import and export activities are leading to increasingly frequent international litigation. For those who are interested in consumers, there is pessimism that they do not have effective means of recourse (Calais-Auloy and Temple 2010), because the problems of the choice of the territorially competent court (respondent or applicant) is complicated with that of determining the applicable law whether national or foreign and then the international execution of the decision. It can take more than 10 years to get justice! For transactions between professionals, things are more convenient because they have appropriate advices and technical means. They may use, *inter alia*, techniques prohibited to consumers (to protect them): mediation or arbitration, or the prerogative clause of jurisdiction which designates the court of a third country (i.e., Switzerland, UK, France, Singapore, Ivory Coast). These questions of private international law and arbitration are regarded by experienced lawyers as the most complex and is therefore only mentioned here in reference.

6. Conclusion

Traceability issues are extremely complex because they involve difficult legal rules that combine international, European and national texts. They are poorly understood, not well communicated and inadequately applied. Yet they presuppose a true revolution in the company's practices, be it its legal, commercial, scientific and technical levels. The challenge of traceability is not in terms of the knowledge of the additional benefits but more radically in the awareness of the terms of heavy losses in the event of a food crisis, reputational damage, even the closure of the company and the condemnation of its leaders. Unfortunately the companies, even in France, have a great ignorance of the implications of the legal risks and an astonishing underestimation of the difficulty of the subject.

Keywords: Food traceability, France and Europe, mandatory rules, sanctions

References

Calais-Auloy, J. and Temple, H. 2015. Droit de la consommation, Dalloz edition 9th edition. 722 pages.

Depincé, M. 2013. Le point de vue juridique in Multon, J.L, Temple, H., Viruega, J.L. Traité pratique de droit alimentaire, Lavoisier edition, Tec et Doc, pp. 745–765.

Directive 2001/95/EC of the European Parliament and of the Council of 3 December 2001 on general product safety (OJ L 11 of 15 January 2002).

EN WS/86 - Authenticity in the feed and food chain – General principles and basic requirements.

Fourgoux-Jeannin, M.-V. 2013. L'étiquetage alimentaire in Multon, J.L, Temple, H., Viruega, J.L. Traité pratique de droit alimentaire, Lavoisier edition, Tec et Doc, pp. 477–493.

Godard, O. 2013. Le point de vue scientifique : entre doctrine sociopolitique et théorie de la décision in Multon, J.L, Temple, H., Viruega, J.L. Traité pratique de droit alimentaire, Lavoisier edition, Tec et Doc, pp. 709–742.

Ilhe, L. 2013. L'organisation de l'entreprise pour se prémunir contre la survenance d'une crise alimentaire in Multon, J.L, Temple, H., Viruega, J.L. Traité pratique de droit alimentaire, Lavoisier edition, Tec et Doc, pp. 1077–1089.

Marchalant, K. 2013. Les obligations générales et leurs applications techniques, in Multon, J.L, Temple, H., Viruega, J.L. Traité pratique de droit alimentaire, Lavoisier edition, Tec et Doc, pp. 295–298.

Marchalant, K. 2013. L'hygiène des aliments, in Multon, J.L, Temple, H., Viruega, J.L. Traité pratique de droit alimentaire, Lavoisier edition, Tec et Doc, pp. 303–326.

Mouillet, L. 2013. Concepts relatifs à la qualité et à l'assurance qualité, in Multon, J.L, Temple, H., Viruega, J.L. Traité pratique de droit alimentaire, Lavoisier edition, Tec et Doc, pp. 343–363.

Regulation EC No 178/2002 of the European Parliament and of the Council of 28 January 2002 laying down the general principles and requirements of food law, establishing the European Food Safety Authority and laying down procedures in matters of food safety.

Temple, H. La loi du 1er août 1905 sur les fraudes et falsifications. Protection des consommateurs. Doc.fr.2007, p.184

Temple, H. 2008. La traçabilité des produits alimentaires et non alimentaires, Revue des techniques de l'ingénieur.

Temple, note under ECJ 2 Dec. 2009, Lamy Droit des Affaires, 2010 p. 40.

Temple, H. 2013. Les règles juridiques de fond du droit alimentaire in Multon, J.L, Temple, H., Viruega, J.L. Traité pratique de droit alimentaire, Lavoisier edition, Tec et Doc, pp. 75–84.

The Hague Convention of 2 October 1973 on the Recognition and Enforcement of Decisions Relating to Maintenance Obligations.

Verger, P. 2013. Analyse des risques et anticipation des crises in Multon, J.L, Temple, H., Viruega, J.L. Traité pratique de droit alimentaire, Lavoisier edition, Tec et Doc, pp. 987–999.

Viruega, J.-L. 2004. Le nouveau droit alimentaire, colloque CIRAD, Faculté de pharmacie, CDCM, Faculté de droit de Montpellier 2005, t.1 et 2007, t.2. Traçabilité, Editions d'organisation.

Viruega, J.-L. 2013. La traçabilité, in Multon, J.L, Temple, H., Viruega, J.L. Traité pratique de droit alimentaire, Lavoisier edition, Tec et Doc, pp. 377–428.

Emerging Trends in Traceability Techniques in Food Systems

Anil Kumar Anal, Muhammad Bilal Sadiq* and
Manisha Singh*

1. Introduction

In the entire food supply chain, the major concern of consumers is increasingly focused on safety levels and transparency. At the same time, trade of food is rapidly getting internationalized by food industries. The production, distribution and trade of food products have greatly been influenced during the past decade due to consumer and industrial concerns (Donnelly et al. 2013). Public sectors and concerned policy makers revise the regulations for food safety in response to these changes, based on the resources provided by international regulatory bodies such as World Organization for Animal Health and *Codex Alimentarius* (Trienekens and Zuurbier 2008). In this scenario, traceability is an important tool, because the traceability practices differ among the countries. This can be elaborated by the following example, the trade, production and distribution of food products in Europe need to comply with the "European General Food Law" (European Commission 2002). European General Food Law defines the traceability as "the ability to trace and follow a food, feed, food-producing animal or substance intended to be, or expected to be incorporated into a

Food Engineering and Bioprocess Technology, Asian Institute of Technology, Bangkok, Thailand; E-mail: m.bilalsadiq@hotmail.com; manshu_123@hotmail.com
* Corresponding author: anilkumar@ait.asia

food or feed, through all stages of production and distribution", however this law does not specify the methods for keeping and validating the records (Canavari et al. 2010). This fact offers a business opportunity to tailor the methods of traceability based on demands and resources that leads to the development of multiple traceability guidelines and standards (Karlsen et al. 2013). The nature of traceability is determined by food business in practice, particular stage in supply chain and applicable legislation based on the origin (ECR Europe 2004). In an organization dimensions of traceability can be summarized in three steps. First, precision that presents the bulk of a traceable batch that is a unique identity ranging from a single product to whole a batch or lot. This step is named as granularity and serves as a key to traceability (Golan et al. 2004, Karlsen et al. 2013).

The collected information that is linked with the product lot or batch is referred as breadth and is the second dimension in the traceability. Depth is the third dimension that determines how regularly the system traces the information upstream and downstream in a food supply chain.

A detailed level of traceability is not confined to a single firm, but the effective tracking and tracing techniques in a food supply chain relies on the mutual consensus among the group of industries. The lack of transparency at any stage affects the whole food supply chain management system. The automation in data collection and record management elevates the precision level and consistency of identification of the individual product or lot to be traced at any stage of the food supply chain. The devices and technologies are continuously improved to depict the actual traceability. Optical systems (bar code, data matrix, quick response QR code) and radio frequency identification devices (RFID) have been implemented in various food supply chain systems (Costa et al. 2013).

The majority of traceability systems that are currently in practice lack the ability to connect food chain records, inaccuracy and errors in records, resulting in delaying the acquisition of required data, particularly in food-borne disease outbreaks. The recall and withdrawal information of non-consumable products should be addressed by these traceability techniques. The latest traceability techniques such as advanced RFID and DNA sequence analysis raise consumer satisfaction by presenting the actual traceability concepts (Badia-Melis et al. 2015).

2. Framework and Issues Concerning the Implementation of Traceability

The level of traceability capacity, either optimal or adequate, depends on the facilities, objectives and resources of the food processing firm and is mainly associated with cost and benefits (Golan et al. 2003). There might be a deviation from expected cost and benefit analysis in case of uncertainties for implementation of actual traceability capacity. Actual traceability can only be achieved by linking internal and external traceability (Fig. 1).

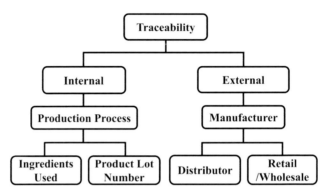

Figure 1. Internal and External Traceability Linked Approach Towards Actual Traceability.

2.1 Bulk products traceability

Various food industries use raw ingredients to formulate the product such as liquids (juices and oils), powders (coffee, dried milk, salt and sugar) or grains that are stored in large storage tanks that are rarely emptied for fresh lots, so that many lots are contemporaneously kept in same storage tanks. In case of liquid food products processing, cleaning between two lots (even the same product) is of prime importance to differentiate between product batches (Cocucci et al. 2002). The cleaning between the batches ensures cleaning and hygienic standards and is the only way to guarantee that different batches cannot contaminate each other. These cleaning processes present cost elevation for industries and are undesirable under certain conditions such as continuous batch production systems. In continuous batch processing of bulk products any interruption for cleaning would result in extra cost and delay in manufacturing (Skoglund and Dejmek 2007). However, RFID (Radio-frequency identification) markers have been developed for online traceability of continuous flows (Kvarnström and Oghazi 2008). The pills-based food grade tracer for processing units dealing with grains by introducing pill sized food grade tracer labeled with edible materials, such as cellulose or sugar particles are proposed to be put into use in grain harvesting (Lee et al. 2010, Liang et al. 2012 and 2013).

The problems concerning the fluid product traceability in case of continuous batch processing were addressed by a dynamic simulation model to elaborate the issues arising from the leftover portions of lots between different batches (Skoglund and Dejmek 2007). The concept of fuzzy traceability arises from the products developed from subsequent mixing of two different lots. The problem can be addressed by considering "composition distance" that measures the differences of products based on the contents of each supply lot. The product can be considered homogeneous (from a single lot) if the composition distance is within defined limits. This concept correlates with the regulations for the traceability of genetically modified (GM) products (European Commission 2003), that defines, the product can be labelled as GM-free if GM content is less than 0.9%.

2.2 Preservation of quality and identity to ensure traceability capacity

The recent advancements in the development of active RFID tags help in the improvement of the food supply chain system with interesting and novel solutions. These RFID tags are encoded with specific sensors (for example, temperature and humidity) for identifying the specific data along with product identification code. This advancement in the traceability system ensures the availability of automated information about product identity and related data (history, storage, etc.), hence providing the complete description of the current status of the food supply chain. This application leads to a dynamic approach to overcome the "fixed life hypothesis" of perishable items by providing real time measured data of the product (Fig. 2).

The data obtained from this traceability system can be utilized to maintain the lot size and the routing of fresh-food supplies by determining the remaining shelf life. Li et al. (2006) have recommended the dynamic planning method for reducing the loss of the product. This method is based on the linear-in-temperature approximation of the food product

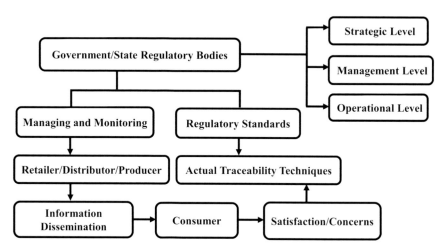

Figure 2. Conceptual Model for Information Dissemination to Achieve Consumer Satisfaction.

deterioration in which temperatures are recorded by the RFID that allows the traceability system to track the product along with its time history. Similarly, an automated traceability system has been established and validated by Abad et al. (2009) which incorporates the online traceability data and monitored cold chain conditions for application to an intercontinental fresh fish logistics chain.

Traceability along with other quality parameters like physical, chemical, microbiological, bio-molecular and organoleptic analysis can be used to achieve the product information of each lot (Fig. 3). Therefore, quality attributes should also be considered while assigning and managing the lot that will allow in variation of price based on quality characteristics (Jang and Olson 2010).

Nature of the product is another parameter that has to be taken into account in the supply chain. So, the design cannot be executed in the supply chain without considering the perishability and variability in different stages of the chain for the fresh food products (Dabbene et al. 2008a, 2008b). During the sizing and preparation of a lot, the quality and residual shelf life should be considered as they are continuously varying. This dynamic nature of the product quality should be embedded in a logistics optimization framework (Rong et al. 2011).

For facilitating the differentiation of product, a concept of identity preservation (IP) has been introduced. IP can preserve the particular traits and/or attributes and enhance the economic value to the product (Bennet 2009). There are some process or credence attributes that contribute to the product value for the consumer/buyer but are tough to be recognized and detected by them for example GMO, halal, country of origin, microbial and allergen contamination, "free-range" livestock, animal welfare, dolphin free, low carbon footprint, etc.

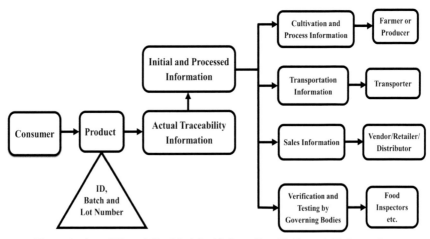

Figure 3. Actual Traceability Model with Stage Specific Essential Information.

These attributes may or may not affect the quality parameters of the product, but they are related with the enhancement of product value as perceived by the consumer (Niederhauser et al. 2008). These attributes should be guaranteed by the producers through the certification along the whole supply chain. The cost analysis and benefits from IP at supply chain level are evaluated by constructing models (Desquilbet and Bullock 2009).

Identity preservation (IP) technique is used to differentiate the genetically modified maize and soybeans varieties (*Bacillus thuringiensis* (BT) maize or Roundup-ready soybean) from that of non-genetically modified (Sobolevsky et al. 2005). Spatial is based on separation of grains in different driers and silos whereas temporal does time scheduling of the grain collection and utilization of the facilities (Coléno 2008). IP can also be implemented for separating different product lots according to their specific traits or known compositions of the ingredients in the mixture for enhancing the properties of the resulting mix. Some of the related cases are balancing protein content in flour, acidity and ethanol content in wine, quantity of lysine, oil, amylose, and extractable starch in maize (Wilson and Dahl 2008).

2.3 Prevention of fraud and anti-counterfeit concerns in the traceability system

There is an increasing trend in the cases of fraud and cheating in the field of food mostly on high value products like wine, ham, cheese, extra virgin oil, etc. These types of activities cause economic loss, unhealthy competition and negative impact on brand reputation. In such cases, a traceability system can be used to avoid and eliminate illegal, unreported and unregulated (IUU) food productions. As the traceability system can trace the history and location of the particular food entity with the help of recorded data and unique identity, it can help in preventing the fraud in food supply chain.

The tracking, tracing and identification of the product through the supply chain can be done to prevent fraud and counterfeiting by both overt and covert technologies (Li 2013, Sun et al. 2014). The data from machine-readable devices such as QR codes, barcodes, datamatrix, etc. permit the enhanced number of checks and also let the data be shared on secured networks. RFID system is one of the most reliable methods in "actual traceability system" due to unique features like automation, object tracking method and non-line-of sight identification. The authentication in RFID can be accomplished by three different ways.

2.3.1 Centralized database checking

Here the authentication of online product in real-time is done by a plausibility check of the unique identity code that is performed over internet. Although this method has higher efficiency but back-end database maintaining cost is expensive and it is tough to keep privacy levels.

2.3.2 Offline object authentication

This system is based on the encrypted tags and cryptographic algorithms carry out the authentication. For few cases, memory cards are also used to store the product information and are transported along with the entity so that the information can be retrieved on-site. Currently this type of system is being used in meat supply chains where weight-by-mass-balance control is accomplished through intelligent selling scales that will release sales receipts along with traceability information only when the carcass weight is reached to desirable value.

2.3.3 Track and trace

Track and trace method is a recent one and has more advantages for authentication of product. For any suspicious activity in the supply chain, the monitoring actions can be increased and extended so that anti-counterfeit system based on the traceability could be shared with different partners along with the customers. Jie et al. (2012) introduced this approach to Chinese authorities to utilize such a system for wine and suggested it for application to other food commodities. Similarly, Borit and Olsen (2012), proposed a traceability system for the prevention of IUU (Illegal, Unreported, and Unregulated) fishing in the supply chain of Nordic fish. While handling any anti-counterfeit case, the information used are mostly collected during the food supply chain. So, the tracking and tracing of anti-counterfeit systems will reduce the cost of methods for protection of fake and fraud activities without losing competitiveness.

3. Trends and Perspectives for Actual Traceability

For the optimized traceability system, there is a need for formulating models based on both operational and financial point of view that will help in evaluation and comparison of procedures in an integrated framework. During the optimization process, the financial gap between the investment cost for establishing and implementation of a traceability system, and the savings from the product recall should be taken into account while designing a strategic operational plan (Wang et al. 2009). Therefore, these different components of cost should be highly considered while developing optimization model so that the solution obtained will be optimal for all the cases. According to Regulation (EC) No. 178/2002 proposed by the European Commission, the main principle of traceability is based on "one-step-back-one-step-forward traceability" in which every actor in the supply chain only handles the information or data that is coming from his supplier and sent to his customer (European Commission 2002). However, to make the traceability system more effective and efficient, it should be considered and implemented at the whole supply chain level. Several companies face problems at the supply chain level in sharing of the data/information due to the lack of widely accepted standards. Maximum benefit from the system can be attained by involving all the stakeholders and by improving traceability in the entire supply chain. Similarly, data exchange can be made faster and efficient by building an inter-organizational communication and information sharing system between all the organizations across the food supply chain (Anica-Popa 2012, Karlsen and Olsen 2011). This will result in:

- the time reduction for identifying all the activities and the food processors that are involved in the food supply chain,
- the identification and elimination of potential critical points of the traceability system, and
- the implementation of more sophisticated management rules that consider the detail history and information of food product.

Extension of optimal lot size and integration of the policies from production to distribution section help in achieving these benefits. This integrated approach of combining production and distribution is a new and promising way that not only characterizes the traceability but also modern management strategies in the food production system (Amorim et al. 2012) (Fig. 4).

Practically, the quality assurance manager of the company conducts the risk analysis and determines the corresponding risk exposure. Risk may come from raw material or during processing steps. The risk exposure in raw materials can be assessed on the basis of the trust worthiness of the ingredient supplier and the possible criticality of the material. Some of the researches that are associated with the risk analysis and risk exposure are stated further. Cassin et al. (1998) developed the process risk model for the quantitative risk assessment for *E. coli* O175:H7 in beef hamburgers. Same model was also revealed by Resende-Filho and Buhr (2010) for estimating the probability of recall in a hamburger supply chain due to *E. coli* O157:H7 contamination of ground meat. The effects of both the traceability system and probable intervention on the quality control system for reduction in the costs of recall were also revealed. A risk rating parameter was proposed by Wang et al. (2009) that accounts for the probability of product recall, which is estimated on the basis of hazard analysis and critical control point (HACCP)-inspired criteria. Similarly, Tamayo et al. (2009) recommended to measure the criticality of product for projecting its current state of risk. Three different parameters: dispersion rate, trust worthiness of supplier and the residual shelf life of product are used for calculating this index.

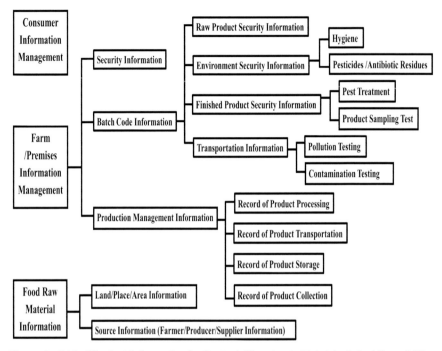

Figure 4. Critical Product Information for Strategic Planning to Maintain Actual Traceability.

Another point that needs to be discussed is the expected nature of the information/data obtained from a traceability system. It is always expected that the information from a traceability system is exact and correct. But in reality, most of the processes in the supply chain are affected by underlying stochastic phenomena that result in deviation from the exact information (Riden and Bollen 2007). The better performance can only be achieved by reducing the absolute certainty constraint and admitting tolerances expressed in probabilistic terms. This will provide a small tolerance (very fine) to the composition of an output lot. Hence, this will follow the same direction as the tolerance-based definition of GM and non-GM products as stated in EC Regulation 1829/2003 (European Commission 2003).

Till now most of the traceability optimization has been carried out as being considered static framework but it is necessary to understand that the production is a dynamic system with continuous change with time. Generally, production is always considered for a fixed time interval (hours, days or weeks) and the batch analysis determines the route of the product. Similar concept was adapted during batch processing of meat for sausage production with different compositions (Dupuy et al. 2005) and productions of cheese (Barge et al. 2014). In both cases, the production manager is the one who plans for daily production and can decide the route of the food supply chain in advance. Nevertheless this may not be applicable in all the cases. In some production lines, there may be continuous introduction of a new ingredient at a specific time after each batch. So each batch must be distinguished appropriately in order to have a better traceability system. In the supply chain, the quantity of products that can be traced at any point and at any time is influenced by the production rate, shelf life and consumption rate of the product. Hence, these variable times should be considered for establishing an optimization and planning framework. The main aim is to track the changes involved in the production line, update and adjust the planning strategies according to the alterations.

In many cases, after getting the initial signal of potential risk in the supply chain, even the modern industries with the best traceability systems are not fully ready to immediately start the recall process and trace back the product. Therefore, the concept of rapidity has been presented by Mgonja et al. (2013) that can assess the speed of a traceability system in response to the primary information of traded injured products. The recalling time should be as minimum as possible because the rise in recall time may increase the possibility of more product injuries/defects. The consumer will also assume it as negligence by the company. However, it cannot be denied that some time is required for the process of product recall. Thus, the relation between the speed of the recalling process and the measure of the product dispersion is also considered as a function of time. Faster the removal of defected product from the production chain, lesser will be the dispersion.

4. Technological Advancements for Traceability Techniques

Any risk associated with the food in the supply chain should be handled seriously due to consciousness among the consumers. Food traceability has become an integral part of the supply chain for controlling the injured products. With globalization, traditional traceability techniques are not adequate and advancements in these methods are desirable for effective an traceability system. Some of the recent technological advancements in the actual traceability techniques are discussed. Figure 5 illustrates the overall strategies of the traceability techniques in food supply chain systems.

4.1 Radio-frequency identification (RFID)

RFID is the wireless traceability technology in which radio waves are used for collecting the data/information that are stored in an electronic tag connected to the product in the supply chain. RFID has been a promising tool for monitoring traceability in food supply chains for more than a decade and its application is becoming more popular day by day (Costa et al. 2013). As compared to the traditional traceability system (barcodes), RFID provides

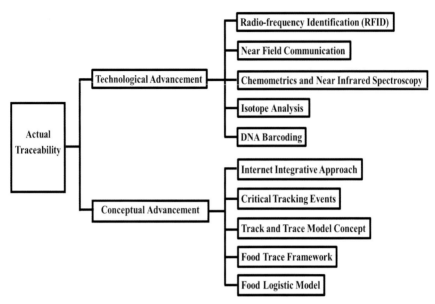

Figure 5. Technological and Conceptual Advancements in Actual Traceability Practice.

higher reading rate, therefore, it is one of the most efficient and reliable methods for food traceability systems (Hong et al. 2011). RFID technology can track and identify the product without any physical connection and offers effective sharing of the information with efficient customization and handling (Zhang and Li 2012).

In the food traceability system, the application of RFID technology is extensive for various purposes due to the recent innovations in its technology and integration of vital tools like sensors and data loggers (Ruiz-Garcia and Lunadei 2011). A framework was developed by Kelepouris et al. (2007) using RFID for comparison of this system with conventional lot sizing and the information collecting method. They have mentioned that RFID possess benefits over traditional system like low investment cost, easy information gathering, automatic identification of the product and identical electronic product code (EPC) for all the participating organizations. RFID traceability system have been used for long time in numerous agro-food logistics and supply chain management processes which is also stated by several authors (Attaran 2007, Ngai et al. 2007, Sugahara 2008).

A RFID system was developed for tracing the temperature for combating the logistics related to feeding by Amador and Emond (2010). A concept of cold traceability was put forward to track the set of products that are temperature sensitive using the different sensors during the transportation in different atmospheric conditions (Ruiz-Garcia et al. 2010). Qian et al. (2012) established a Wheat Flour Milling Traceability System (WFMTS) in

China which is the combination of RFID and 2D barcode technology for validating the wheat flour mill system. In this system, the small packages of wheat flour were identified by the QR Code labels and tags from RFID were used for identifying wheat flour bins and for automatic recording of all the logistics data. The cost analysis indicated that there is an increase in the total cost of the system by 17.2% but in return it is giving 32.5% more sales income. Therefore, this WFMTS has higher scope to be used in both medium and large wheatmill industries for traceability of product. Similarly, a cattle/ beef traceability system has been designed and tested for the application and management of traceability in the production line. This system has combined RFID technology along with bar code printer and personal digital assistant (PDA) to achieve the acquisition and transmission of real-time and accurate data. And the result indicated that the efficiency of the traceability system was improved in the supply chain (Feng et al. 2013). Barge et al. (2014) fixed the RFID tags to the cheese wheels by various methods to track the movement of cheese wheels during production, handling, storage, delivery, packaging and distribution process. RFID was combined with wireless sensor network (WSN) for improving the traceability system in the whole supply chain of white wine starting from the vineyard to consumer (Catarinucci et al. 2011). WSN is a similar wireless technology as RFID that has been used to enhance the effectiveness of traceability system in food and aquaculture industries (Qi et al. 2014).

Besides having so many advantages, RFID technology also has some drawbacks. There are more interferences in some of the Ultra High Frequency tags (bands 860 MHz Europe, 915 MHz USA) that are used in controlling the traceability and monitoring of the cold chain as compared to other frequencies. Also there is a problem in the availability of the bands, as Japan and China restrict these transmissions and only USA and Europe provide these bands (Ruiz-Garcia and Lunadei 2011). In the application of RFID technology in the traceability system, other major barriers are the difficulty in entering data, disorganized sequence of data input and mode of communication with the RFID reader (Feng et al. 2013). RFID technologies have more advantages than disadvantages but still the companies do not prefer to adopt this technology because of its extra cost of investment. But, if the company conducts detail analysis on the overall benefit and gives importance to the safety of product, they will definitely adopt traceability technologies like RFID (Zhang and Li 2012).

4.2 Near field communication (NFC)

Near Field Communication can be considered as part or subgroup or modified version of RFID technology which has similar features as other wireless technologies used at the consumer level (the existing standards for contactless card technologies ISO/IEC 14443 A and B and JIS-X 6319-4). NFC technology offers two better ways of communications between electronic devices for sharing the information in the very short distance of less than 4 cm (Badia-Melis et al. 2015). This technology operates at the frequency of 13.56 MHz and data rates of 106, 212, 424 or 848 Kbit/s are supported by this (Mainetti et al. 2013). For the purpose of specific identity, security and anti-theft of the information, very small sized tags are prepared which can be inserted in the products. The advantage of this technology over the traditional barcode and QR codes is that a laser beam does not have to travel in a solid path in order to travel between two different devices.

The implementation of an integrated system of RFID and NFC in the supply chain can provide the full history of purchased product to the consumer. Mainetti et al. (2013) mentioned that NFC is linked with mobile applications so that the traceability information can be correlated with the products directly. In the recent future, with the use of the mobile, a consumer can get information about the product prior to buying and decide whether to buy or not (Chen et al. 2014). A chemiresistor-based NFC tag has the potential of detecting certain gases. The electronic circuit is disconnected by creating a hole and then reconnected with the carbon nanotube linker. In presence of specific gas, there is change in the conductivity of the nanotubes allowing the retrieval of information from the tag into their smartphone and finally detecting the presence of the targeted gas (Badia-Melis et al. 2015).

4.3 Advances in unique identification and quality of livestock

The main aim of traceability is to improve the production efficiency of food and to minimize the generation of food waste. A model for tracking the quality of vacuum-packed lamb in the supply chain was generated by Mack et al. (2014). This system was provided with a temperature sensor and was capable of recording significant information, analyzing it with the Arrhenius model and making final decision for enhancing the quality of meat during processing and storage.

Shanahan et al. (2009) developed an integrated traceability system involving all the stakeholders along the supply chain which can detect all forms of traceability in beef starting from farm to slaughter. With the accessibility of traceability data to the consumer, their confidence towards the beef products can increase. RFID, biometrics and identifiers can be used as tool for verifying the identity of individual cattle. The ID data of the animal is translated to EPC which is then shared through the EPC global Network for the traceability system.

The traditional ear tagging of calves and printing barcodes in each quarter of the carcass after slaughter provide inadequate data and in harsh condition, the integrity of the label is weak. Hence, RFID tags were recommended instead of barcodes and simple tags (Mc Carthy et al. 2011). Mc Inerney et al. (2010) conducted the *in vivo* study on e-tracking of poultry by printing barcodes with different inks onto the beaks of live broiler chickens. Since the result of this study indicated very poor readability of the data, they further recommended to use various kinds of inks with delayed effect, such as laser printing, etc.

4.4 Isotope analysis and DNA barcoding

Isotope analysis is the study of the relative abundance of stable isotopes (isotopes from same element) and is expressed as isotope ratio which can be used for research purpose. The combination of isotope ratio analysis with other techniques (e.g., chromatography, isotope ratio mass spectroscopy (IRMS), inductively coupled plasma mass (ICP-MS), near infrared spectroscopy (NIRS)) may prove to be a better method for authenticity and traceability of agro-products (Zhao et al. 2014). The analysis of a stable isotope has developed as one of the main techniques for identifying the geographical origin of the agro food products like milk, meat, cereal crops, oil and wine.

Horacek and Min (2010) successfully identified the origin of a beef sample by determining the isotope ratio of carbon, nitrogen and hydrogen in the defatted dry mass of beef samples from different countries. Similarly, Perini et al. (2009) categorized the lambs from seven different regions of Italy on the basis of their feed and by estimating the multi-element stable isotopes. Some of the studies related with the differentiation of the food products using isotope analysis for traceability are listed below:

- Conventional and organic milk by analyzing carbon isotope (Molkentin and Giesemann 2007)
- Rice samples from various areas of Australia, Japan and USA according to the elemental and isotopic composition (Suzuki et al. 2008)
- Wines from different places of Brazil on the basis of isotopic ratios of oxygen of wine water and carbon of ethanol (Dutra et al. 2011)
- Olive oil samples from eight European regions by estimating the stable isotope ratios of H, C and O and 14 other elements (Camin et al. 2009).

The validation of food authenticity mainly depends upon protein analysis and DNA sequencing. Immunological assays, chromatography and electrophoretical methods come under the protein-based analysis methods (Galimberti et al. 2013).

DNA barcoding system has been established as an effective and reliable approach in the actual traceability system of seafood (Becker et al. 2011), mammalian meat (Cai et al. 2011), raw milk (Arcuri et al. 2013), edible plants and also for chocolates, cookies, processed food, purees, fruit residues in juices, etc.

4.5 Chemometrics and near-infrared spectroscopy (NIRS)

Chemometrics is the method which is based on the mathematical and statistical model used for analytical data obtained from different points for getting the necessary chemical information in the system. As compared to the general statistical model, chemometric models are simple, easy to maintain, have better updatability, give accurate data and can be transformed easily into a set of specifications. This will aid in generating the guidelines for decision making on the authenticity of the origin of food, which is vital in the traceability system (Vandeginste 2013). Chemometrics models can be used in integration with other instrumental analytical techniques for developing the authenticity and traceability models that can be used for enhancing the quality and safety of food (Bertacchini et al. 2013).

The combination of near infrared (NIR) spectroscopy and mid infrared (MIR) spectroscopy with multivariate data analysis (like principal component analysis—PCA), partial least squares discriminant analysis (PLS-DA), linear discriminant analysis (LDA), etc. for authentication and traceability of cereals has been studied by Cozzolino (2014) and revealed that the new spectroscopy technologies combined with new algorithms can be the promising tools for traceability and authentication of cereals. Bevilacqua et al. (2012) applied mid-and near-infrared spectroscopic techniques along with chemometric models for the traceability system of extra virgin olive oil and found that the method is inexpensive, fast and non-destructive for the traceability of olive oil.

Versari et al. (2014) found that the integration of chemometrics with other analytical tools can be used for authentication and classification of wines that will later help the traceability process. Nuclear Magnetic Resonance (NMR) and chemometrics was used to track the geographical origin and quality of traditional food (Consonni and Cagliani 2010). Likewise, Near-infrared spectroscopy (NIRS) was combined with chemometrics for evaluating the quality and identifying the geographical origins of black tea (Ren et al. 2013). González-Martín et al. (2014) successfully predicted the geographical origin of wheat grain by combining near-infrared spectroscopy (NIRS) with chemometric.

5. Conceptual Advancements in Actual Traceability

Apart from technological advancements, conceptual approaches have also been developed to validate the actual traceability capacity in food systems.

5.1 Critical tracking events (CTEs) and key data elements

Critical Tracking Event (CTE) is one of the most flexible, efficient, scalable, adaptive and interconnected concepts that can recognize the smallest and significant units in the food logistics supply chain. It combines the modern technologies for tracking the unit of interest and offers fast, accurate and easily understandable information while recalling in the traceability system. The approach of CTE is not to collect the information of a specific product but of the process/events that is responsible for manipulating products in the supply chain. The process/events are monitored, corresponding data are recorded according to time, date and specific locations, and are stored for future recall process (Miller and Welt 2014).

CTEs are classified into four different groups on the basis of the main categories of supply chain (Miller and Welt 2014).

a. **Terminal CTEs:** events present at the terminal end of the supply chain that include the perishable nature and the variability of the products travelling in the chain;
b. **Aggregation/Disaggregation CTEs:** events that come after terminal CTEs, e.g., packaging;
c. **Transfer CTEs:** events involving any type of product movement in the supply chain such as shipping, receiving, and loading;
d. **Commingling CTEs:** Commingling CTEs are the events which arise when a new product is formed by using different products from different sources.

CTEs provide freedom to the operator for selecting methods, tools and technologies to maintain and control the CTE data which is one of the advantages of this concept over other food traceability concepts.

5.2 Track and trace model concept

In the Food Track and Trace Ontology (FTTO) model, a unique body of knowledge is managed through the combination of different concepts and terms that are coming from heterogenous sources of information and users involved in the supply chain. The main objective of FTTO ontology is to involve the most representative food concepts involved in a supply chain, integrate and link them with the key features of the food traceability domain. For this purpose, the information is generally collected from books, dictionary, food databases and internet. Many authors have stated the definition of ontologies based on the particular product or class of products which can be appropriately used again for implementing Global Food Track and Trace Ontology (FTTO) in the traceability system. The major classes that are involved in FTTO ontology are agent, food product, service product and process (Pizzuti et al. 2014).

5.3 Trace food framework

The Trace Food Framework concept is developed so as to formulate international, non-proprietary standards for exchanging the food traceability data. Storøy et al. (2013) have mentioned that for unique identification of products and good traceability practice, a general standard has been formed for the electronic exchange of traceability information. It defines how to construct, send and receive the information and also explains how to analyze and interpret the data elements in the messages. The components that are comprised in this framework are the principle of unique identifications, documentation of transformations of products, generic language for electronic exchange of information, sector-specific language, generic guidelines for implementation of traceability and sector-specific guidelines.

5.4 Internet integrative approach

The Internet of Things (IoT) is a network which integrates, interacts and controls normal objects so that they can be linked with the internet to ease the exchange of information in the system (Giusto et al. 2010). IoT is an emerging technology that can create a world where machines can communicate with each other and permit people to interact with the digital world for enabling the exchange of product information (Atzori et al. 2010, Amaral et al. 2011). This system has three different dimensions which include information items (for identifying and collecting data), independent networks (networks capable of self-configuration, self-healing, self-optimization and self-protection) and intelligent applications (intelligent control and processing).

5.5 Food logistics based actual traceability

Food traceability is a vital part of logistics management as it can store the information which can be retrieved later at any point of time (Bosona and Gebresenbet 2013). Jedermann et al. (2014) have mentioned that intelligent food logistics can minimize the perishable waste in the food supply chain through the reduction of the deviation from the optimal cold chain. Scheer (2006) described that in the traditional tracking and tracing (T and T) systems, the obtained data are only used for recall management. However, they proposed a quality oriented tracking and tracing system (QTT) that utilizes the information retrieved from intelligent logistic tools like wireless sensors. This QTT concept maintains a perfect association between logistics and traceability for improving the products in the supply chain. Jedermann et al. (2014) revealed that shelf life models can be used for combining QTT with FEFO (first expire first out) which will then improve the demand and the supply chain. Van der Vorst et al. (2007) highlighted that if the quality control logistics is implemented in the supply chain networks, the quality of the product can be projected in advance and there will be better flow of product and improved chain design.

5.6 Legal advancements towards actual traceability

As per the Directorate General Health and Consumers Affairs (DG SANCO) of the European Commission (2010), traceability is "the ability to trace and follow a food, feed, food-producing animal or substance through all stages of production and distribution". EU has made tracing and tracking mandatory in 2005 but the specifications in this regulation are low, lacks monitoring of batches and restrict the documentation process (Regattieri et al. 2007). U.S. Food and Drug Administration (FDA) assigned Institute of Food Technologists (IFT) to design pilot product tracing projects to meet the growing requirements of agriculture and food traceability (Bhatt et al. 2013). The main goal of IFT's Traceability Improvement Initiative is to work in the tracing and tracking of food product through several means, pilots and implementation studies in the products like dairy, seafood and other food industries. Some recommendations have been given by IFT for fast and effective investigations during food borne illness outbreaks for improving the protection of public health (Bhatt et al. 2013). A study was carried out by Bhatt and Zhang (2013) on information tracing to identify the contaminated ingredient and probable source, along with the distribution of the product.

6. Conclusion

The increased concern of food safety has raised the consumers and producers attention towards an effective traceability system. The traceability system can be improved by implementing actual traceability techniques and managing the important food product information during the entire food supply chain. The combination of traceability system with planning and logistics can significantly improve the actual traceability practices.

Keywords: Actual traceability, RFID (Radio-frequency identification), technological advancements, conceptual advancements, isotope analysis and DNA barcoding, bulk product traceability

References

Abad, E., Palacio, F., Nuin, M., de Zárate, A.G., Juarros, A., Gómez, J.M. and Marco, S. 2009. RFID smart tag for traceability and cold chain monitoring of foods: demonstration in an intercontinental fresh fish logistic chain. Journal of Food Engineering 93: 394–399.
Amador, C. and Emond, J.P. 2010. Development of RFID temperature tracking systems for combat feeding logistics. In XVIIth World Congress of the International Commission of Agricultural and Biosystems Engineering (CIGR), CSBE/SCGAB. Québec City, Canada.

Amaral, L.A., Hessel, F.P., Bezerra, E.A., Corrêa, J.C., Longhi, O.B. and Dias, T.F. 2011. eCloudRFID—A mobile software framework architecture for pervasive RFID-based applications. Journal of Network and Computer Applications 34(3): 972–979. http://dx.doi.org/10.1016/j.jnca.2010.04.005.

Amorim, P., Günther, H.O. and Almada-Lobo, B. 2012. Multi-objective integrated production and distribution planning of perishable products. International Journal of Production Economics 138(1): 89–101. http://dx.doi.org/10.1016/j.ijpe.2012.03.005.

Anica-Popa, I. 2012. Food traceability systems and information sharing in food supply chain. Management and Marketing 7(4): 749–758.

Arcuri, E.F., El Sheikha, A.F., Rychlik, T., Piro-Metayer, I. and Montet, D. 2013. Determination of cheese origin by using 16S rDNA fingerprinting of bacteria communities by PCR-DGGE: preliminary application to traditional Minas cheese. Food Control 30: 1–6.

Attaran, M. 2007. RFID: an enabler of supply chain operations. Supply Chain Management: An International Journal 12(4): 249–257. http://dx.doi.org/10.1108/13598540710759763.

Atzori, L., Iera, A. and Morabito, G. 2010. The internet of things: a survey. Computer Networks 54(15): 2787–2805.

Badia-Melis, R., Mishra, P. and Ruiz-García, L. 2015. Food traceability: new trends and recent advances. A review. Food Control 57: 393–401. http://dx.doi.org/10.1016/j.foodcont.2015.05.005.

Barge, P., Gay, P., Merlino, V. and Tortia, C. 2014. Item-level Radio-Frequency Identification for the traceability of food products: application on a dairy product. Journal of Food Engineering 125: 119–130. http://dx.doi.org/10.1016/j.jfoodeng.2013.10.019.

Becker, S., Hanner, R. and Steinke, D. 2011. Five years of FISH-BOL: brief status report. Mitochondrial DNA 22: 3–9. http://dx.doi.org/10.3109/19401736.2010.535528.

Bennet, G.S. 2009. Food identity preservation and traceability: Safer grains. Florida, US: CRC Press, 686 pp. ISBN 978-1-4398-0486-5.

Bertacchini, L., Cocchi, M., Vigni, M.L., Marchetti, A., Salvatore, E., Sighinolfi, S. and Durante, C. 2013. The impact of chemometrics on food traceability. pp. 371–410. *In*: Marini, F. (ed.). Chemometrics in Food Chemistry, Elsevier, Oxford, UK, ISBN 978-0-444-59528-7; ISSN 0922-3487.

Bevilacqua, M., Bucci, R., Magrì, A.D., Magrì, A.L. and Marini, F. 2012. Tracing the origin of extra virgin olive oils by infrared spectroscopy and chemometrics: A case study. Analytica Chimica Acta 717: 39–51. http://dx.doi.org/10.1016/j.aca.2011.12.035.

Bhatt, T., Buckley, G., McEntire, J.C., Lothian, P., Sterling, B. and Hickey, C. 2013. Making traceability work across the entire food supply chain. Journal of Food Science 78: B21–B27. http://dx.doi.org/10.1111/1750-3841.12278.

Bhatt, T. and Zhang, J.J. 2013. Food product tracing technology capabilities and interoperability. Journal of Food Science 78: B28–B33. http://dx.doi.org/10.1111/1750-3841.12299.

Borit, M. and Olsen, P. 2012. Evaluation framework for regulatory requirements related to data recording and traceability designed to prevent illegal, unreported and unregulated fishing. Marine Policy 36(1): 96–102. http://dx.doi.org/10.1016/j.marpol.2011.03.012.

Bosona, T. and Gebresenbet, G. 2013. Food traceability as an integral part of logistics management in food and agricultural supply chain. Food Control 33(1): 32–48. http://dx.doi.org/10.1016/j.foodcont.2013.02.004.

Cai, Y., Zhang, L., Shen, F., Zhang, W., Hou, R., Yue, B. and Zhang, Z. 2011. DNA barcoding of 18 species of Bovidae. Chinese Science Bulletin 56(2): 164–168. http://dx.doi.org/10.1007/s11434-010-4302-1.

Canavari, M., Centonze, R., Hingley, M. and Spadoni, R. 2010. Traceability as part of competitive strategy in the fruit supply chain. British Food Journal 112(2): 171–186. http://dx.doi.org/10.1108/00070701011018851.

Camin, F., Larcher, R., Nicolini, G., Bontempo, L., Bertoldi, D., Perini, M., Voerkelius, S., Horacek, M., Ueckermann, H., Froeschl, H., Wimmer, B., Heiss, G., Baxter, M., Rossmann, A. and Hoogewerff, J. 2009. Isotopic and elemental data for tracing the origin of European olive oils. Journal of Agricultural and Food Chemistry 58(1): 570–577. http://dx.doi.org/10.1021/jf902814s.

Cassin, M.H., Lammerding, A.M., Todd, E.C., Ross, W. and McColl, R.S. 1998. Quantitative risk assessment for *Escherichia coli* O157: H7 in ground beef hamburgers. International Journal of Food Microbiology 41(1): 21–44. http://dx.doi.org/10.1016/S0168-1605(98)00028-2.

Catarinucci, L., Cuinas, I., Exposito, I., Colella, R., Fernández, J.A.G. and Tarricone, L. 2011. RFID and WSNs for traceability of agricultural goods from Farm to Fork: electromagnetic and deployment aspects on wine test-cases. In software, telecommunications and computer networks (SoftCOM), 2011 19th international Conference on Software, Telecommunications and Computer Networks, IEEE, pp. 1–4.

Chen, Y.Y., Wang, Y.J. and Jan, J.K. 2014. A novel deployment of smart cold chain system using 2G-RFID-Sys. Journal of Food Engineering 141: 113–121. http://dx.doi.org/10.1016/j.jfoodeng.2014.05.014.

Cocucci, M., Ferrari, E. and Martello, S. 2002. MILK: Milk traceability: from farm to distribution. Italian Food and Beverage Technology 29: 15–20.

Coléno, F.C. 2008. Simulation and evaluation of GM and non-GM segregation management strategies among European grain merchants. Journal of Food Engineering 88(3): 306–314. http://dx.doi.org/10.1016/j.jfoodeng.2008.02.013.

Consonni, R. and Cagliani, L.R. 2010. Nuclear magnetic resonance and chemometrics to assess geographical origin and quality of traditional food products. Advances in Food and Nutrition Research 59: 87–165. http://dx.doi.org/10.1016/S1043-4526(10)59004-1.

Costa, C., Antonucci, F., Pallottino, F., Aguzzi, J., Sarriá, D. and Menesatti, P. 2013. A review on agri-food supply chain traceability by means of RFID technology. Food and Bioprocess Technology 6(2): 353–366. http://dx.doi.org/10.1007/s11947-012-0958-7.

Cozzolino, D. 2014. An overview of the use of infrared spectroscopy and chemometrics in authenticity and traceability of cereals. Food Research International 60: 262–265. http://dx.doi.org/10.1016/j.foodres.2013.08.034.

Dabbene, F., Gay, P. and Sacco, N. 2008a. Optimization of fresh-food supply chains in uncertain environments, Part I: Background and methodology. Biosystems Engineering 99(3): 348–359. http://dx.doi.org/10.1016/j.biosystemseng.2007.11.011.

Dabbene, F., Gay, P. and Sacco, N. 2008b. Optimisation of fresh-food supply chains in uncertain environments, part II: a case study. Biosystems Engineering 99(3): 360–371. http://dx.doi.org/10.1016/j.biosystemseng.2007.11.012.

Donnelly, K.A.-M., Thakur, M. and Sakai, J. 2013. Following the mackerel–Cost and benefits of improved information exchange in food supply chains. Food Control 33(1): 25–31. http://dx.doi.org/10.1016/j.foodcont.2013.01.021.

Desquilbet, M. and Bullock, D.S. 2009. Who pays the costs of non-GMO segregation and identity preservation? American Journal of Agricultural Economics 91(3): 656–672. https://doi.org/10.1111/j.1467-8276.2009.01262.x.

Dupuy, C., Botta-Genoulaz, V. and Guinet, A. 2005. Batch dispersion model to optimise traceability in food industry. Journal of Food Engineering 70(3): 333–339. http://dx.doi.org/10.1016/j.jfoodeng.2004.05.074.

Dutra, S.V., Adami, L., Marcon, A.R., Carnieli, G.J., Roani, C.A., Spinelli, F.R., Leonardelli, S., Ducatti, C., Moreira, M.Z. and Vanderlinde, R. 2011. Determination of the geographical origin of Brazilian wines by isotope and mineral analysis. Analytical and Bioanalytical Chemistry 401: 1571–1576. http://dx.doi.org/10.1007/s00216-011-5181-2.

ECR EUROPE. 2004. L'impiego della rintracciabilità all'interno della Supply Chain al fine di soddisfare le aspettative del consumatore in materia di sicurezza.

European Commission. 2002. Regulation (EC) No. 178/2002 of the European Parliament and of the Council of 28 January 2002 laying down the general principles and requirements of food law, establishing the European Food Safety Authority and laying down procedures in matters of food safety. Official Journal of the European Union, L31, pp. 1–24, 1/02/2002.

European Commission. 2003. Regulation (EC) No. 1829/2003 of the European Parliament and of the Council of 22 September 2003 on genetically modified food and feed. Official Journal of the European Union, L268, pp. 1–23, 18/10/2003.

European Commission—Directorate General Health and Consumers Affairs (DG SANCO). 2010. Guidance on the implementation of articles 11, 12, 14, 17, 18, 19 and 20 of Regulation (EC) No. 178/2002 on general food law. Conclusions of the Standing Committee on the Food Chain and Animal Health.

Feng, J., Fu, Z., Wang, Z., Xu, M. and Zhang, X. 2013. Development and evaluation on a RFID-based traceability system for cattle/beef quality safety in China. Food Control 31(2): 314–325. http://dx.doi.org/10.1016/j.foodcont.2012.10.016.

Galimberti, A., De Mattia, F., Losa, A., Bruni, I., Federici, S., Casiraghi, M., Martellos, S. and Labra, M. 2013. DNA barcoding as a new tool for food traceability. Food Research International 50(1): 55–63. http://dx.doi.org/10.1016/j.foodres.2012.09.036.

Golan, E., Krissoff, B., Kuchler, F., Nelson, K., Price, G. and Calvin, L. 2003. Traceability in the US food supply: dead end or superhighway. Choices 18: 17–20.

Golan, E., Krissoff, B., Kuchler, F., Calvin, L., Nelson, K. and Price, G. 2004. Traceability in the US food supply: economic theory and industry studies. Agricultural Economic Report 830: 183–185.

González-Martín, M.I., Moncada, G.W., González-Pérez, C., San Martín, N.Z., López-González, F., Ortega, I.L. and Hernández-Hierro, J.M. 2014. Chilean flour and wheat grain: tracing their origin using near infrared spectroscopy and chemometrics. Food Chemistry 145: 802–806. http://dx.doi.org/10.1016/j.foodchem.2013.08.103.

Giusto, D., Iera, A., Morabito, G. and Atzori, L. (eds.). 2010. The internet of things: 20th Tyrrhenian workshop on digital communications. Springer Science and Business Media.

Hong, I.H., Dang, J.F., Tsai, Y.H., Liu, C.S., Lee, W.T., Wang, M.L. and Chen, P.C. 2011. An RFID application in the food supply chain: A case study of convenience stores in Taiwan. Journal of Food Engineering 106(2): 119–126. http://dx.doi.org/10.1016/j.jfoodeng.2011.04.014.

Horacek, M. and Min, J.S. 2010. Discrimination of Korean beef from beef of other origin by stable isotope measurements. Food Chemistry 121(2): 517–520. http://dx.doi.org/10.1016/j.foodchem.2009.12.018.

Jang, J. and Olson, F. 2010. The role of product differentiation for contract choice in the agro-food sector. European Review of Agricultural Economics 37: 251–273.

Jedermann, R., Nicometo, M., Uysal, I. and Lang, W. 2014. Reducing food losses by intelligent food logistics. Philosophical Transactions of the Royal Society of London A: Mathematical, Physical and Engineering Sciences 372: 20130302. http://dx.doi.org/10.1098/rsta.2013.0302.

Jie, Y., Jun, L., Xin, C., ZiJiang, Y. and Wei, X. 2012. Study of traceability and anti-counterfeiting of imported wines. In 2012 International Conference on Information Management, Innovation Management and Industrial Engineering 2: 95–97. IEEE. http://dx.doi.org/10.1109/ICIII.2012.6339786.

Karlsen, K.M. and Olsen, P. 2011. Validity of method for analysing critical traceability points. Food Control 22(8): 1209–1215. http://dx.doi.org/10.1016/j.foodcont.2011.01.020.

Karlsen, K.M., Dreyer, B., Olsen, P. and Elvevoll, E.O. 2013. Literature review: does a common theoretical framework to implement food traceability exist? Food Control 32(2): 409–417. http://dx.doi.org/10.1016/j.foodcont.2012.12.011.

Kelepouris, T., Pramatari, K. and Doukidis, G. 2007. RFID-enabled traceability in the food supply chain. Industrial Management and Data Systems 107(2): 183–200. http://dx.doi.org/10.1108/02635570710723804.

Kvarnström, B. and Oghazi, P. 2008. Methods for traceability in continuous processes–Experience from an iron ore refinement process. Minerals Engineering 21(10): 720–730. http://dx.doi.org/10.1016/j.mineng.2008.02.002.

Lee, K.M., Armstrong, P.R., Thomasson, J.A., Sui, R., Casada, M. and Herrman, T.J. 2010. Development and characterization of food-grade tracers for the global grain tracing and recall system. Journal of Agricultural and Food Chemistry 58(20): 10945–10957. http://dx.doi.org/10.1021/jf101370k.

Liang, K., Thomasson, J., Lee, K., Shen, M., Ge, Y. and Herrman, T.J. 2012. Printing data matrix code on food-grade tracers for grain traceability. Biosystems Engineering 113: 395–401. http://dx.doi.org/10.1016/j.biosystemseng.2012.09.012.

Liang, K., Thomasson, J.A., Shen, M.X., Armstrong, P.R., Ge, Y., Lee, K.M. and Herrman, T.J. 2013. Ruggedness of 2D code printed on grain tracers for implementing a prospective grain traceability system to the bulk grain delivery system. Food Control 33(2): 359–365. http://dx.doi.org/10.1016/j.foodcont.2013.03.029.

Li, D., Kehoe, D. and Drake, P. 2006. Dynamic planning with a wireless product identification technology in food supply chains. The International Journal of Advanced Manufacturing Technology 30(9): 938–944. http://dx.doi.org/10.1007/s00170-005-0066-1.

Li, L. 2013. Technology designed to combat fakes in the global supply chain. Business Horizons 56(2): 167–177. http://dx.doi.org/10.1016/j.bushor.2012.11.010.

Mack, M., Dittmer, P., Veigt, M., Kus, M., Nehmiz, U. and Kreyenschmidt, J. 2014. Quality tracing in meat supply chains. Philosophical Transactions of the Royal Society of London A: Mathematical, Physical and Engineering Sciences 372(2017), 20130308. http://dx.doi.org/10.1098/rsta.2013.0308.

Mainetti, L., Patrono, L., Stefanizzi, M.L. and Vergallo, R. 2013. An innovative and low-cost gapless traceability system of fresh vegetable products using RF technologies and EPC global standard. Computers and Electronics in Agriculture 98: 146–157. http://dx.doi.org/10.1016/j.compag.2013.07.015.

Mc Carthy, U.M., Ayalew, G., Mc Donnell, K., Butler, F. and Ward, S. 2011. The Case for UHF RFID application in the meat supply chain in the Irish context: a review perspective. Agricultural Engineering International: CIGR Journal 13(3): 1–9.

Mc Inerney, B., Corkery, G., Ayalew, G., Ward, S. and Mc Donnell, K. 2010. A preliminary *in vivo* study on the potential application of e-tracking in poultry using ink printed 2D barcodes. Computers and Electronics in Agriculture 73: 112–117. http://dx.doi.org/10.1016/j.compag.2010.06.004.

Mgonja, J.T., Luning, P. and Van der Vorst, J.G. 2013. Diagnostic model for assessing traceability system performance in fish processing plants. Journal of Food Engineering 118(2): 188–197. http://dx.doi.org/10.1016/j.jfoodeng.2013.04.009.

Miller, B.D. and Welt, B.A. 2014. Critical tracking events approach to food traceability. Encyclopaedia of Agriculture and Food Systems 387–398.

Molkentin, J. and Giesemann, A. 2007. Differentiation of organically and conventionally produced milk by stable isotope and fatty acid analysis. Analytical and Bioanalytical Chemistry 388(1): 297–305. http://dx.doi.org/10.1007/s00216-007-1222-2.

Ngai, E.W., Cheng, T.E., Au, S. and Lai, K.H. 2007. Mobile commerce integrated with RFID technology in a container depot. Decision Support Systems 43(1): 62–76. http://dx.doi.org/10.1016/j.dss.2005.05.006.

Niederhauser, N., Oberthür, T., Kattnig, S. and Cock, J. 2008. Information and its management for differentiation of agricultural products: The example of specialty coffee. Computers and Electronics in Agriculture 61(2): 241–253. http://dx.doi.org/10.1016/j.compag.2007.12.001.

Perini, M., Camin, F., Bontempo, L., Rossmann, A. and Piasentier, E. 2009. Multielement (H, C, N, O, S) stable isotope characteristics of lamb meat from different Italian regions. Rapid

Communications in Mass Spectrometry 23(16): 2573–2585. http://dx.doi.org/10.1002/rcm.4140.

Pizzuti, T., Mirabelli, G., Sanz-Bobi, M.A. and Goméz-Gonzaléz, F. 2014. Food track and trace ontology for helping the food traceability control. Journal of Food Engineering 120: 17–30. http://dx.doi.org/10.1016/j.jfoodeng.2013.07.017.

Qian, J.P., Yang, X.T., Wu, X.M., Zhao, L., Fan, B.L. and Xing, B. 2012. A traceability system incorporating 2D barcode and RFID technology for wheat flour mills. Computers and Electronics in Agriculture 89: 76–85. http://dx.doi.org/10.1016/j.compag.2012.08.004.

Qi, L., Xu, M., Fu, Z., Mira, T. and Zhang, X. 2014. C²SLDS: A WSN-based perishable food shelf-life prediction and LSFO strategy decision support system in cold chain logistics. Food Control 38: 19–29. http://dx.doi.org/10.1016/j.foodcont.2013.09.023.

Regattieri, A., Gamberi, M. and Manzini, R. 2007. Traceability of food products: general framework and experimental evidence. Journal of Food Engineering 81(2): 347–356. http://dx.doi.org/10.1016/j.jfoodeng.2006.10.032.

Ren, G., Wang, S., Ning, J., Xu, R., Wang, Y., Xing, Z., Wan, X. and Zhang, Z. 2013. Quantitative analysis and geographical traceability of black tea using Fourier transform near-infrared spectroscopy (FT-NIRS). Food Research International 53(2): 822–826. http://dx.doi.org/10.1016/j.foodres.2012.10.032.

Resende-Filho, M.A. and Buhr, B.L. 2010. Economics of traceability for mitigation of food recall costs. Available at SSRN https://ssrn.com/abstract=995335 or http://dx.doi.org/10.2139/ssrn.995335.

Riden, C.P. and Bollen, A.F. 2007. Agricultural supply system traceability, Part II: Implications of packhouse processing transformations. Biosystems Engineering 98(4): 401–410. http://dx.doi.org/10.1016/j.biosystemseng.2007.07.004.

Rong, A., Akkerman, R. and Grunow, M. 2011. An optimization approach for managing fresh food quality throughout the supply chain. International Journal of Production Economics 131(1): 421–429. http://dx.doi.org/10.1016/j.ijpe.2009.11.026.

Ruiz-Garcia, L., Steinberger, G. and Rothmund, M. 2010. A model and prototype implementation for tracking and tracing agricultural batch products along the food chain. Food Control 21(2): 112–121. http://dx.doi.org/10.1016/j.foodcont.2008.12.003.

Ruiz-Garcia, L. and Lunadei, L. 2011. The role of RFID in agriculture: applications, limitations and challenges. Computers and Electronics in Agriculture 79(1): 42–50. http://dx.doi.org/10.1016/j.compag.2011.08.010.

Scheer, F.P. 2006. Optimising supply chains using traceability systems. Improving Traceability in Food Processing and Distribution, pp. 52–64.

Shanahan, C., Kernan, B., Ayalew, G., McDonnell, K., Butler, F. and Ward, S. 2009. A framework for beef traceability from farm to slaughter using global standards: an Irish perspective. Computers and Electronics in Agriculture 66(1): 62–69. http://dx.doi.org/10.1016/j.compag.2008.12.002.

Skoglund, T. and Dejmek, P. 2007. Fuzzy traceability: a process simulation derived extension of the traceability concept in continuous food processing. Food and Bioproducts Processing 85(4): 354–359. http://dx.doi.org/10.1205/fbp07044.

Sobolevsky, A., Moschini, G. and Lapan, H. 2005. Genetically modified crops and product differentiation: Trade and welfare effects in the soybean complex. American Journal of Agricultural Economics 87(3): 621–644.

Storøy, J., Thakur, M. and Olsen, P. 2013. The trace food framework–principles and guidelines for implementing traceability in food value chains. Journal of Food Engineering 115(1): 41–48. http://dx.doi.org/10.1016/j.jfoodeng.2012.09.018.

Sugahara, K. 2008. Traceability system for agricultural products based on RFID and mobile technology. pp. 2293–2301. *In*: International Conference on Computer and Computing Technologies in Agriculture, Springer US.

Sun, C.H., Li, W.Y., Zhou, C., Li, M., Ji, Z.T. and Yang, X.T. 2014. Anti-counterfeit code for aquatic product identification for traceability and supervision in China. Food Control 37: 126–134. http://dx.doi.org/10.1016/j.foodcont.2013.08.013.

Suzuki, Y., Chikaraishi, Y., Ogawa, N.O., Ohkouchi, N. and Korenaga, T. 2008. Geographical origin of polished rice based on multiple element and stable isotope analyses. Food Chemistry 109(2): 470–475. http://dx.doi.org/10.1016/j.foodchem.2007.12.063.

Tamayo, S., Monteiro, T. and Sauer, N. 2009. Deliveries optimization by exploiting production traceability information. Engineering Applications of Artificial Intelligence 22(4-5): 557–568. http://dx.doi.org/10.1016/j.engappai.2009.02.007.

Trienekens, J. and Zuurbier, P. 2008. Quality and safety standards in the food industry, developments and challenges. International Journal of Production Economics 113(1): 107–122. http://dx.doi.org/10.1016/j.ijpe.2007.02.050.

Vandeginste, B. and Brereton, P. 2013. Chemometrics in studies of food origin. New Analytical Approaches for Verifying the Origin of Food, pp. 117–145.

Van der Vorst, J.G.A.J., Van Kooten, O., Marcelis, W., Luning, P. and Beulens, A.J. 2007. Quality controlled logistics in food supply chain networks: integrated decision-making on quality and logistics to meet advanced customer demands. In Proceedings of the Euroma 2007 conference, Ankara, Turkey, pp. 18–20.

Versari, A., Laurie, V.F., Ricci, A., Laghi, L. and Parpinello, G.P. 2014. Progress in authentication, typification and traceability of grapes and wines by chemometric approaches. Food Research International 60: 2–18. http://dx.doi.org/10.1016/j.foodres.2014.02.007.

Wang, X., Li, D. and O'brien, C. 2009. Optimisation of traceability and operations planning: an integrated model for perishable food production. International Journal of Production Research 47(11): 2865–2886. http://dx.doi.org/10.1080/00207540701725075.

Wilson, W.W. and Dahl, B.L. 2008. Procurement strategies to improve quality consistency in wheat shipments. Journal of Agricultural and Resource Economics 33(1): 69–86.

Zhang, M. and Li, P. 2012. RFID application strategy in agri-food supply chain based on safety and benefit analysis. Physics Procedia 25: 636–642. http://dx.doi.org/10.1016/j.phpro.2012.03.137.

Zhao, Y., Zhang, B., Chen, G., Chen, A., Yang, S. and Ye, Z. 2014. Recent developments in application of stable isotope analysis on agro-product authenticity and traceability. Food Chemistry 145: 300–305. http://dx.doi.org/10.1016/j.foodchem.2013.08.062.

Chemical and Biological 'Barcodes' or Markers in Food Traceability
A Case Study on Wines
Gargi Dey[1],* and *Didier Montet*[2]

1. Introduction

Wine is a complex beverage which is influenced by many factors amongst which the geographical origin, vintage, grape variety, agronomic conditions, vinification process and aging process play the major roles. Throughout the world wines are produced from cultivars of a single grape species, *Vitis vinifera* L. However, each wine producing country uses grape varieties that differ from region to region. For instance, France has ten prominent regions—Bordeaux, Burgundy/Beaujolais, Champagne, Alsace, Provence, The Loire valley, The Rhône valley, Jura and Savoie, Southwest, Languedoc-Roussillon, Provence; Italy has twelve prominent regions—Piedmont, Alto Adige, Veneto, Tuscany, Emilia-Romagna, Marche, Abruzzo, Molise, Puglia, Campania, Sicily, Sardinia; Spain has sixteen prominent regions—Rioja and Navarra, catalonia, Jerez, Cava, Ribera del Duero, La Mancha, Valdepeñas, Valencia, Montilla, Bullas, Vinos de Madrid, Campo de Borja, Chacoli de

[1] School of Biotechnology, KIIT University, Bhubaneswar, Pin 751024, India.
[2] Cirad, UMR 95 Qualisud TA B-95/16 73, rue Jean-François Breton 34398 Montpellier cedex 5, France; E-mail: didier.montet@cirad.fr
* Corresponding author: drgargi.dey@gmail.com

Bizkaia, Calatayud, Cariñena; Portugal has ten prominent regions—Altenjo, Algarve, Dăo, Estremadura, Ribatejo, Vinho Verde, Třas-os-Montes, Terras do Sado, Douro, Beiras; Germany has ten prominent regions—Mosel-Saar-Ruwer, Rheingau, Rheinhassen, Nahe, pfalz, Württemberg, Baden, Franken, Mittelrhein, and Ahr. Other countries like England and Wales, Eastern Europe, Middle East and North Africa, South Africa, Australia, New Zealand, Argentina, Chile, California (USA), Canada, Greece, India, Turkey, China, Japan and Brazil are also involved in Industrial scale wine production.

According to the EU Regulation (EU) No. 1308 (2013) wine is defined as a product obtained exclusively from the alcoholic fermentation of fresh grapes, whether crushed or not, or grape must. The geographic location, the climatic, and the edaphic and orographic factors influence the process of vine growth, which in turn influences the compositional and sensory parameters of wines (Ballabio et al. 2006). This in turn gave rise to the concept of *"terroir"* referring to the specific characteristics of a wine that are induced mainly by geographical location and characteristics of production in the concerned areas. The "denomination of origin" (DOC) or "Denomination of Origin Certified and Guaranteed" (DOCG), regions are areas within designated traditional wines, which have typical quality features.

Since wine is an easily adulterated product due to its chemical composition and its availability throughout the world, the wine supply chain requires traceability from grape production to processing and wine distribution, mainly aimed at preserving the identity of unique quality traits against frauds or commercial disputes. This has led to a growing demand of new traceability systems for the collection of information related to units/batches of food ingredients and products. A method that allows verification of the origin and composition of each batch of wine, its condition of production, storage and all products that were in contact with wine post-production may be defined as traceability of wine. A traceability information system on the wine supply chain must provide a better management of all representative events which are part of this chain, from grape production to wine selling.

The practice of wine fraudulence since ancient times has made it more and more sophisticated. It is a daunting task for consumers, manufacturers and regulatory bodies to detect and prevent wine frauds. Over the decade, these studies have been aided by the introduction of advanced analytical techniques, used for the measurement of trace compounds or the simultaneous determination of a wide array of markers, emergence of newer chemical and biological marker along with the development of more comprehensive approaches for data mining and data analysis. The following sections of the chapter discuss the state-of-art techniques that are being use to detect conventional 'barcodes' or markers. The chapter also describes the non-conventional biological markers that have recently emerged in the field of wine traceability and evaluates its applicability for larger sample sets.

2. Chemical 'Barcodes'/Markers

In earlier years, a few volatile markers were used to characterize varieties and a few elemental markers were used to trace geographical differentiation. With time, however, several new chemical markers have been established. This section reviews the emerging chemical markers which allow not only the grape varieties to be traced but also allow classification of wines, detection of adulteration and determination of wine age (Table 1).

2.1 Aromatic markers

Aroma of wine is usually classified into primary (varietal and pre-fermentative), secondary (fermentative) and tertiary (wine aging). The knowledge that the grape variety, climate, fermentation conditions, yeast strain, wine making techniques and aging conditions have a strong impact on aromatic profile of wines (Ebeler and Thorngate 2009, Styger et al. 2011, Cordente et al. 2012), has given rise to many volatile markers. Among the volatile compounds, amino acid (arginine, alanine, tyrosine, valine and leucine) profiles have proved helpful in identification of geographical origins of wines (Flamini and De Rosso 2006, Chambery et al. 2009). In past, these amino acids have been used to differentiate port wines from port imitation (Herbert et al. 2000).

Other than amino acids, French red wines and Spanish white wines could be geographically differentiated by use of volatile compounds like, ethyl esters, isoamyl esters, aldehydes, and acetals (Rapeanu et al. 2009). A very recent study with 54 Brazilian wine samples from grape variety of Cabernet Sauvignon, Merlot, Chardonnay, Sauvignon blanc and Pinot Noir was conducted with volatile markers nerol, β-santalol and 4-carene, diethyl malonate and ethyl-9-decenoate and tetrahydro-2-(2H) pyranone and 3-methyl-2(SH)-furanone. The two esters (diethyl malonate and ethyl-9-decenoate) were distinguishing markers for Chardonnay. The furanone markers (tetrahydro-2-(2H) pyranone and 3-methyl-2(5H)-furanone) were distinguishing markers for Cabernet Sauvignon (Welke et al. 2013). Earlier on, it was found that volatile markers like 2-phenylethanol, butyric acid, ethyl acetate, iso-amyl acetate, iso-amyl alcohol and iso-butyric acid, differentiated Pinotage wines from other varieties like Sauvignon blanc, Chardonnay, Merlot, Shiraz, Cabernet Sauvignon. The work which included a sample set of 496 single varietal South African young wines was conducted by Louw et al. (2009).

Table 1. Chemical and Biological "barcodes" for Wine Traceability.

Type	Chemical class	Markers	Application	References
Chemical "barcodes"	Amino acids	Arg, Ala, Tyr, Val, Leu	Classification based on geographical origin	Flamini and De Rosso 2006, Chambery et al. 2009
	Aromatic compounds	Ala, Tyr, Val, Leu	Identification of port imitations	Herbert et al. 2000
		Tetra-hydro-2(2H)-pyranone, 3-methyl-2(5H)-furranone	Distinctive for Cabernet Sauvignon from Brazil	Welke et al. 2013
		2-phenyl ethanol, butyric acid, ethyl acetate, isoamyl acetate, isoamyl alcohol and isobutyric acid	Differentiation of pinotage wine from other varieties	Louw et al. 2009
		Ethyl-hex-4-enoate, Pentyl-ethanoate, Propyl decanoate, 1-hexanol, *trans*-2-hexanyl-butyrate	Differentiation of French, Australian and Tasmanian varieties	Berna et al. 2009
		3-methyl-1-butanol-decanoic acid, isoamyl acetate, 1-dodecene-ethyl-1-octanoate, 2-phenethyl acetate, ethyl decanoate, benzophenone, 2-propoanone, 1-hydroxy-ethyl-butanate, 2(5H) furranone, ethyl hexanoate, octanal, hexyl hexanoate, 1-hexadecene, isobutyl pthalate	Differentiation of wines fermented from starter cultures and indigenous microflora	Vararu et al. 2016
	Anthocyanins and Flavonoids	Acyl-anthocyanin, Coumaryl-anthocyanin	Authentication of red wines	Von Baer et al. 2008

Table 1 contd. ....

...Table 1 contd.

Type	Chemical class	Markers	Application	References
		Catechin, quercitin, trans-resveratrol, p-coumaric acid, trans-caffeic acid	Categorization of Greek red and white wines	Anastasiadi et al. 2009
		Di-glucoside anthocyanin	Segregation of wines from hybrid grape varieties	Baldi et al. 1995
		Acylated anthocyanin	Absence confirms Pinot Noir grape variety	Ferrandino et al. 2012
		Quercitin-3-O-glucoside, myrecitin-3-O-glucoside	Varietal markers	Mattivi et al. 2006, Castillo-Munoz et al. 2007
		Procyanidin B1 and B2	Deterministic for wine variety and provenance	Makris et al. 2006
		Malvidin-3-O-acetyl glucoside	Differentiation between Cabernet sauvignon, Merlot, Syrah and Monastrell	Ortega-Regules et al. 2006
	Elements	As, B, Be, Cs, Li, Mg, Pb, Si, Sn, Sr, Ti, W, Y	Geographical origin (Germany)	Thiel et al. 2004
		Li, B, Mg, Fe, Zn, Sr, Cs, Pb	Geographical origin (Germany)	Gomez et al. 2004
		Y, U, Cr	Geographical origin (Hungary, Romania, Czech Republic and S. Africa)	Capron et al. 2007
		Li, Si, Mn, Na, Y, V, U, Cd	Distinctive for white wines from Romania, Hungary, Czech Republic	Romisch et al. 2009
		Cd, S, Cr, La, V, Cl, Pb	Distinctive for red wines from Romania, Hungary, Czech Republic	Romisch et al. 2009
		Na, K, Mg, Ca, Fe, Mn, Zn, Cu, Pb	Geographical origin (Serbia, Montenegro and Macedonia)	Ražić and Onjia 2010
		Mg, Rb	Differentiation of Merlot and Cabernet Sauvignon from same geographical location	Dutra et al. 2011
		Sc, V, Cr, Ga, Se, Sr, Pd, Sn, Tl and U	Distinction of soil, grape and wines (China)	Versari et al. 2014

Category	Markers	Application	Reference
	Cd, Cr, Cs, Er, Ga, Mn and Sr	Differentiation of geographical locations (Spain and England)	Versari et al. 2014
	Ca, K, Sr	Markers for Italian and French wines	Serapinas et al. 2008
	Cr, K, Rb and Sr	Markers for Bulgarian and Hungarian wines	Serapinas et al. 2008
	Al, Ca, Na	Markers for white wines	Rodrigues et al. 2011
	B, Ba, Fe, K, Mg, Mn, Ni, Sr	Markers for red wines	Rodrigues et al. 2011
	Ba, As, Pb, Mo, Co	Differentiation of Argentinean wines	Azcarate et al. 2015
	Co, Ba, Pb	Classification of Argentinean wines	Di Paolo-Naranjo et al. 2011
Isotopes	$^{13}C/^{12}C$	Geographical traceability	Christoph et al. 2006
	$^{18}O/^{16}O$	Geographical traceability (Italy)	Dordevic et al. 2013
	$^{13}C/^{12}C$	Markers of wine adulteration	Dordevic et al. 2013
	$^{13}C/^{12}C$ and $^{18}O/^{16}O$	Markers of adulteration	Geana et al. 2016
	$^{13}C/^{12}C$ and $^{18}O/^{16}O$	Determination of wine age	Palade and Popa 2014
	^{210}Pb, ^{210}Po, ^{239}Pu, ^{240}Pu and ^{137}Cs; $^{87}Sr/^{86}Sr$	Geographical traceability	Geana et al. 2016
Biological "Barcodes" SSRs (Nuclear)	VVS 2, VVS 5, VVMD5, VVMD 7, VrZag79, VrZag 62, VrZag 47	Monovarietal classification of musts	Garcia-Beneytez et al. 2002a
	VrZag 7, VrZag 12, VrZag 21, 47, VrZag 62, VVs 2, VVS 4, VVIC 05, VVMD 27	Cultivar classification	Muccillo et al. 2014
	VVMD 5, VVMD 6, VVMD 7, VVS 2	Varietal classification	Faria et al. 2000

Table 1 contd. ....

...*Table 1 contd.*

Type	Chemical class	Markers	Application	References
		VVMD 5, VVMD 6, VVMD 7, VVS 2	Monovarietal classification of musts	Bigliazzi et al. 2012
		VrZag 62, VrZag 79	Tracing adulteration in industrial musts	di Rienzo et al. 2016
		VVS2, VVMD 5, VVMD 7, VVMD 25, VVMD 27, VVMD 28, VVMD 32, VrZag 62, VrZag 79, VMC1B11	Characterization of Italian must varieties	Catalano et al. 2016
	SSRs (Chloroplast)	ccmp 10, ccSSR 14	Authentication of cultivars from must	Catalano et al. 2016
		ccmp 3, ccmp 10, NTcp8, NTcp 12, CCSSR 14	For Asti spumanate wines	Boccacci et al. 2012
	Proteins & Peptides	Thaumatin-like-proteins isoforms VVTL1	Classification of Muscat	Tattersall et al. 1997
		VVTL1, VVTL2 and VVTL 3	Characterization of Sauvignon Blanc	Peng et al. 1997
		Arg-Ile, Ile-Arg, Ile-Val, Lys-Phe, Lys-Tyr, Phe-Tyr, tyr-Gln	Differentiation of champagne wines	de Person et al. 2004

With advances in GC analyzers and development of newer data interpretation tools, several more volatile markers have emerged in wine traceability studies. For instance, recently, 30 samples of Australian Cabernet Sauvignon wines from different regions were categorized on the basis of their geographic origins based on volatile markers 2-isobutyl-3-methoxypyrazine, menthone, iso-menthone, carvacrol, δ-octalactone, p-methylacetophenone, and m-dimethoxybenzene. Instead of conventional GC, a GCxGC-time–of-the-flight mass analyzer was used and data interpretation was carried out by CVA and PLS-DA (Robinson et al. 2012). Volatile compound profiling with conventional GC method and other current methods coupling MS and other detectors have given rise to the popular 'electronic nose' or E-nose. An elegant work performed by Berna et al. (2009), applied GC-MS E-nose to analyze 34 Sauvignon blanc wines from different regions of France, Australia and New Zealand. The workers reported specific volatile markers which could be used as 'barcodes' for different types of wines, like, high levels of ethyl-hex-4-enoate, pentyl ethanoate and propyl decanoate were found in French, Tasmanian and West Australian wines while high contents of 1-hexanol and trans-2-hexanyl butyrate content were found in wines from New Zealand and Malborough and teh Australian region of Victoria. South Australian wines contained distinctively high contents of ethyl acetate and nerol oxide.

Very recently Vararu et al. (2016) showed that a chemometric analysis of volatile markers could be used to differentiate wines fermented from eight different yeast cultures. A sample set of nine Romanian wines obtained by fermenting Aligotè musts with eight individual starter cultures and also with an indigenous microbiota were classified based on 49 aroma markers. The compounds were arranged into six chemical family viz., acids, alcohols, aldehydes and ketones, benzene compounds, esters, terpenes and non-isoprenoids. Among the markers which had useful discriminative powers were 3-methyl-1-butanol, decanoic acid, isoamyl acetate, 1-dodecene, ethyl octanoate, 2-phenethyl acetate, ethyl decanoate, benzophenone, 2-propanone-1-hydroxy ethyl butanate, 2(5H)-furanone, ethyl hexanoate, octanal, hexyl hexanoate, 1-hexadecene, and isobutyl phthalate. A stir-bar-sorptive absorption GC-MS coupled with chemometric analysis established that wines obtained from indigenous microbiota have different aroma marker profile as compared to those obtained from yeast starter cultures in spite of the belonging to the same PDO-grape growing area.

2.2 Non aromatic markers

Non-aromatic compounds including polyphenols, anthocyanins and tannins also contribute to the characterisation of grapes, musts and wines (Carpentieri et al. 2007).

Polyphenols and anthocyanin are usually present in the free form, as anthocyanins/anthocyanidins (malvidol, delfinidol, cyanidol, peonidol and petunidol), as well as acyl and coumaryl compounds. Since, during the grapes processing and wine preservation acyl and coumaryl anthocyanins are more stable, acyl anthocyanins + coumaryl anthocyanins and a ratio of acyl anthocyanins/coumaryl anthocyanins has been proposed as a dependable chemical marker for authentication of grape varieties and red wines (von Baer et al. 2008).

Polyphenol profiling was applied successfully to categorize between Greek red wines (Agiorgitiko and Mandalaria) from whites (Moschofilero and Asyrtiko) over two vintage years of 2005 and 2006 (Anastasiadi et al. 2009) confirming that the most useful phenolics for separation of the four wines were (+) catechin, quercitin, trans-resveratrol, p-coumaric acid and trans-caffeic acid.

Several past researchers have established the usefulness of polyphenolic markers in differentiation of grape varieties. Baldi et al. (1995) used a diglucoside anthocyanin marker. Interestingly this derivative of anthocyanin is not present in *Vitis vinifera* L. varieties; hence, it could be used to segregate wines made from hybrid grape varieties. More recently, Ferrandino et al. (2012) established that absence of acylated anthocyanin could be applied as markers to characterize the Pinot Noir grape variety.

Among polyphenols, flavanols have also proved as successful marker in wine typification studies. Past reports confirm that quercitin-3-O-glucoside and myrecetin-3-O-glucoside are suitable varietal markers (Mattivi et al. 2006, Castillo-Munoz et al. 2007). Similarly, flavanols with procyanidin B1 and B2 were found to be deterministic markers for characterizing wine variety and provenance (Makris et al. 2006).

Along with single class of compounds, combinations of phenolics (including anthocyanin and flavanols) as markers have also been applied in traceability studies. There are several reports where anthocyanin fingerprint has been used for varietal classification (Arozarena et al. 2002, Garcia-Beneytez et al. 2002b, Romero-Cascales et al. 2005). In 2006, Ortega-Regules et al. established an anthocyanin based characterization of varieties based on eight markers (malvidin-3-O-acetylglucoside, malvidin-3-o-trans-p-coumarylglucoside, delphinidin-3-O-glucoside, peonidin-3-O-acetyl glucoside, malvidin-3-O-glucoside and cyanidin-3-O-glucoside). The most useful discriminating marker was malvidin 3-O-acetylglucoside which could differentiate between Cabernet Sauvignon, Merlot, Syrah, and Monastrell. Anthocyanin profiles generated using HPLC-DAD was used to categorize Tannet, Cabernet Sauvignon and Merlot (Gonzalez-Neves et al. 2007). In the subsequent years, Jaitz et al. (2010) reported classification of Australian wines using a group of phenolics analyzed by LC-MS/MS.

Similarly, Spanish wines were also classified based combination of phenolic profiles generated by HPLC-UV/VS FLD.

While there is abundant literature on classification and grouping using phenolic markers, one must exercise caution when it comes to accreditation of authenticity of wines. The typical anthocyanin compounds in *Vitis vinifera* L. varieties are 3-O-glucosides of malvidin, petunidin, cyaniding, peomidin and delphimidin along with their acetyl coumaryl and caffeoyl derivatives. The phenolic composition and concentration are usually said to reflect cultivar, geographic location, and seasonal conditions. However, wine making techniques, e.g., aging conditions may result in variation of phenolic concentration and composition which may be a perceived as a potential disadvantage in case of phenol markers.

However, the consensus is that use of large set of target compounds and upgradation of data analysis models performed on a relatively large number of samples could definitely solve the problem of conditional variations and provide useful information on typification and traceability. Apart from anthocyanins, shikimic acid has also been recommended for white wines authentication and as the additional indicator for red wines authentication (Makris et al. 2006). Mazzei et al. (2013) used metabolites like α-glucose, succinic acid, glycerol and amino acids leucine/isolucine ratio and differentiated two wines groups from 'Fiano di Avellino' grapes that were produced: (a) from commercial grape starters, (b) autochthonous *S. cerevisiae* yeast starters. In the same year, Papotti et al. (2013), used 2,3 butanediol, threonine, lactic acid, succinic acid and malic acid as markers to differentiate between Italian sweet and dry lambrusco red wines from Modena.

Over the years research on wine traceability has brought forth many new chemical markers and combination markers which would not have been possible without advanced analytical techniques combined with conventional liquid chromatography approaches (including HPLC and FPLC) and mass spectrometry (mainly GC/MS and ESI/MS) which have high resolution and detection power (Wang and Sporns 1999, Flamini and De Rosso 2006, Flamini and Panighel 2006). ESI–MS in negative ion-mode, a technique that is particularly suitable for ionisation and identification of acidic compounds such as phenols, flavonoids and sugars was used for characterisation of must samples derived from six grapes varieties by (Catharino et al. 2006).

2.3 Elemental markers

The applicability of elemental/mineral markers has been assessed in past, for authentication of geographical origin of wine. It was found that major changes are brought about in macro- and micro-nutrients, Na, K, Ca, Fe,

Cu, Zn and others, during wine processing. However, some elements like Li, Rb, Cr, Co, Sb, Cs, Sc, Eu, Hf, and Ta showed very small variations during technological process, although they are found in small quantities. Fortunately modern methods like flame atomic emission spectrometry has made it possible to quantify very small quantities, in the range of 1–200 ppm for lithium and 0.5–5.0 ppm of rubidium. Quantification of rare earth elements like lanthanides using mass spectrometry with inductively coupled plasma (ICP-MS) have also provided information on detection of growing areas of the vine (Dutra et al. 2011). However, wine fining with bentonites could modify the element distribution causing fractionation and, thus, modifying the original fingerprint coming from soil (Nicolini et al. 2004, Catarino et al. 2007). A recent study on the traceability of Moscato authentic wines based on the distribution of lanthanides as chemical markers, confirmed that the passage from soil-grapes-must does not affect the elements fingerprint, whereas wine fining with bentonites causes elements fractionation (Aceto et al. 2013). Gomez et al. (2004) reported that only a few elements such as Li, B, Mg, Ca, Rb, Cs, and Pb remain nearly constant (changes in concentration of < 50%) throughout the winemaking processes, being independent of the time of bentonite addition.

A study of 88 German commercial wines from four sub region areas showed that 13 elemental markers viz., As, B, Be, Cs, Li, Mg, Pb, Si, Sn, Sr, Ti, W and Y, analyzed using ICP-MS and ICP-OES were useful for tracing geographical origin of wines, with overall prediction rate by cross validation of 88.6% (Thiel et al. 2004). Another investigation in 127 German wines with elemental markers (Li, B, Mg, Fe, Zn, Sr, Cs and Pb), when measured by ICP-MS using quadratic discriminant analysis, allowed the distinction between four regions. Interestingly, when the same data were analyzed on the basis of non parametric classification and decision tree (CART) the classification of samples could be carried out using only four elemental markers (Li, Zn, Mg and Sr) (Gomez et al. 2004).

Studies and research on geographical traceability of wines using elemental composition as markers have given a new insight that while experimental procedures or high throughput analytical techniques like ICP-MS and ICP-OES have taken data mining abilities further. However, it is equally important to apply a robust data analysis model to increase accuracy and precision of the prediction abilities. As mentioned earlier, the CART method proved superior to the QDA model. This was further confirmed by a study performed on 786 wines collected from Hungary, Romania, Czech Republic and South Africa. The data set analyzed by CART method gave accurate classification based on fewer trace element markers, viz., Y, U, Cr (Capron et al. 2007).

A brief mention of the WINES-DB project is appropriate here. In total 1188 wine samples were collected for three consecutive years. The data were analyzed using partial least square discriminant analysis (PLS-DA) and classified the wines from Romania, Hungary, and Czech Republic. The most useful discriminant markers were Li, Si, Mn, Na, Y, V, U and Cd for white wines and Cd, S, Cr, La, V, Cl, Pb for red wines (Romisch et al. 2009). Other data analysis tools like κ-nearest neighbour (KNN), soft independent modelling of class analogy (SIMCA) and artificial neural network (ANN) have also been evaluated for application in traceability studies using elemental markers. For instance, Ražić and Onjia (2010) reported that 41 commercial wines from Serbia, Montenegro and Macedonia could be classified based on markers Na, K, Mg, Ca, Fe, Mn, Zn, Cu and Pb.

The elemental profile of wine samples is representative of (a) mobility of elements from soil to grape, (b) agronomic practices, (c) uptake rate of the grape variety, (d) wine making process and (e) packaging and storage conditions (Almeida and Vasconcelos 2003, Hopfer et al. 2013). Some interesting studies have been performed related to the effect of grape varieties and provenance on the elemental profiles. Dutra et al. (2011) reported that the Cabernet Sauvignon variety showed higher Mg and Rb content compared to Merlot in spite of being produced in the same region. Similarly, differences in the uptake rate was reported by Vystavna et al. (2014) in case of Muscat white and Chardonnay grapes for selected markers like Cu, Cr, Ni, Pb and Zn. Versari et al. (2014) reported that in order to distinguish soils, grapes and authentic wines (56 samples) from seven Chinese wine producing regions by using SLDA, 10 markers (Sc, V, Cr, Ga, Se, Sr, Pd, Sn, Tl, U) were needed for 100% accuracy.

ICP-MS data analyzed by LDA was useful in separating 55 Spanish wines and 57 wines from England using 7 elemental markers-Cd, Cr, Cs, Er, Ga, Mn, Sr (Versari et al. 2014). Studies in the past have also reported that there may be selected markers for distinction of wines from specific regions. For example, Serapinas et al. (2008) showed that concentration of Ca, K and Sr were better markers for differentiation of Italian and French wine and Cr, K, Rb and Sr were more suitable markers for Bulgarian and Hungarian wines. Similarly, Rodrigues et al. (2011) reported that white and red wines differ in their major elemental composition, i.e., white wines are high on Al, Ca and Na and red wines are high on B, Ba, Fe, K, Mg, Mn, Ni and Sr. In 2015, Azcarate et al. (2015) worked with 57 monovarietal white wine samples. The samples were from four representative wine producing areas of Argentina, i.e., from Mendoza (Torrontès, Chardonnay, Sauvignon blanc) from Rio Negro (Torrontès, Sauvignon blanc) from San Juan (Torrontès) and from Salta (Chardonnay, Sauvignon blanc). Confirmation of provenance genuineness was achieved by a combination of five elemental markers (Ba, As, Pb, Mo, and Co). Similar results of successful Argentinean wine classification using Co, Ba and Pb markers has also reported by Di Paolo-Naranjo et al. (2011). However, other studies conducted on Argentinean wine have reported application of different set of descriptor elemental markers (Fabani et al. 2010).

The soil, the grape variety, the state of grape maturity, climatic conditions, wine making techniques, impurities, all contribute to the elemental or metal composition of a finished wine product and hence, elemental markers are in fact not only identifiers of geographical origin but also indicators of quality assurance and quality control.

2.4 Isotopes markers

Isotopes of D, H, O and C (extracted from water and alcohol of wine), and their isotopic ratios (D/H, $^{18}O/^{16}O$, $^{13}C/^{12}C$) have been used as stable markers, which show trace variations caused mainly by climatic factors and as such may be used to locate areas of growing, especially depending on their climate (cold and dry, cold and wet, hot and humid, and hot and dry). For example, there is a decrease in content of oxygen isotopes (^{18}O) and of deuterium (2H) of wine water, when geographic location shifts from warmer areas to areas with a temperate climate and from West to East.

Similarly, the isotopic ratio value of $^{13}C/^{12}C$ is also influenced by weather conditions, particularly the temperatures at which grapes are ripening. $^{13}C/^{12}C$ values generate reliable information for pinpointing the area of production of a wine. In fact, $^{13}C/^{12}C$ ratio, also referred to as δ^{13}, remains constant in foods and beverages and this serves as an important isotopic marker for tracing wine adulteration, since the δ^{13} diverts from reference value depending on whether carbon source is from atmospheric carbon dioxide, C4 and C3 plants, or animal carbon (Christoph et al. 2006). In past, several traceability and authentication methods based on isotope markers have evolved which couples determination of isotopes and subsequent interpretation of experimental data using chemometric methods like PCA, LDA. The isotopic δ^{13} marker continues to prove useful for characterization of geographic origin even in present times. Di Paolo-Naranjo et al. (2011), characterized 51 red wine samples from Argentinean regions of Mendoza, San Juan and Cordoba. More recently, in 2013, δ^{13} was applied for geographic distinction of 5220 Italian wines that were collected from EU wines database during 2000–2010 (Dordevic et al. 2013).

Apart from tracing geographical location, determination of radioactivity of 14C isotope of wine ethanol is also frequently used for authentication of wine age. The presence of water in grapes decrease during ripening because of water evaporation which in turn results in higher $^{18}O/^{16}O$ ratio in grapes as compared to that of ground water transported through roots. The $^{18}O/^{16}O$ ratio has emerged as another stable chemical 'barcode' which mainly aids in identification of adulteration by added ground and/or tap water in grape must. A recent work by Dordevic et al. (2013) further confirmed this fact when they found a $\delta^{18}O$ in the range of (–1.3 to 8.9) for Italian wine samples collected for Italian Wine Databank (collected from 2000–2010) and for that of ground water in the range of –5 and –9 for southern Italy and northern Italy, respectively.

However, it is important to mention here that while considering the $^{18}O/^{16}O$ ratio, one has to also consider the climatic conditions during grape ripening since that directly influences the evaporation rates of water. For instance, hotter regions like South Africa, south of France Australia, the reference δ^{18} value is 8.5, 6.0 and 7.5 respectively, while for colder regions like Argentina it is –1.0 (Fauhl-Hassek 2009).

A very recent application of the $\delta^{13}C$ and $\delta^{18}O$ has been for detection of adulteration in 29 Romanian sweet and medium sweet wines sold commercially (Geana et al. 2016). The $^{18}O/^{16}O$ and $^{13}C/^{12}C$ isotopic ratios were measured for water and ethanol (fractions extracted from wine), respectively to detect adulterations like addition of water and sugar. Wine samples with $\delta^{13}C$ in the range of –27.31‰ to –24.9‰, $\delta^{18}O$ between –0.8‰ to 7.4‰ and alcoholic content of 10.48% vol to 13.67% vol were classified as good wines. Commercial wines with $\delta^{13}C$ from –27.00‰ to –26.05‰, $\delta^{18}O$ between –2.9‰ to –1.10‰ and alcoholic content of 10.69% vol to 13.09% vol were classified as adulterated wine with addition of water. And wines with $\delta^{13}C$ in the range of –24.25‰ to –17.10‰, $\delta^{18}O$ between –5.4‰ to –2.20‰ and alcohol content from 9.81% vol to 12.46% vol were classified as wines which were chaptalized with C4-sugar along with addition of water.

Apart from adulteration, traceability studies also include determination of wine age. Traditionally wine aging has been done by isotopic measurement of carbon and oxygen markers. Currently new isotopic markers such as ^{210}Pb, ^{210}Po, ^{239}Pu, ^{240}Pu, and ^{13}Cs have also emerged. The measurement of these isotopic markers in wine sediment is another way of knowing the wine age. The validity of the method was verified by appropriate dating of wine during 1850–1968, for every six year (Palade and Popa 2014). Very recent work by Geana et al. (2016) evaluated the suitability of another isotopic marker $^{87}Sr/^{86}Sr$ for wine geographical traceability. The rational of using $^{87}Sr/^{86}Sr$ ratio is that it is not influenced by plant metabolic process and hence is assumed to be characteristic of the region's soil. A sample set of 21 red wines from two Romanian wine regions of Dealurile Vrancei and Terasele Dunarii were included in the study. A collection of 20 elemental markers (Li, ga, Se, Ag, TL, Pb, Ni, Cr, ba, Zn, mn, Sr, Rb, Fe, Ca, Mg, Na, K, Al and Cu) were measured. There was distinctive difference in the elemental content based on geographical locations. For instance, Sr, Al, Ca and Mg were higher in wines from Dealurile Vrancei region while concentration of Ag marker was higher in Insuratei region and Ga, Ni, Fe, Cu and Na were higher in wines from Cuza Voda areas. Both Insratei and Cuza Voda areas were representatives of Terasele Dunarii region. Wine samples from Terasele

Dunarii had a lower $^{87}Sr/^{86}Sr$ ratio compared to Dealurile Vrancei region, with an average of 0.70917 and 0.71679 respectively. The study concluded that a combination of elemental marker and $^{87}Sr/^{86}Sr$ ratio proved to be a robust geographical marker for wine traceability studies. However, one must also take into consideration that, in order to use $^{87}Sr/^{86}Sr$ isotopic marker, there is a requirement of a pre-step for separation of Sr from Rb and the requirement of advanced analytical instrumentation thus limiting the usefulness of this marker in routine traceability analyses.

3. Biological Markers

Biomolecules like DNA and proteins represent a relatively newer breed of markers with potential application in wine traceability. This section reviews the reports of certain established biological markers and their suitability for future large scale application (Table 1).

3.1 DNA-based markers

Authentication of grape variety, geographic origin, wine types have been successfully accomplished by profiling or fingerprinting and chemical markers like volatile compounds, polyphenol, carbohydrates and finally mineral or elements. Arguably, chemical markers are influenced by environmental conditions and wine making techniques. This has led to a shift towards biological markers; because of their unique property genotype remains uninfluenced by environmental conditions. A thorough survey of literature strongly indicates that DNA-based methodologies are indeed becoming popular.

3.1.1 Nuclear SSR markers

Currently, two classes of DNA markers, viz., SSRs and SNPs, have emerged. The application of microsatellite or SSRs was reported much earlier in 1990s and the Italian VITIS Database, which is the collection of SSR grapevine genotypes, has helped greatly in establishing SSR as a vital authentication marker. The SSR markers which are generally used are VMD34, VrZag 83, VrZag 64, VrZag 21, VVMD 31, VrZag 47, and VVMD 21. In past, genotyping of 7 SSr loci (VVS 2, VVS5, VVMD 5, VVMD 7, VrZag 47, VrZag 62, and VrZag 79) coupled with fluorescence detection yielded mixed results. While the extraction and detection method was suitable for monovarietal classification, the same showed limited applicability with multivarietal musts (Garcia-Beneytez et al. 2002).

An interesting work was reported by Muccillo et al. (2014). A sample set of 72 wines from 7 cultivars Piedirosso (P), Cabernet Sauvignon (CS),

Merlot (M), Aglianico di Taurisi (T), Aglianico del Taburno (A), Aglianico del vulture (V), Lingua di Femmina (L) was investigated for polymorphisms by the SSR markers viz., VrZag 7, VrZag 12, VrZag 21, VrZag 47, VrZag 62, VVS2, VVS 4, VVIC 05, and VVMD 27. The power of discrimination values were in the range of 0.53 to 0.81, lowest being for VVIC05, highest for VrZag 21. In addition, the authors also performed a cluster analysis of seven cultivars and as per the dendrogram they grouped them into two clusters. Based on their genetic relatedness the first cluster included CS, M, P and L and the second cluster included A, T and V. In past Faria et al. (2000) attempted a varietal characterization of five varieties using microsatellite markers VVMD 5, VVMD 6, VVMD 7 and VVS2. Among the markers, polymorphism was found at locus VVMD 5 for all the five varieties making this a more useful marker compared to VVS2, VVMD 6 and VVMD 7. Recently Bigliazzi et al. (2012) classified seven monovarietal wines, Merlot, Pinot Noir, Zinfereld, Riesling, Sauvignon blanc, Sangiovese and Alicante based on the above mentioned SSR loci.

A very recent work by di Rienzo et al. (2016) demonstrated that SSR markers VrZag 62 and VrZag 79 could be suitably applied for tracing adulteration in industrial musts. Another important point that this recent work highlighted, was the need for modern techniques that would allow not only adequate DNA isolation from complex media like must or wine but also facilitate fine genotyping or providing allele size. In di Rienzo's method they compared two methods viz., high resolution melting analysis and capillary electrophoresis; and found that the combined information collected from both the methods was useful for detection of adulteration.

Popularity of SSR markers has grown since several studies have confirmed that it is a reliable technique for varietal characterization and authentication. However, strong reliance on *Vitis vinifera* SSR database has somewhat limited its application to characterization of musts originating from well established regions which have authorization for only few varieties for production of high quality wines. Another point of concern is the complications in processing of the DNA from heterogenous matrix like wine, which comes not only from the grape but also from yeast and bacterial strains used for fermentation. It has also been observed that the DNase from yeast fragments denatures the grape DNA during fermentation. Additional steps involved in wine making like decantation, filtration, treatment with fining agents, further decrease the final DNA concentration. Very recently in a study of Catalano et al. (2016), it was demonstrated that the maceration step also influences the amount of DNA recovered. Two commercial wines, Lambrusco and Brunello 2008, when used for DNA extraction, showed that while the Brunello 2008 sample yielded a large DNA pellet, the former sample yielded no pellet. The reason being that Lambrusco production is characterized by limited period of maceration, where as Brunello has a particularly long maceration period allowing greater skin contact and higher content of suspended solids in wine.

Recently, Catalano et al. (2016) tried to address and overcome the problems associated with wine matrix for DNA based techniques. Certain ground rules were established through this study, for instance, (a) the solid or the cellular part of the must is the main source of DNA, the aqueous part does not yield much DNA; (b) the DNA content is generally high at the pre- and middle stage of fermentation and it is lowest at the end of the fermentation process; (c) a proteinase K treatment step during the DNA extraction process enhances the yield of DNA since this enzyme denatures proteins and deactivates DNase; and (d) modified Qiagen DNeasy kit (QDEK) method improved the DNA yield extracted from wine. The authors also attempted SSR genotyping using nuclear SSR markers VVS2, VVMD5, VVMD7, VVMD25, VVMD27, VVMD28, VVMD32, VrZag62, VrZag 79, and VMC1B11 for Italian varieties Brunello di Montalcino, Lambruschi Modenesi and Trento DOC. They reported amplification of the earlier mentioned SSR markers and matching of must samples with the reference. Unfortunately the results were not reproducible for a larger sample set.

Apart from employing microsatellite markers, several studies have also reported the employment of SNP markers which have the advantage of being less prone to mutation than SSRs (Syvanen 2001, Crespen 2004). Spaniolas et al. (2008) investigated on SNP markers that could differentiate between Greek grape varieties of Agiorgitiko from Cabernet Sauvignon. Here a cleaved amplified polymorphic sequence (CAPS) assay was used for must mixtures of two grape varieties during microvinification. A randomly selected EST from NCBI database was used. In the PCR amplicons, one of the SNP had a Bpu10I restriction site in Cabernet sauvignon and it was absent in Agiorgitiko, thus differentiating between the two. Despite these reports, presently, SNP-based methods have not gained the popularity as SSRs. For this approach to be used extensively, exhaustive grapevine genotyping has to be carried out to generate a comprehensive grapevine SNP database.

3.1.2 Chloroplast SSR (cpSSRs) markers

Due to the fact that chloroplast DNA is more abundant and it is more resistant to hydrolysis by exonuclease because of its circular form, this has led to exploration of chloroplast markers for wine authentication studies. Catalano et al. (2016) reported using eight cpSSRs to evaluate wine authentication for cultivars, Sangiovese, Lambrusco and Teroldego. Among them four SSR markers (ccmp2, ccmp4, ccmp6 and NTcp12) were not polymorphic. Two markers ccmp10 and ccSSR14 were polymorphic. Four haplotypes (A, B, C, D) were identified; Sangiovese cultivar had all the four, Lambrusco cultivars had A and B; and Teroldego cultivar showed only haplotype B. A similar investigation was carried out by Bocacci et al. (2012) with ccmp 3, ccmp10, NTcp8, NTcp12 and CCSSR14 chloroplast SSR markers in Asti Spumante wines.

It must be clarified that amplification results of chloroplast markers have not been reproducible across the labs. While Boccacci et al. (2012) were able to amplify the SSR markers from both must and commercial wines; more recently, Catalano et al. (2016) were successful in amplification of these markers from only the must. Though chloroplast markers have a few advantages over nuclear markers, however, its application may be limited because of its low mutation rates and the fact that it is inherited from single parent. Currently, the numbers of reports that have explored the full potential of its application is still very low. Much needs to done with respect to method standardization and validation before chloroplast SSR markers can be exploited as a useful marker.

3.2 Protein and peptide markers

Wine proteins, mainly in the range of 20–30 kDa, contribute to the taste, clarity and stability of the product. Like DNA, proteins and peptides, though present in small quantities, have also been explored as biological markers for varietal identification of wine, musts and grape (Sarmento et al. 2001). Furthermore, many studies have been recently focused on proteins and polypeptides that despite the minimal contribution to nutritive values seem to play important roles in various technological and enological issues (Ferreira et al. 2002). Kwon (2004) had profiled protein isolated from Sauvignon blanc wine samples using a nano-HPLC-Tandem MS. Five grape proteins were identified, i.e., basic extracellular β-1,3-glucanase precursor, thaumatin-like protein (VVTL1), C class IV endochitinases and vacuolar invertase 1. Among these, the pathogenesis-related (PR) proteins like chitinases and thaumatin-like proteins, though present in all grape varieties, show marked variations in their isoforms within the varieties, which has formed the basis for varietal differentiation based on wine protein profiles. Moreover, the thaumatin-like proteins persist till the end of the vinification process. Because of these reasons, thaumatin-like proteins and chitinases have emerged as robust protein markers for wine typification and authenticity studies.

Past studies have established the occurrence of thaumatin-like protein isoform VVTL-1 in Muscat of Alexandria variety (Tattersall et al. 1997) and VVTL-I, VVTL-2, VVTL-3 in Sauvignon blanc wine (Peng et al. 1997). Similarly, consistent difference in molecular weights of same PR proteins, isolated from grape juice was the basis of harvest year and varietal differentiation of grape juices from different locations (Peng et al. 1997).

LC/ESI-MS analysis was performed on low molecular weight peptide markers isolated from champagne wine (de Person et al. 2004). The champagne wines were differentiated based on vintage using 9 dipeptide markers Arg-Ile, Ile-Arg, Ile-Val, Lys-Phe, Lys-Tyr, Phe-Lys, tyr-Gln, Tyr-Lys and Val-Ile. In past, MALDI-TOFF has also been applied to characterize proteins isolated from Chardonnay and Suavignon Blanc wines (Szilágyi et al. 1996). Comparative analysis of MALDI spectra showed that the m/Z 25520 peak was distinctive of Sauvignon Blanc. Similar studies on proteins and peptide fingerprinting were performed by Weiss et al. (1998) to differentiate between Chardonnay, Sauvignon Blanc, Muscat of Alexandria grape varieties. More recently, Chambery et al. (2009) isolated whole wine protein, subjected to tryptic digests and obtained peptide profiles which were analyzed by MALDI-TOFF. These MALDI-TOFF fingerprints were obtained from Campania white wines. Though the authors reported that MALDI fingerprints were able to differentiate between grape varieties, the number of representative samples analyzed was very few.

Though there is existing literature on application of protein profiles and thaumatin-like protein and chitinases have been established as useful marker, it must be confessed that these studies are more of academic interest. Therefore, as of now, among the biological 'barcode', protein and peptide fingerprints is not yet a popular SSR markers.

4. Newer Approach: Metabolomic Fingerprinting

Conventional wine classification has been based on specific chemical markers. Considering the importance of wine traceability, several methods have emerged. Over the years, biological markers, especially DNA based SSR markers have also been established. The latest trend is the metabolomic approach using combination of markers. Oliver et al. (1998) defined metabolome as the whole set of metabolites with molecular weight less than 1500 daltons that are found within the grape, the must or the wine obtained from it. Pools of metabolites involved in the same metabolic cycle results in metabolic profile which have been used for wine typification, authenticity and traceability. It can either be targeted pre-defined group of markers like organic acids, phenolics, volatile compounds, or it may be untargeted metabolomics involving partial quantification of many markers. For instance, metabolomic finger printing using NMR entire spectra and multivariate techniques. The fingerprinting technique provides more information from all molecules that give NMR signals without having to quantify or identify them.

In the wine world, one of the important aspects is to be able to distinguish between blends, identification of the composition. By employing the fingerprinting approach using H'-NMR and data analysis with LDA differentiated between monovarietal and blended wines along with ANN, the blend percentage could be identified. Imparato et al. (2011) performed this fingerprinting study with 4 wines, i.e., Montepulciano d'Abruzzo, Sangiovese, Cabernet and Merlot which were used in blend ratio from 10–70%. Untargeted approaches gives rise to distinct signatures or fingerprints which are useful in differentiating wines based on origin, however, further down streaming and identification of markers is necessary to enable comparison of fingerprint data generated from different labs. The identification of markers is even more important because metabolomic-based approaches apply different techniques like LC-HRMS or Q-TOF, etc. which contribute further variance with respect to sample treatment, stationary and mobile phases, and data processing techniques. Adequate normalization would also be required to nullify the experimental variations. Diaz et al. (2016) recently reported application of untargeted approach for classification of 42 Spanish wines samples belonging to three different protected denominations of origin (PDO). A comparative analysis of fingerprints from two platforms, LC-HRMS and Q-TOF was made. Since they restricted the metabolome to only polyphenols and plant related molecules, they were able to identify some discriminating markers. For instance, catechin, epicatechin flavanols were higher in Pendes wines while, derivatives like gallocaechin and epicatechin were higher in Ribera del Duero wines. The results were confirmed by both the techniques (LC-HRMS and Q-TOF). The observations made by Diaz et al. (2016) further emphasizes the bottlenecks of non-targeted metabolomic platforms, viz., identification of metabolites and complication arising from direct comparison of data using different techniques and finally the expected complications arising from higher coverage of metabolomes.

A refined literature survey does indicate that the non-targeted metabolomic method with limited coverage of metabolites is more popular. Arbulu et al. (2015) worked with non-volatile metabolomic profile involving 411 metabolites like amino acids, biogenic amines, fatty acids, organic acids and secondary metabolites such as phenols and esters. The profile was established with Spanish Graciano *Vitis vinifera* wine variety using LC-ESI-QTOF. This study focused more on identification and at least 15 chemical markers analyzed were identified and found to be able to be differentiated between the Tempranillo and Graciano varieties. It is noteworthy to mention that from the data generated by this work, the authors were able to develop a WinMet database (http://databases-metabolomips.rhdound.co/) especially adapted for QTOF-MS analyses of metabolite profiles. The database is said to include 2030 putative markers usually found in wine matrices belonging to 10 chemical classes. Another database of putative wine compounds which deserve a mention here is METLIN (http://metlin.scripps.edu).

Several hundreds of metabolites found in the oenological matrices depict the panoramic picture of the wine making process starting from the geographic conditions, grape varieties, yeast strain used, fermentation conditions, storage, aging and other down streaming processes. Thus metabolic approaches are very promising for global analysis of metabolites. Despite advances in instrumentation and multivariate data manipulation techniques, roadblocks still exists. In order to be effective in industrial scale, it requires multiple expertises of biologists, biochemists, chemists and chemometry.

5. Conclusion and Future Perspectives

The field of wine traceability has seen enormous progress over the decade. Beginning with a few elemental markers for determination of geographical locations to carbon dating for detection of wine age, we now have a huge number of markers supported with advanced analytical tools that are available for traceability, typification and classification of wines. Currently, the scale is tilted in favour of chemical markers especially with chemometric analysis of polyphenols, which give useful information concerning all aspects of the wine traceability chain. Exploration with DNA and peptide based biological markers have not been that extensive. However, SSR markers seem definitely poised to be more popular in the coming years. The most current flavour in wine traceability is the metabolomic approach. Untargeted metabolome, because of its complicated nature and possible misinterpretation of data collected across the labs, is not likely to be applied for routine traceability analyses. However, application of restricted or targeted metabolome may become more visible, as it can provide important support information for confirmation of data.

Abbreviations

CDA	:	Canonical Discriminant Analysis
CVA	:	Canonical Variate Analysis
CART	:	Classification and Decision Tree
DOC	:	Controlled Designation of Origin
CA	:	Cluster Analysis
DAD	:	Diode Array Detector Electronic Nose (Enose)
FT-ICR-MS	:	Fourier Transform Ion Cyclotron Resonance Mass Spectrometry
FTIR	:	Fourier Transform Infrared
FT-MS	:	Fourier Transform Ion Cyclotron Mass Spectrometry
GC	:	Gas Chromatography
HPLC	:	High Performance Liquid Chromatography

ICPAES	:	Inductively Coupled Plasma Atomic Emission Spectrometry
ICP-MS	:	Inductively Coupled Plasma Mass Spectrometry
PDO	:	Protected Designation of Origin
MS	:	Mass Spectrometer
QDA	:	Quadratic Discriminant Analysis
SLDA	:	Stepwise Linear Discriminant Analysis
SBSE	:	Stir Bar Sorptive Extraction
GC×GC-TOFMS	:	Two-Dimensional Gas Chromatography With Time-of-The-Flight Mass Analyzer
SIMCA	:	Soft Independent Modelling of Class Analogy

Keywords: Barcodes, chemometry, SSR, peptide fingerprints, METLIN, WinMet

References

Aceto, M., Robotti, E., Oddone, M., Baldizzone, M., Bonifacino, G., Bezzo, G., Di Stefano, R., Gosetti, F., Mazzucco, E., Manfredi, M. and Marengo, E. 2013. A traceability study on the Moscato wine chain. Food Chemistry 138(2-3): 1914–1922.

Almeida, C.M.R. and Vasconcelos, M.T.S.D. 2003. Multielement composition of wines and their precursors including provenance soil and their potentialities as fingerprints of wine origin. Journal of Agricultural and Food Chemistry 51: 4788–4798.

Anastasiadi, M., Zira, A., Magiatis, P., Haroutounian, S.A., Skaltsounis, A.L. and Mikros, E. 2009. HNMR based metabonomics for the classification of Greek wines according to variety, region, and vintage. Comparison with HPLC data. Journal of Agricultural and Food Chemistry 57(23): 11067–11074.

Arbulu, M., Sampedro, M.C., Gómez-Caballero, A., Goicolea, M.A. and Barrio, R.J. 2015. Untargeted metabolomic analysis using liquid chromatography quadrupole time-of-flight mass spectrometry for non-volatile profiling of wines. Analytica Chimica Acta 858: 32–41.

Arozarena, I., Ayestaran, B., Cantalejo, M.J., Navarro, M., Vera, M. and Abril, I. 2002. Anthocyanin composition of tempranillo garnacha and Cabernet Sauvignon grapes from high-and-low-quality vineyards over two years. European Food Research Technology 214: 303–309.

Azcarate, S.M., Martinez, L.D., Savio, M., Camiña, J.M. and Gil, R.A. 2015. Classification of monovarietal Argentinean white wines by their elemental profile. Food Control 57: 268–274.

Baldi, A., Romani, A., Mulinacci, N., Vincieri, F.F. and Casetta, B. 1995. HPLC/MS application to anthocyanins of *Vitis-vinifera* L. Journal of Agricultural and Food Chemistry 43: 2104–2109.

Ballabio, D., Mauri, A., Todeschini, R. and Buratti, S. 2006. Geographical classification of wine and olive oil by means of classification and influence matrix analysis (CAIMAN). Analytica Chimica Acta 570(2): 249–258.

Berna, A.Z., Trowell, S., Clifford, D., Cynkar, W. and Cozzolino, D. 2009. Geographical origin of Sauvignon Blanc wines predicted by mass spectrometry and metal oxide based electronic nose. Analytica Chimica Acta 648(2): 146–152.

Bigliazzi, J., Scali, M., Paolucci, E., Cresti, M. and Vignani, R. 2012. DNA extracted with optimized protocols can be genotyped to reconstruct the varietal composition of monovarietal wines. American Journal of Enology and Viticulture 63(4): 568–573.

Boccacci, P., Akkak, A., Marinoni, D.T., Gerbi, V. and Schneider, A. 2012. Genetic traceability of Asti Spumante and Moscato d'Asti musts and wines using nuclear and chloroplast microsatellite markers. European Food Research and Technology 235(3): 439–446.

Capron, X., Smeyers-Verbeke, J. and Massart, D.L. 2007. Multivariate determination of the geographical origin of wines from four different countries. Food Chemistry 101: 1585–1597.

Carpentieri, A., Marino, G. and Amoresano, A. 2007. Rapid fingerprinting of red wines by MALDI mass spectrometry. Analytical and Bioanalytical Chemistry 389(3): 969–982.

Castillo-Munoz, N., Gomez-Alonso, S., Garcia-Romero, E. and Hermosin-Gutierrez, I. 2007. Flavonol profiles of *Vitis vinifera* red grapes and their single-cultivar wines. Journal of Agricultural and Food Chemistry 55: 992–1002.

Catalano, V., Moreno-Sanz, P., Lorenzi, S. and Grando, M.S. 2016. Experimental review of DNA-based methods for wine traceability and development of a single-nucleotide polymorphism (SNP) genotyping assay for quantitative varietal authentication. Journal of Agricultural and Food Chemistry 64(37): 6969–6984.

Catarino, S., Madeira, M., Monteiro, F., Rocha, F., Curvelo-Garcia, A.S. and de Sousa, R.B. 2007. Effect of bentonite characteristics on the elemental composition of wine. Journal of Agricultural and Food Chemistry 56(1): 158–165.

Catharino, R.R., Cunha, I.B., Fogaca, A.O., Facco, E.M., Godoy, H.T. and Daudt, C.E. 2006. Characterization of must and wine of six varieties of grapes by direct infusion electrospray ionization mass spectrometry. Journal of Mass Spectrometry 41(2): 185–190.

Chambery, A., del Monaco, G., Di Maro, A. and Parente, A. 2009. Peptide fingerprint of high quality Campania white wines by MALDI-TOF mass spectrometry. Food Chemistry 113(4): 1283–1289.

Christoph, N., Rossmann, A., Schlicht, C. and Voerkelius, S. 2006. Wine Authentication Using Stable Isotope Ratio Analysis: Significance of Geographic Origin, Climate, and Viticultural Parameters. *In*: Ebeler, S.E., Takeoka, G. and Winterhalter, P.R. (eds.). Authentication of Food and Wine, Washington, DC: American Chemical Society 952: 166–179.

Cordente, A.G., Curtin, C.D., Varela, C. and Pretorius, I.S. 2012. Flavour-active wine yeasts. Applied Microbiology and Biotechnology 96(3): 601–618.

Crespan, M. 2004. Evidence on the evolution of polymorphism of microsatellite markers in varieties of *Vitis vinifera* L. Theoretical and Applied Genetics 108(2): 231–237.

de Person, M., Sevestre, A., Chaimbault, P., Perrot, L., Duchiron, F. and Elfakir, C. 2004. Characterization of low-molecular weight peptides in champagne wine by liquid chromatography/tandem mass spectrometry. Analytica Chimica Acta 520(1): 149–158.

Díaz, R., Gallart-Ayala, H., Sancho, J.V., Nuñez, O., Zamora, T., Martins, C.P., Hernández, F., Hernández-Cassou, S., Saurina, J. and Checa, A. 2016. Told through the wine: a liquid chromatography–mass spectrometry interplatform comparison reveals the influence of the global approach on the final annotated metabolites in non-targeted metabolomics. Journal of Chromatography A 1433: 90–97.

Di Paola-Naranjo, R.D., Baroni, M.V., Podio, N.S., Rubinstein, H.R., Fabani, M.P., Badini, R.G., Inga, M., Ostera, H.A., Cagnoni, M., Gallegos, E., Gautier, E., Peral-Garcia, P., Hoogewerff, J. and Wunderlin, D.A. 2011. Fingerprints for main varieties of Argentinean wines: terroir differentiation by inorganic, organic, and stable isotopic analyses coupled to chemometrics. Journal of Agricultural and Food Chemistry 59(14): 7854–7865.

di Rienzo, V., Miazzi, M.M., Fanelli, V., Savino, V., Pollastro, S., Colucci, F., Miccolupo, A., Blanco, A., Pasqualone, A. and Montemurro, C. 2016. An enhanced analytical procedure to discover table grape DNA adulteration in industrial musts. Food Control 60: 124–130.

Dordevic, N., Camin, F., Marianella, R.M., Postma, G.J., Buydens, L.M.C. and Wehrens, R. 2013. Detecting the addition of sugar and water to wine. Australian Journal of Grape and Wine Research 19(3): 324–330.

Dutra, S.V., Adami, L., Marcon, A.R., Carnieli, G.J., Roani, C.A., Spinelli, F.R., Leonardelli, S., Ducatti, C., Moreira, M.Z. and Vanderlinde, R. 2011. Determination of the geographical origin of Brazilian wines by isotope and mineral analysis. Analytical Bioanalytical Chemistry 401: 1571–1576.

Ebeler, S.E. and Thorngate, J.H. 2009. Wine chemistry and flavor: looking into the crystal glass. Journal of Agricultural and Food Chemistry 57(18): 8098–8108.

Fabani, M.P., Arrúa, R.C., Vázquez, F., Diaz, M.P., Baroni, M.V. and Wunderlin, D.A. 2010. Evaluation of elemental profile coupled to chemometrics to assess the geographical origin of Argentinean wines. Food Chemistry 119(1): 372–379.

Faria, M.A., Magalhaes, R., Ferreira, M.A., Meredith, C.P. and Monteiro, F.F. 2000. *Vitis vinifera* must varietal authentication using microsatellite DNA analysis (SSR). Journal of Agricultural and Food Chemistry 48(4): 1096–1100.

Fauhl-Hassek, C. 2009. Trends in wine authentication. Bulletin de l'OIV, 82(935-936-937), 94–100.

Ferrandino, A., Carra, A., Rolle, L., Schneider, A. and Schubert, A. 2012. Profiling of hydroxycinnamoyl tartrates and acylated anthocyanins in the skin of 34 *Vitis vinifera* genotypes. Journal of Agricultural and Food Chemistry 60: 4931–4945.

Ferreira, R.B., Picarra-Pereira, M.A., Monteiro, S., Loureiro, V.B. and Teixeira, A.R. 2002. The wine proteins. Trends in Food Science and Technology 12: 230–239.

Flamini, R. and De Rosso, M. 2006. Mass spectrometry in the analysis of grape and wine proteins. Expert Review of Proteomics 3(3): 321–331.

Flamini, R. and Panighel, A. 2006. Mass spectrometry in grape and wine chemistry. Part II: The consumer protection. Mass Spectrometry Reviews 25(5): 741–774.

García-Beneytez, E., Moreno-Arribas, M.V., Borrego, J., Polo, M.C. and Ibáñez, J. 2002a. Application of a DNA analysis method for the cultivar identification of grape musts and experimental and commercial wines of *Vitis vinifera* L. using microsatellite markers. Journal of Agricultural and Food Chemistry 50(21): 6090–6096.

Garcia-Beneytez, E., Revilla, E. and Cabello, F. 2002b. Anthocyanin pattern of several red grape cultivars and wines made from them. European Research Technology. European Food Research Technology 215: 32–37.

Geana, E.I., Popescu, R., Costinel, D., Dinca, O.R., Stefanescu, I., Ionete, R.E. and Bala, C. 2016. Verifying the red wines adulteration through isotopic and chromatographic investigations coupled with multivariate statistic interpretation of the data. Food Control 62: 1–9.

Gomez, C.M.D.M., Brandt, R., Jakubowski, N. and Andersson, J.T. 2004. Changes of the metal composition in German white wines through the winemaking process. A study of 63 elements by inductively coupled plasma-mass spectrometry. Journal of Agricultural and Food Chemistry 52(10): 2953–2961.

Gonzalez-Neves, G., Franco, J., Barreiro, L., Gil, G., Moutounet, M. and Carbonneau, A. 2007. Varietal differentiation of Tannat, Cabernet-Sauvignon and Merlot grapes and wines according to their anthocyanic composition. European Food Research and Technology 225: 111–117.

Herbert, P., Barros, P. and Alves, A. 2000. Detection of port wine imitations by discriminant analysis using free amino acids profiles. American Journal of Enology and Viticulture 51: 262–268.

Hopfer, H., Nelson, J., Mitchell, A.E., Heymann, H. and Ebeler, S.E. 2013. Profiling the trace metal composition of wine as a function of storage temperature and packaging type. Journal of Analytical Atomic Spectrometry 28(8): 1288–1291.

http://metlin.scripps.edu.

Imparato, G., Paolo, E.D., Braca, A. and Lamanna, R. 2011. Nuclear magnetic resonance profiling of wine blends. Journal of Agricultural and Food Chemistry 59(9): 4429–4434.

Jaitz, L., Siegl, K., Eder, R., Rak, G., Abranko, L., Koellensperger, G. and Hann, S. 2010. LC-MS/MS analysis of phenols for classification of red wine according to geographic origin, grape variety and vintage. Food Chemistry 122: 366–372.

Kwon, S.W. 2004. Profiling of soluble proteins in wine by nano-high-performance liquid chromatography/tandem mass spectrometry. Journal of Agricultural and Food Chemistry 52(24): 7258–7263.

Louw, L., Roux, K., Tredoux, A., Tomic, O., Naes, T., Nieuwoudt, E. and van Rensburg, P. 2009. Characterization of selected South African young cultivar wines using FTMIR spectroscopy, Gas Chromatography, and multivariate data analysis. Journal of Agricultural and Food Chemistry 57(7): 2623–2632.

Makris, D.P., Kallithraka, S. and Mamalos, A. 2006. Differentiation of young red wines based on cultivar and geographical origin with application of chemometrics of principal polyphenolic constituents. Talanta 70: 1143–1152.

Mattivi, F., Guzzon, R., Vrhovsek, U., Stefanini, M. and Velasco, R. 2006. Metabolite profiling of grape: flavonols and anthocyanins. Journal of Agricultural and Food Chemistry 54: 7692–7702.

Mazzei, P., Spaccini, R., Francesca, N., Moschetti, G. and Piccolo, A. 2013. Metabolomic by ^{1}H-NMR spectroscopy differentiates "Fiano di Avellino" white wines obtained with different yeast strains. Journal of Agricultural and Food Chemistry 61(8): 1741–1746.

Muccillo, L., Gambuti, A., Frusciante, L., Iorizzo, M., Moio, L., Raieta, K., Rinaldi, A., Colantuoni, V. and Aversano, R. 2014. Biochemical features of native red wines and genetic diversity of the corresponding grape varieties from Campania region. Food Chemistry 143: 506–513.

Nicolini, G., Larcher, R., Pangrazzi, P. and Bontempo, L. 2004. Changes in the contents of micro- and trace elements in wine due to winemaking treatments. VITIS-Journal of Grapevine Research 43(1): 41–45.

Oliver, S.G., Winson, M.K., Kell, D.B. and Baganz, F. 1998. Systematic functional analysis of the yeast genome. Trends in Biotechnology 16(9): 373–378.

Ortega-Regules, A., Romero-Cascales, I., Lopez-Roca, J.M., Ros-Garcıa, J.M. and Gomez-Plaza, E. 2006. Anthocyanin fingerprint of grapes: environmental and genetic variations. Journal of the Science of Food and Agriculture 86: 1460–1467.

Peng, Z., Pockoc, K.F., Water, E.J., Francis, I.L. and Williams P.J. 1997. Taste properties of grape (*Vitis vinifera*) pathogenesis-related proteins isolated from wine. Journal of Agricultural and Food Chemistry 45: 4639–4643.

Palade, M. and Mona-Elena, P.O.P.A. 2014. Wine traceability and authenticity–A literature review. Scientific Bulletin. Series F. Biotechnologies 18: 226–233.

Papotti, G., Bertelli, D., Graziosi, R., Silvestri, M., Bertacchini, L., Durante, C. and Plessi, M. 2013. Application of one- and two-dimensional NMR spectroscopy for the characterization of Protected Designation of Origin Lambrusco wines of Modena. Journal of Agricultural and Food Chemistry 61(8): 1741–1746.

Rapeanu, G., Vicol, C. and Bichescu, C. 2009. Possibilities to asses the wines authenticity. Innovative Romanian Food Biotechnology 5(12): 1–9.

Ražić, S. and Onjia, A. 2010. Trace element analysis and pattern recognition techniques in classification of wine from central Balkan countries. American Journal of Enology and Viticulure 61(4): 506–511.

Regulation (EU) No 1308/2013 of the European Parliament and of the Council of 17 December 2013 establishing a common organization of the markets in agricultural products and repealing Council Regulations (EEC) No 922/72. (EEC) No. 234/79, (EC) No 1037/2001. Official Journal of the European Union, L 347, (2013) 671–854.

Robinson, A.L., Adams, D.O., Boss, P.K., Heymann, H., Solomon, P.S. and Trengove, R.D. 2012. Influence of geographic origin on the sensory characteristics and wine composition of *Vitis vinifera* cv. Cabernet Sauvignon wines from Australia. American Journal of Enology and Viticulture: ajev-2012.

Rodrigues, S.M., Otero, M., Alves, A.A., Coimbra, J., Coimbra, M.A., Pereira, E. and Duarte, A.C. 2011. Elemental analysis for categorization of wines and authentication of their certified brand of origin. Journal of Food Composition and Analysis 24(4): 548–562.

Romero-Cascales, I., Ortega-Regules, A., López-Roca, J.M., Fernández-Fernández, J.I. and Gómez-Plaza, E. 2005. Differences in anthocyanin extractability from grapes to wines according to variety. American Journal of Enology and Viticulture 56(3): 212–219.

Romisch, U., Jager, H., Capron, X., Lanteri, S., Forina, M. and Smeyers-Verbeke, J. 2009. Characterization and determination of the geographical origin of wines. Part III: multivariate discrimination and classification methods. European Food Research Technology 230: 31–45.

Sarmento, M.R., Oliveira, J.C., Slatner, M. and Boulton, R.B. 2001. Comparative quantitative analysis of the effect of cultivar, wine growing region and vinification method on the protein profiles of some white wines. International Journal of Food Science & Technology 36(7): 759–766.

Serapinas, P., Venskutonis, P.R., Aninkevicius, V., Ezerinskis, Z., Galdikas, A. and Juzikiene, V. 2008. Step by step approach to multi-element data analysis in testing the provenance of wines. Food Chemistry 107: 1652–1660.

Spaniolas, S., Tsachaki, M., Bennett, M.J. and Tucker, G.A. 2008. Toward the authentication of wines of Nemea denomination of origin through cleaved amplified polymorphic sequence (CAPS)-based assay. Journal of Agricultural and Food Chemistry 56(17): 7667–7671.

Styger, G., Prior, B. and Bauer, F.F. 2011. Wine flavor and aroma. Journal of Industrial Microbiology and Biotechnology 38(9): 1145–59.

Syvänen, A.C. 2001. Accessing genetic variation: genotyping single nucleotide polymorphisms. Nature Reviews Genetics 2(12): 930–942.

Szilagyi, Z., Vas, G., Mady, G. and Vekey, K. 1996. Investigation of macromolecules in wines by matrix-assisted laser desoption/ionization time-of-flight mass spectrometry. Rapid Communication in Mass Spectrometry 10: 1141–1143.

Tattersall, D.B., Van Heeswijck, R. and Hoj, P.B. 1997. Identification and characterization of a fruit-specific, thaumatin-like protein that accumulates at very high levels in conjunction with the onset of sugar accumulation and berry softening in grapes. Plant Physiology 114(3): 759–769.

Thiel, G., Geisler, G., Blechschmidt, I. and Danzer, K. 2004. Determination of trace elements in wines and classification according to their provenance. Analytical and Bioanalytical Chemistry 378: 1630–1636.

Vararu, F., Moreno-García, J., Zamfir, C.I., Cotea, V.V. and Moreno, J. 2016. Selection of aroma compounds for the differentiation of wines obtained by fermenting musts with starter cultures of commercial yeast strains. Food Chemistry 197: 373–381.

Versari, A., Laurie, V.F., Ricci, A., Laghi, L. and Parpinello, G.P. 2014. Progress in authentication, typification and traceability of grapes and wines by chemometric approaches. Food Research International 60: 2–18.

Von Baer, D., Rentzsch, M., Hitschfeld, M.A., Mardones, C., Vergara, C. and Winterhalter, P. 2008. Relevance of chromatographic efficiency in varietal authenticity verification of red wines based on their anthocyanin profiles: Interference of pyranoanthocyanins formed during wine ageing. Analytica Chimica Acta 621(1): 52–6.

Vystavna, Y., Rushenko, L., Diadin, D., Klymenko, O. and Klymenko, M. 2014. Trace metals in wine and vineyard environment in southern Ukraine. Food Chemistry 146: 339–344.

Wang, J. and Sporns, P. 1999. Analysis of anthocyanins in red wine and fruit juice using MALDI-MS. Journal of Agricultural and Food Chemistry 47(5): 2009–2015.

Weiss, K.C., Yip, T.T., Hutchens, T.W. and Bisson, L.F. 1998. Rapid and sensitive fingerprinting of wine proteins by matrix-assisted laser desorption/ionization time-of-flight (MALDI-TOF) mass spectrometry. American Journal of Enology and Viticulture 49(3): 231–239.

Welke, J.E., Manfroi, V., Zanus, M., Lazzarotto, M. and Alcaraz Zini, C. 2013. Differentiation of wines according to grape variety using multivariate analysis of comprehensive two-dimensional gas chromatography with time-of-flight mass spectrometric detection data. Food Chemistry 141(4): 3897.

WinMet database (http://databases-metabolomips.rhdound.co/).

Traceability and Authentication of Organic Foodstuffs

Céline Bigot[1,2],* Romain Métivier,[4] Didier Montet[1] and
Jean-Christophe Meile[3]

1. Introduction

The existing consumer expectations concerning food authenticity and traceability have made them increasingly interested in food labels, such as the signs of origin (e.g., Protected Geographical Indication or PGI, Protected Designation of Origin or PDO) and agricultural labels (Traditional Specialties Guaranteed or TSG, organic label, etc. …), because they are considered as guarantee of quality and safety. However, despite the strict regulations and controls realized on these types of products, they may be subjected to frauds. In addition, the excessive media exposure of various food crises over the past few years, such as those linked to "organic cucumbers" (fenugreek sprouts were actually implicated) in 2011 or the beef meat adulteration scandal in 2013, have led consumers to be more focused on what they eat. Food

[1] CIRAD-UMR Qualisud, TA B-95/16, 73, rue Jean-François Breton, 34398 Montpellier Cedex 5, France.
[2] CEA/IG/CNG/LMPD-DT, Evry, 2 rue Gaston Crémieux, CP 5721, 91057 Evry Cedex, France;
[3] CIRAD-UMR Qualisud, Station de Ligne-Paradis, 7 chemin de l'IRAT 97410 Saint-Pierre, Réunion, France; E-mail: jean-christophe.meile@cirad.fr
[4] LMBA – UMR CBMN 5248, Bordeaux Sciences Agro, 1 cours du Gal De Gaulle - CS 40201, 33175 Gradignan Cedex, France; E-mail: rom.metivier@gmail.com
* Corresponding author: celine.bigot83@hotmail.fr; celine.bigot@cea.fr

traceability and food authenticity have thus become major challenges for inspection bodies, food industries and scientific community. Many studies, in the recent years, have been conducted around these two concepts with the development of new tools that mainly use the physico-chemical and molecular biology fields (notably through European Union funded projects such as the Food Integrity, CoreOrganic or Authent-Net projects).

Food authenticity is in line with the current expectations of consumers for natural and local products. Assessing food authenticity means being able to detect (i) if food description is correct or incorrect and does not meet the requirements of a legal name, (ii) if ingredients have been replaced by others less expensive, (iii) if processes are not reported (e.g., irradiation, freezing …) and (iv) if the origin (geographic, species …) is false. Industrialization and globalization have made the authentication of foods particularly difficult given the fact that a large part of food products available on the market are transformed and therefore, more susceptible to falsifications (Johnson 2014). In addition, frauds are constantly innovating to bypass controls and thus, rendering them more difficult to detect by classical analyses. These concerns have become major, not only for consumers but also for producers, distributors and authorities. These illicit activities involve not only significant economic losses worldwide, but also affect the confidence of consumers who call into question the credibility of food labels. So, it is necessary to develop advanced analytical methods to detect non-compliant products, notably false organic products which are among commodities the most subjected to fraud in Europe (European Commission 2013/2091(INI)).

Organic farming is a method of sustainable production which contributes to the environmental and animal protection by a set of specific agricultural practices. The label "organic", like other official labels (such as TSG, PDO, PGI …), attests to the quality of a product. Indeed, it is subjected to specific regulations, whose applications are controlled by certification bodies agreed by public authorities. Farmers having opted for this mode of production have to comply with the organic specifications and standards associated to be able to sell products under the "organic" label. These specifications are different from one country to another but have been harmonized at the European level (EEC No. 2092/91, repealed by EC No. 834/2007 since the 1st January 2009). Consumer demands for fresh local foods or cultivated according to organic methods is increasing (Willer and Kilcher 2012). This growing interest, related to the rapid development of organic production systems, has made more difficult the certification and the guarantee of organic products from around the world (considering also that organic foods are part of the most difficult products to control because it is dependent on the type of product: meats, fruits, vegetables and others). Although the "added value" of these products is guaranteed by a certification system (EC No. 834/2007 notably), current systems are mainly administrative and can be falsified. Risks levels also differ from one country to another: very few cases of fraud were reported in France, while frauds are regularly reported on organic product originating from Spain or Romania. Food frauds are mostly motivated by the price of organic products that are significantly higher than conventional ones (Everstine et al. 2013).

Thus, to ensure the authenticity of foods commercialized as "organic", there is an existing need for robust, accurate and inexpensive methods to support the certification, control and traceability systems. This chapter firstly gives an overall presentation of organic farming and its guarantees in terms of regulation. The second part of this chapter reviews some of the most recent analytical approaches applied to authenticate or compare and the biomarkers identified to ensure the authenticity of organic food products.

A. Part 1: Generalities on Organic Farming

1.1 What is organic farming?

Organic farming is an alternative agricultural method for vegetable and livestock that uses natural sources and nutrients from compost and crop residues. It excludes the use of synthetic chemical molecules, Genetic Modified Organisms (GMOs), irradiation and genetic engineering. The Food and Agriculture Organization (FAO) and the World Health Organization (WHO) defined, in *Codex Alimentarius,* organic agriculture as "a holistic production management system which promotes and enhances agro-ecosystem health, including biodiversity, biological cycles, and soil biological activity. It emphasizes the use of management practices in preference to the use of off-farm inputs, taking into account that regional conditions require locally adapted systems. This is accomplished by using, where possible, cultural, biological and mechanical methods, as opposed to using synthetic materials, to fulfil any specific function within the system" (Joint FAO/WHO *Codex Alimentarius* Commission, Food and Agriculture Organization of the United Nations, et World Health Organization 2007).

1.1.1 History of organic farming

Until the 1920s, agriculture was generally "organic". Farmers used natural means to feed the soil and to control pests (The Organics Institute 2016). The concept of organic farming started to emerge in the early part of the twentieth century. The pioneers of the organic movement desired to reverse the problems of agriculture such as soil erosion and depletion, loss of quality of foods, livestock feeds and crop varieties. This movement embodied a commitment to sustainability through soil regeneration and sought to avoid wasteful exploitation of natural resources (Kuepper 2010). The concept of organic farming started through four main currents of thought based on ethical and ecological principles (Besson 2007): the works of the Austrian philosopher Steiner, conducted on the biodynamic farming in the 1920s, were based on the use of bio-stimulant (such as vegetable and mineral substances) and composting with consideration of the telluric and cosmic strengths in the agricultural practices. Howard's works, a precursor of organic farming in United Kingdom, also contributed to this concept by developing a system of composting that became widely adopted. His concept of soil fertility (defined in his book, *An Agricultural Testament* (Hall 1940)) was focused on building soil humus with an emphasis on how soil life was connected to the health of crops, livestock, and mankind (Heckman 2006). Then, in the 1960s, the Swiss biologist Rusch developed a test to determine the level of soil fertility. He collaborated with Müller and his wife to promote the organic agriculture in Switzerland. In Japan, the notions of natural or savage agriculture were used and introduced by the farmer and philosopher Fukuoka. His system took place in 1930 and was based on the recognition of the complexity of living organisms that shape an ecosystem and deliberately exploiting it. These four currents of thought were in contrast to most of agricultural systems in the world with mismanagement of resources (Lowdermilk 1975).

However, after the Second World War, farming methods changed dramatically and became "industrials". Indeed, it was essential and urgent to reconstruct the industries and the millions of hectares of farmlands devastated by years of war. In this period, simultaneous advances in engineering and biochemistry rapidly and profoundly changed farming practices. It was amplified by the need to feed the populations. Scientific and technological innovations during the Second World War were mainly directed towards industrial applications to intensify the production. Thus, there were large advances in mechanization (tractors, large-scale irrigation…) accompanied by the massive arrival of synthetic molecules as pesticides and fertilizers. In particular, the ammonium nitrate, used for munitions during wartime, was a good source of nitrate in abundant quantity. Moreover, this agricultural type was encouraged by an international campaign called the Green revolution in 1944. This industrial agriculture, based on the combination of technological and scientific progress, allowed to considerably increase yields and to avoid famine of a growing population from 1960 to 1970. This intensive agriculture was also considered as being a solution against food insecurity. But, very quickly, the unregulated use of fertilizers led to an imbalance: deficiencies of soil, reduction of the biodiversity, loss of soil permeability.

At the beginning of the 60s, there was a growing awareness about the consequences of intensive farming and the effects of pesticides on health and environment. Some social movements in opposition to this farming type appeared. At this time, "productivist" agriculture was deeply criticized for its strong fossil energy consumption and for its polluting nature. In this context, organic farming appeared as an interesting alternative. For example, Rachel Carson, a marine biologist and conservationist, published her book "Silent Spring" in 1962 which showed the effects of DDT and pesticides on the environment (Carson 2002). This book helped the US government to decide the exclusion of DTT use for agriculture in 1972.

At the beginning of the 70s the notions of specifications, guarantees and controls were developed to ensure a quality defined for the consumer. In 1972, Nature and Progress (A French federation of consumers and professionals engaged in the agro-ecology since 1964) impulsed the creation of the International Federation of Organic Agriculture Movements (IFOAM) in Versailles (France). Its objective was to coordinate the active organizations in the organic sector. IFOAM is a world reference for minimal standards that all organizations members have to respect. During this period, the awareness in favor of the environmental protection and the wish of a new quality of life had strengthened. But it took a long time for the methods of organic farming to be adopted as they were often marginalized.

In 1975, Fukuoka wrote "The One-Straw Revolution". This book had a strong impact in many areas of agriculture in the world. His approach to small-scale grain production showed equilibrium of local farming ecosystems and human actions.

In 1979, a law project on the fertilizing materials allowed to approach for the first time the question of the organic farming in France. One year after, the agricultural guidance law of July 4th, 1980 recognized "Organic" without quoting it in particular (Law No. 80-502 of July 4th, 1980). In the mean time in the U.S., Oregon Tilth started an organic certification program. It created the Western Alliance of Certification Organizations with California Certified Organic Farmers (CCOF) and the WA State Dept. of Agriculture's Organic Program to formulate material list standards and further align the three western certification programs (1984) (Oregon Tilth 2016). This led to legislation and certification standards from 1990. In the U.S., the Organic Foods Production Act (1990) ended in the creation of the National Organic Program, published in Federal register in 2000 (Huber 2005).

1.1.2 World-wide organic agriculture

The results of the latest survey on organic agriculture world-wide showed that organic farming is present in 172 countries (IFOAM–Organics International 2016). Approximately, 2.3 million of producers cultivate 43.7 million hectares of agricultural land that are managed organically (including in conversion areas). The global sales of organic foods and drinks reached 80 billion US dollars in 2014 (FiBL 2016). Oceania has the largest areas of organic agricultural lands with hosting of about 40% of the total world organic agricultural land. Europe represents 27%. Latin America comes third with 15%, followed by Asia (8%), North America (7%), and Africa (3%). In 2014, almost 500 thousand supplementary hectares of organic agricultural lands were reported compared to the previous year. This increase of organic agricultural land concerns all regions, except Latin America. In Europe, organic areas increased by almost 2%. In France, organic areas increased by 24% in 2014 and 2015 (Agence Bio 2016). In Africa, the total surface devoted to organic agriculture increased by almost 5.5%.

Apart from agricultural land, there are areas for organic wild species collection. Forests, grazing and aquaculture represent other areas of non-agricultural lands. In total, agricultural (43.7 million hectares) and non-agricultural (37.6 million hectares) areas represent 81.2 million hectares in the organic sector. Table 1 gives key numbers and top countries in the organic sector (IFOAM – Organics International 2016).

Table 1. Organic Agriculture 2016: Key Indicators and Top Countries (IFOAM–Organics International 2016).

Indicator	Worlds	Top countries
Countries with organic activities	2014: 172 countries	New countries: Kiribati, Puerto Rico, Suriname, United States Virgin Islands
Organic agricultural land	2014: 43.7 million hectares (1999: 11 million hectares)	Australia (17.2 million hectares; 2013) Argentina (3.1 million hectares) US (2.2 million hectares, 2011)
Wild collection and further, non-agricultural areas	2014: 37.6 m hectares (1999: 4.1 million hectares)	Finland (9.1 million hectares) Zamia (6.8 million hectares) India (4million hectares)
Producers	2014: 2.3 million producers (1999: 200 thousand producers)	India 650,000 producers (2013) Uganda (190,552 producers) Mexico 160,703 producers (2013)
Organic market size	2014: 80 billion US Dollars (1999: 15.2 billion US Dollars)	US (35.9 billion USD; 27.1 billion euros) Germany (10.5 billion USD; 7.9 billion euros) France (6.8 billion USD; 4.8 billion euros)
Per capita consumption	2014: 11 US dollars (14 euros)	Switzerland (221 euros) Luxemburg (164 euros) Denmark (162 euros)
Number of countries with organic regulations	2015: 87 countries	
Number of IFOAM affiliates	2015: 784 affiliates from 117 countries	Germany – 92 affiliates China – 57 affiliates India – 44 affiliates USA – 40 affiliates

So, the market of organic farming is rapidly growing. Worldwide retail sales of organic foods and drinks reached 80 billion US dollars in 2014 against 28.7 billion US dollars in 2004. North America and Europe are the first consumers of organic products and drain 90% of sales, respectively 38.5 and 35 billion US dollars (IFOAM–Organics International 2016). The major import markets for organic products are the European Union, the United States, Canada, and Japan. In these countries, products may only be imported if the certifying agency has approved them. However, the legislation can be different between the importer and exporting countries. These confusions make the existence of control and certification bodies to guarantee the organic origin of products necessary.

1.2 Guarantees of organic farming

Organic farming is a multi-scale regulation that appears to be very complex. The first European regulation on organic plant production was published in 1991 and the term "organic" was officially recognized (EC No. 2092/91 of 24th June 1991) as part of the reform of the Common Agricultural Policy (CAP). For organic animal production, the legal basis is ruled by the EC Regulation No. 1804/99 dating back to 1999. Until the implementation of these regulations, the quality defined for the consumer was provided by specifications managed by private organizations, which served as guarantee and control, until the rules of production were harmonized.

In the FIBL survey (IFOAM–Organics International 2016), 87 countries in the world have an organic standard and 18 are in the process of drafting legislation. In these countries, organic standards are overseen by the government but also public or private organizations like the "United States Department of Agriculture" (USDA) in the US, "Japanese Agricultural Standard" (JAS) in Japan and "AB—agriculture biologique" in France (see below). These rules have led to the harmonization of regulations on organic farming, with essentially the help of the International Federation of Organic Agriculture Movements (IFOAM) and the Codex Alimentarius.

1.2.1 Codex alimentarius

Today in Europe, and since 1st January 2009, organic farming is governed by the Regulation (EC) No. 834/2007 which repeals and replaces the Regulation (EC) No 2092/91. This new regulation specifies the set of rules to follow for the conditions and principles of production, processing, distribution, import, control or labeling of organic products. There are complementing regulations, notably the Regulation (EC) No. 889/2008 (for the organic production, the labeling rules and the controls). These rules are regularly reviewed to be adapted and to enable the development of organic agriculture. But this harmonization has brought some debates and raised questions about the quality of European organic products. Indeed, some national measures are stricter than the European regulation, for example those established in France concerning the Genetically Modified Organisms (GMOs). This has led to reluctances for the implementation of harmonized rules. Despite that, the harmonization of rules in a European regulatory framework has been beneficial to the organic sector, as it allowed to this agriculture to become one of the most dynamic sectors in agriculture in Europe (see Section 1.1).

Currently, public and private national specifications still exist and may complete the European Regulation 834/2007 to provide additional guarantees, even if they do not replace it. Besides, some operators choose to follow the two types of specifications, national and European.

Concerning the organic certification and control systems, in most European countries, the certification and control bodies are private (http://ec.europa.eu). Each Member State has to designate the control authority(ies), which mostly correspond to a department of the Ministry of Agriculture or Public Health (e.g., in France it is notably the Directorate General for Competition, Consumer Affairs and Prevention of Fraud, or DGCCRF). This competent authority may delegate all or a part of its inspection tasks to one or more private/public inspection bodies. The private inspection bodies must be approved and supervised by the competent authority. It is also possible to make use of both systems (public and private inspection bodies). A list of these European certification and control bodies is available at: https://ec.europa.eu/agriculture/organic/index_en (European Commission 2017). Each year, the Member States have to submit to the European Commission their report about the controls that have been taken on organic operators and about the measures taken in case of non-compliance. Concurrently, the European Commission supervises the Member States to fulfill their responsibilities, according to the information provided in their annual reports and audits performed.

Thus, the rules that define and control the organic farming are on three scales: national, European, and global scales (with notably the standard rules and definitions of the *Codex Alimentarius*). All of these regulatory and control systems can be found behind the organic logo. The official European organic logo (named "Euro-leaf") became mandatory since 2010 (EU Regulation No. 271/2010 of the Commission of 24 March 2010). Widely broadcasted by the media, the logo allows strengthening the value of organic products and also their protection against imitation. But the main purpose is to allow consumers to quickly identify with certainty an organic product and to enhance consumer confidence. Indeed, the logo complies with the organic rules under the Regulation which frames them (EC No. 834/2007) and ensures that the products contain at least 95% of ingredients certified as "organic". The remaining 5% must be mentioned on a positive list annexed to the framework regulation. If an ingredient is not available in the organic agriculture, a temporary authorization for use can be given. The place of production and the code of the certifying body must systematically be included. It is also possible to mention the country (when at least 98% in weight of the raw materials come from this country). The European logo cannot be used if (i) the products do not lie within the scope of EU legislation (such as hunting or fishing of wild animals, cosmetics and textiles), (ii) the products contain less than 95% of organic ingredients, and if (iii) the products are transitioning toward organic agriculture.

National and private logos may also be included optionally with the European logo. They may be affixed alone in case the product only complies to national level but only after certification by an accredited certifier body. So, the logo will always be accompanied by the mention "controlled or Certified by" followed by the name of the certifying body. These various guarantee levels of the organic sector show that the rules are very strict but also very complex. The products are guaranteed by a certification system which should justify their traceability/authenticity at all stages of the production, processing and commercialization (EC 834/2007 notably). However, current systems are mainly administrative and may be non-compliant, voluntarily or not: involuntarily because, as noted above, the regulation of organic agriculture is divided between several European and national texts, which evolve frequently. So, the producer has difficulties for understanding what could be applied to his production and his farm. And voluntarily, because an important economic issue is hidden behind the "organic" mention, which is mainly due to the certification costs and the lower yield of organic farming compared to conventional farming (Capuano et al. 2012). Thus, frauds are mostly motivated by the prospect of financial gain. These non-compliances are mainly related to the non-respect of organic regulations concerning pesticide residues, GMOs or heavy metals. The administrative systems are not self-sufficient to detect and identify the non-compliances. They must be accompanied by analytical tools to be more effective. But the current and only analytical tool applied by control bodies is based on the analysis of pesticide residues, which is expensive and not effective enough (Laursen et al. 2014).

To conclude with this section, consumer interest in organic food products is increasing, and so is the risk for more and more motivated organic food frauds. Nowadays, the authentication of organic products is a complex challenge which is important to be able to raise in order to limit frauds and meet the needs of consumers in terms of organic food traceability and authenticity and thus, in a broader perspective, to contribute to food safety. The next part of this chapter provides an overview of the different approaches proposed by the scientific community for ensuring the authentication of organic products.

Part 2: How to Prevent Organic Food Fraud? Different Trends in Food Authentication

2.1 Analytical methods to authenticate organic foods

Regulations and controls of organic food production systems mainly focus on the methods of production but not necessarily on the quality of the final products. Therefore, there are no specific legal expectations for organic food products. This it renders very difficult to define what makes organic food products different from conventional ones. From the analytical point of view, one could expect less or a total absence of specific pesticides residues (vegetal products) or antibiotics (animal products). But so far, there are no criteria defining organic foods products *per se* and no recognized worldwide standard. Hence, finding a way to differentiate or authenticate organic products constitutes a real challenge.

The general assumption is that different agricultural practices will impact the chemical composition of a given food product (mineral composition, nitrogen content, micronutrients, etc. …). However, different parameters can also influence this chemical content and be source of variations such as crop variety (or cultivar), climate, soil, human practices and post-harvest treatments or storage conditions.

The main issue is to make sure that significant variations or differences observed when analyzing food products can be attributed to the agricultural practices only. To this end, there is still data and knowledge to be produced in order to better understand the contributions of each parameter to food chemical composition variations. In this way, one can think of seeking for specific organic or discriminating markers.

Various techniques based on chemical composition variations, have been employed in trying to discriminate between organic and conventional food products. Up to now, classical approaches to food composition analysis revealed no significant differences between organic and conventional products especially for carbohydrates and minerals (for a review, see Vallverdú-Queralt and Lamuela-Raventós 2016). Therefore, there is a need to develop innovative tools able to rapidly and efficiently analyze simultaneously multiple (profiling) and/or minor markers that could be specific to a type of food product.

In the following lines, different examples of studies using various analytical methods to discriminate (or authenticate) organic food products from non-organic ones will be browsed. [Since most of the following techniques are presented in other chapters, their principle will not be developed here.]

2.1.1 Pesticides/antibiotics residues

The first expected difference between organic and conventional commodities would be the number and amount of pesticides (or antibiotics) residues that can be detected in vegetal (or animal) products. Considering pesticides residues in organic vegetables, in practice, no significant difference could be observed from various contexts, and most importantly pesticide residues analysis could not be seen as discriminant markers. Indeed, various studies showed that pesticides residues could be detected in organic food products, rendering interpretations uneasy. This is due to the tolerance of pesticide use in specific conditions for certain type of crops and non-homogenous regulation on authorized molecules in organic production systems. Large public-funded studies in the US (see USDA survey) and Canada lead to similar conclusions, detecting pesticides residues in a significant number of organic samples, sometimes reaching amounts comparable to conventional counterparts. Several hypotheses could explain these observations, apart from a deliberate fraud, including contamination of water, soil, or drift during treatment spraying, or contamination during post-harvest storage. In all, pesticides residues analysis appears inadequate to accurately discriminate between organic and conventional foodstuffs.

In the case of antibiotics, their utilization in animals used to produce organic food is strictly prohibited in most countries. Hence, in an organic farm, animals that are treated with antibiotics must be clearly identified and sold separately into the non-organic market. In some countries, the regulation is more flexible, for example in Canada, a withholding period of 30 days is allowed to produce organic milk again. In all, no antibiotics residues are expected to be found in organic animal products.

2.1.2 Stable isotope analysis

This method relies on the detection and measure of stable isotopes of main elements (C, N, O and S) in food products. Stable isotopes ratios (SIR) are determined by isotope ratio mass spectrometry (IRMS) and have proved to be a very useful and reliable method to detect adulteration and certify the geographical origin of various added value products, especially wine (Christoph et al. 2015).

Considering organic products, this stable isotope analysis approach (essentially using 15N/14N ratio or $\delta^{15}N$) allows getting insight about the fertilization history of the plant. Food products grown in conventional agriculture with synthetic fertilizers tend to have lower $\delta^{15}N$ values than organic products grown with manure and/or compost. Even if in theory the $\delta^{15}N$ value is a very efficient marker it cannot be used solely for differentiating between organic and conventional, because in some cases, other parameters can induce variations in N isotopes ratios (Kelly and

Batesman 2010). To overcome this limitation, it seems that the combined analysis of two or more isotopes ratios gives more accurate results. A recent study shows that $\delta^{15}N$ and $\delta^{13}C$ values analyzed on specific compounds rather than bulk tissue can help improving the discrimination between organic and conventional food products. In this case, $\delta^{15}N$ and $\delta^{13}C$ values were determined on amino acids after protein hydrolysis and derivatization, using gas chromatography–combustion–isotope ratio mass spectrometry (GC-C-IRMS). In particular, the combination of $\delta^{15}N$ and $\delta^{13}C$ of 10 amino-acids could improve the discrimination between conventional and organic wheat compared to stable isotope bulk tissue analysis (Paolini et al. 2015).

Stable isotope analysis is also widely used in animal products authentication. The combination of stable isotopes and trace elements permits to get more accurate results and discrimination (for a review see Camin et al. 2016).

The main limitation of isotope analysis in authentication is the need of authentic reference samples that serve to build a reference database. Therefore, very large datasets are required to validate tested samples.

2.1.3 Chemical compound family-targeted analysis

Certain families of chemical compounds can be targeted to differentiate between types of food as they constitute discriminant markers of production mode. For example, carotenoids profiling (using HPLC) of egg yolks can help authenticate organic eggs. Total carotenoids composition in organic and conventional eggs differs depending upon hen-feed additives used. The combined analysis of two carotenoids (carotenes and xanthophylls) provided a set of markers that allowed the multivariate authentication of organic eggs (Van Ruth et al. 2011).

In dairy products, specific compounds can be searched for their prevalence in organic farming systems. For example, organic cheeses are prepared from milk originating from grass-fed cows. Vetter and Schröder (2010) showed that two compounds produced by chlorophyll bacterial rumen metabolism were more abundant in German organic cheese. By using Gas Chromatography coupled to Mass Spectrometry (GC-MS) they observed that organic cheese contain on average 50% more phytanic acid and 30% more pristanic acid than conventional cheese (Vetter and Schröder 2010).

Combinations of methods appear to be very reliable and accurate to discriminate and authenticate organic food products. For example, in the UK, a comparative study of lipid profiles in Lamb from various origin and quality was performed in combination with sensory analysis (Angood et al. 2008). In Netherlands, Volatile Organic Compounds (VOC) fingerprints of organic and conventional tomatoes were analyzed using Proton-Transfer

Reaction coupled to Mass Spectrometry (PTR-MS). In parallel, sensory analysis was performed (Muilwijk et al. 2015).

Compound family-target or multi marker fingerprints reveal to be powerful to authenticate organic food products even if these approaches have to be adapted to each food matrix studied.

2.1.4 Spectral signatures (NIR – MIR – NMR – spectroscopy) – hyperspectral imaging

Spectroscopic methods have been greatly utilized for assessing food quality. These methods are very convenient for food analysis as they usually require minimal sample preparation, provide rapid and on-line analysis, and have the potential to be applied to virtually any type of food samples (Nawrocka and Lamorska 2013).

Global spectral approaches provide a snapshot of the chemical composition of a sample. These rapid methods provide complex spectral signatures from which specific features can be assigned (using adequate statistical treatments) to chemical components (water, fat content, etc. …). For example, Mid-Infrared spectroscopy (MIR) was used to discriminate between organic and non-organic wines grown in Australia. Collected spectra were analyzed using chemometrics and showed the potential of MIR spectroscopy as a rapid tool for the wine industry (Cozzolino et al. 2009).

Tres et al. (2012) published a feasibility study using Near-Infrared Spectroscopy (NIR) combined with chemometrics to authenticate organic feed. Nuclear Magnetic Resonance (NMR) spectroscopy was used to differentiate conventionally and organically Grown Tomatoes (Hohmann et al. 2014). And more recently, ^{1}H NMR foodomics approaches were used to show that grapes berries produced from biodynamic and organic farming exhibited different metabolomes. Therefore it was possible to differentiate the two types of grapes according to their production systems (Picone et al. 2016).

Although the following method has not been (to our knowledge) applied up to now to organic food, we believe it has to be mentioned here. Hyperspectral imaging combines both spectroscopy and imaging techniques to perform direct identification of chemical components and their spatial distribution in samples. The combination of spatial and spectral data has proved to be a promising technology with great potential for application in food quality and authentication (Elmasry et al. 2012).

Spectroscopic methods combined with chemometric analyses are a very powerful tool to differentiate groups of samples that have very similar properties but consistent overall differences. In addition, they provide rapid and cost-effective analysis (price per sample). However, calibration samples and application of multivariate calibration techniques are required for spectral analytical methods to extract the sought chemical information and potential markers for organic products authentication.

2.1.5 Foodomics/big data

The term "Foodomics" was first employed in 2009 by Aljandro Cifuentes. Foodomics is a discipline that integrates untargeted "OMICS" analyses in a holistic approach to food and nutrition. This includes metabolomics, proteomics, ionomics and any other type of high-throughput approaches (Cifuentes 2009).

These type of approaches generated large amounts of data that required proper storage, interpretation and adequate statistical analyses to be able to extract the desired information. Therefore, tools like Multivariate Analysis, Chemometrics, Data mining and Machine learning will co-evolve fast together with food analysis methods. One can imagine that organic products authentication will get benefit from the current and future advances in that field. Another promising approach focusing on comparative analysis of microbial ecology in organic and conventional foods showed that molecular microbial markers could be detected. These examples will be developed in the following section.

2.2 Are there discriminant organic markers? Applications for authentication purpose

The recent development of modern analytical tools led to the adjustment of methods for ensuring the genuine character of organic food products. The overall current research works that have been conducted by applying these tools allowed the identification of some biomarkers for ensuring the authentication of organic foods. The publication of Capuano et al. (2012) offers a broad overview of approaches that have been investigated, and especially the biomarkers that have been proposed by the scientific community as being potentials to discriminate organic foods. It appears that the measurement of the isotopic composition of organic/non-organic products is mostly used and depends on the type of food studied. Notably nitrogen isotopes (^{15}N/^{14}N) are used for the authentication of organic crops (as corns, tomatoes, zucchini, potatoes or lettuces), and the carbon-isotopic composition (^{13}C/^{12}C) for the authentication of organic meats (e.g., beef, lamb or pork). Indeed, the emergence of recent studies that have been conducted on this topic showed the potential of these biomarkers (Camin et al. 2011, Sturm and Lojen 2011, Capuano et al. 2012, Flores et al. 2013, Laursen et al. 2014). This interest is based on the fact that fertilization and husbandry practices are very different between organic and conventional production modes. This leads to a measurable and significant variation of the multi-elemental and isotopic composition of plants and meat (Laursen et al. 2014). However, the comparative study of these various works by Capuano et al. (2012) and Laursen et al. (2014) allowed to raise the issue that the values of these ratios were dependent on the food studied, but also on local practices that may vary from one producer to another ("terroir effect"). These differences seem to be even more important in meat industry (apart from season, geographical origin, type of meat or agricultural practices, feed ingredients must also be taken into account). So, there is a certain degree of variability that could lead to bias organic food authentication if only these types of biomarkers are taken into account. The mono-marker analysis may therefore be less useful and effective than multi-markers analysis given the fact that the result can lead to the non-detection of fraudulent products. For example, Laursen et al. (2014) have demonstrated that multi-isotope analysis has the potential to detect frauds in organic products in a more efficient way than the targeted analysis of specific stable isotopes. This is also underlined by Capuano et al. (2012). Indeed, analysis strategies based on the measurement of multi-markers would be more promising for ensuring the authentication of organic foods.

Emerging «omics» technologies, such as metabolomics and proteomics tools, have been applied for the discrimination of organic foods (Nawrocki et al. 2011, Novotná et al. 2012, Mie et al. 2014). Although these tools are not yet sufficiently exploited, interesting results were obtained on several types of foods which demonstrated their potential. Besides, these studies lead to the same conclusion as it was done by applying physicochemical methods: a combination of multiple organic biomarkers, metabolomics/proteomics with those derived from other data analysis tools (such as stable isotopes), could improve the reliability of the authentication of organic foods (Laursen et al. 2014).

Other strategies were considered to achieve this objective, such as the possibility to use discriminating markers which are of the microbial type. Very few studies have investigated the possibility of using microorganisms as potential biomarkers for the discrimination of foods according to their mode of production. Indeed, the microorganisms sought in organic foods (as in conventional ones) are rather human pathogens (Hoogenboom et al. 2008, Ottesen et al. 2009, Maffei et al. 2013), or from organic and conventional soils (Shannon et al. 2002, Wallis et al. 2010, Reilly et al. 2013). However, agricultural practices are very different between organic farming and the others (for examples, pesticides and GMOs are prohibited in organic farming while the use of pest controls is favoured), so it can be assumed that farming types will have a significant impact on overall microbial flora of food. Besides, this has been demonstrated on fruits and vegetables (Ottesen et al. 2009, Leff et al. 2013, Bigot et al. 2015). These studies showed that the "microbial signature", related to a given mode of production, could be used for authentication purposes. Its efficacy has been demonstrated to link the microbial fingerprint of foods to their geographical origin (Le Nguyen 2008, El Sheikha et al. 2009, Tatsadjieu et al. 2010, Dufossé et al. 2013). In addition, some studies have already highlighted and identified some organic discriminant microbial groups (Bigot et al. unpublished). The latter have shown for example, by both profiling (DGGE) and high-throughput sequencing (HTS) of microbial DNA extracted from apple surface that discriminant bacterial species can be identified from different phyla (*Proteobacteriea*, *Bacteroidetes*, *Actinobacteria* and *Firmicutes*). Discriminant fungal groups belonged to the division of Ascomycota but also to the division of Basidiomycota. Chemometrics was used to extract information from DNA profiling and sequencing and permitted to confirm the importance of these microbial groups to discriminate apples according to their farming types, and thus the interest to conduct a more targeted analysis of food microbiota. Indeed, microbial species that do not contribute to the discrimination of organic foods cover the information given by those that are necessary for this purpose. The combined analysis of PLS-DA and VIP (Variable Important in the Projection) showed that about 20 microbial species, some belonging to the "organic class" and the others to the "conventional class", are essential to discriminate organic fruits (data being published). This study remains at an exploratory stage: further research would be needed to verify the phylotypes associated to a farming type and to ensure that these discriminant microbial markers are present over time, in other varieties of apples and maybe on other type of fruits, from different varieties and geographical origins. However, the results obtained remain interesting and are correlated with previous studies (Ottesen et al. 2009, Leff et al. 2013). In addition, following the recent advances in molecular biology, DNA analysis has become a very effective and powerful tool to meet the needs of various areas (medical, food…). It provides additional information that could bring the methods based on protein or chemical analysis (Madesis et al. 2014). DNA is a relatively thermostable molecule, and so offers a great power of preservation compared to other molecule types. DNA would be more resistant to physical and chemical industrial processes (Galimberti et al. 2013). So this molecule, and especially methods based on DNA amplification by PCR (Polymerase Chain Reaction), has quickly become the basis for the development of tools that are widely used in the food control context (Restriction Fragment Length Polymorphisms, RLFP; Random Amplified Polymorphic DNA, RAPD; Simple Sequence Repeat, SSR; Single Nucleotide Polymorphism, SNP and others…) (Galimberti et al. 2013, Scarano and Rao 2014). The results of different studies applying tools based on DNA analysis clearly show the interest to develop methodologies that appeal to molecular biology for their applicability in agri-food sector (Palmieri et al. 2009, Galimberti et al. 2015) notably to validate food authenticity and traceability for both fresh and processed foods. It shows their relevance and encourages continuing on this path to develop, in a longer-term perspective, an authentication tool of organic foods based on the use of these discriminant DNA-microbial markers.

3. Conclusion

Finally, the authentication of foods from organic farming is a difficult challenge since most foods, notably organic foods, are complex because they are processed and are from various production modes and many varied parameters (geographical origin, variety). However, studies that have been conducted until now have demonstrated that potential markers exist and one can imagine their utilization in the near future. The type of statistical analyses to apply to identify markers is critical. The multi-markers analysis (identification and quantification), by fingerprinting/profiling approaches combined with multivariate statistics and chemometrics (e.g., prediction models regularly updated) (Tres et al. 2012), seems to be the preferred path to develop an authentication tool based on its rapid, reliable and cost-effective detection.

Once specific markers will be identified from research study, the next step is method development for application. Method development is strongly linked to regulation that can impose controls and authentication analyses. Therefore, applications of new methods for organic food authentication in the supply chain will depend on how regulation on organic food products will evolve and who will be in charge of the tests. Certifiers? Farmers? Traders? NGO's? Researchers? Industry? Control laboratories?

Keywords: Traceability, organic foods, authenticity, food safety, CE 178/2002, biomarkers

References

Agence Bio. 2016. «Communiqués et dossiers de presse - Agence Française pour le Développement et la Promotion de l'Agriculture Biologique - Agence BIO». http://www.agencebio.org/communiques-et-dossiers-de-presse.

Angood, K.M., Wood, J.D., Nute, G.R, Whittington, F.M., Hughes, S.I. and Sheard, P.R. 2008. A comparison of organic and conventionally-produced lamb purchased from three major UK supermarkets: Price, eating quality and fatty acid composition. Meat Science 78(3): 176–84.

Besson, Y. 2007. Histoire de l'agriculture biologique: une introduction aux fondateurs, Sir Albert Howard, Rudolf Steiner, le couple Müller et Hans Peter Rusch, Masanobu Fukuoka. Troyes. http://www.theses.fr/2007TROY0003.

Bigot, C., Meile, J.C., Kapitan, A. and Montet, D. 2015. Discriminating organic and conventional foods by analysis of their microbial ecology: An application on fruits. Food Control 48: 123–129.

Camin, F., Perini, M., Bontempo, L., Fabroni, S., Faedi, W., Magnani, S., Baruzzi, G., Bonoli, M., Tabilio, M.R., Musmeci, S., Rossmann, A., Kelly, S.D. and Rapisard, P. 2011. Potential isotopic and chemical markers for characterising organic fruits. Food Chemistry 125: 1072–1082.

Camin, F., Bontempo, L., Perini, M. and Piasentier, E. 2016. stable isotope ratio analysis for assessing the authenticity of food of animal origin. Comprehensive Reviews in Food Science and Food Safety 15: 868–877.

Capuano, E., Boerrigter-Eenling, R., Van der Veer, G. and Van Ruth, S.M. 2012. Analytical authentication of organic products: an overview of markers. Journal of the Science of Food and Agriculture 1: 12–28.

Carson, R. 2002. Silent Spring. Houghton Mifflin Harcourt.

Christoph, N., Hermann, A. and Wachter, H. 2015. 25 Years authentication of wine with stable isotope analysis in the European Union–Review and outlook. In BIO Web of Conferences (Vol. 5, p. 02020). EDP Sciences.

Cifuentes, A. 2009. Food analysis and foodomics. Journal of Chromatography A 1216(43): 7109.

Cozzolino, D., Holdstock, M., Dambergs, R.G., Cynkar, W.U. and Smith, P.A. 2009. Mid infrared spectroscopy and multivariate analysis: A tool to discriminate between organic and non-organic wines grown in Australia. Food Chemistry 116(3): 761–765.

Dufossé, L., Donadio, C., Valla, A., Meile, J.C. and Montet, D. 2013. Determination of speciality food salt origin by using 16S rDNA fingerprinting of bacterial communities by PCR-DGGE: an application on marine salts produced in solar salterns from the French Atlantic Ocean. Food Control 32: 644–649.

EC (1991) Council Regulation (EEC) No 2092/91 of 24 June 1991 on organic production of agricultural products and indications referring thereto on agricultural products and foodstuffs. Official Journal of the European Communities, L198 (22.7.91), 1–15.

EC (1999) Council Regulation (EC) No 1804/1999 of 19 July 1999 supplementing Regulation (EEC) No 2092/91 on organic production of agricultural products and indications referring thereto on agricultural products and foodstuffs to include livestock production. OJ L 222, 24.8.1999, p. 1–28.

EC (2007) Council Regulation (EC) No 834/2007 of 28 June 2007 on organic production and labelling of organic products and repealing Regulation (EEC) No 2092/91. Official Journal of the European Communities, L189/1 (20.7.2007), 1–23.

EC (2008) Commission Regulation (EC) No 889/2008 of 5 September 2008 laying down detailed rules for the implementation of Council Regulation (EC) No 834/2007 on organic production and labelling of organic products with regard to organic production, labelling and control. OJ L 250, 18.9.2008, p. 1–84.

EU Regulation (2010) Commission Regulation (EU) No 271/2010 of 24 March 2010 amending Regulation (EC) No 889/2008 laying down detailed rules for the implementation of Council Regulation (EC) No 834/2007, as regards the organic production logo of the European Union. OJ L 84/19, 31.3.2010, p.1.

European Commission. 2017. European Commission. http://ec.europa.eu/.

European Commission. 2017. Organic Farming - European Commission. https://ec.europa.eu/agriculture/organic/index_en.

European Parliament, Committee on the Environment, Public Health and Food Safety. 2013. Draft Report on the food crisis, fraud in the food chain and the control thereof (2013/2091(INI)).

Everstine, K., Spink, J. and Kennedy, S. 2013. Economically motivated adulteration (EMA) of food: common characteristics of EMA incidents. Journal of Food Protection 76: 723–735.

Elmasry, G., Kamruzzaman, M., Sun, D.W. and Allen, P. 2012. Principles and applications of hyperspectral imaging in quality evaluation of agro-food products: A review. Critical Reviews in Food Science and Nutrition 52(11).

El Sheikha, A.F., Condur, A., Métayer, I., Nguyen, D.D., Loiseau, G. and Montet, D. 2009. Determination of fruit origin by using 26S rDNA fingerprinting of yeast communities by PCR-DGGE: preliminary application to *Physalis* fruits from Egypt. Yeast 26: 567–573.

FiBL. 2016. «FiBL-Statistics». http://www.fibl.org/en/themes/organic-farming-statistics.html.

Flores, P., Lopez, A., Fenoll, J., Hellin, P. and Kelly, S. 2013. Classification of organic and conventional sweet peppers and lettuce using a combination of isotopic and bio-markers with multivariate analysis. Journal of Food Composition and Analysis 31: 217–225.

Galimberti, A., De Mattia, F., Losa, A., Bruni, I., Federici, S., Casiraghi, M., Martellos, S. and Labra, M. 2013. DNA barcoding as a new tool for food traceability. Food Research International 50: 55–63.

Galimberti, A., Brunoa, A., Mezzasalma, V., De Mattiab, F., Brunia, I. and Labra, M. 2015. Emerging DNA-based technologies to characterize food ecosystems. Food Research International 69: 424–433.

Hall, A.D. 1940. Review of an agricultural testament, par Albert Howard. Journal of the Royal Society of Arts 88(4571): 885–87.

Heckman, J. 2006. A history of organic farming: Transitions from Sir Albert Howard's War in the soil to USDA national organic program. Renewable Agriculture and Food Systems 21: 143–150.

Hohmann, M., Christoph, N. , Wachter, H. and Holzgrabe, U. 2014. 1H NMR Profiling as an approach to differentiate conventionally and organically grown tomatoes. J. Agric. Food Chem. 62(33): 8530–8540.

Hoogenboom, L.A.P., Bokhorst, J.G., Northolt, M.D., Van de Vijver, L.P.L., Broex, N.J.G., Mevius, D.J., Meijs, J.A.C. and Van der Roes, J. 2008. Contaminants and microorganisms in Dutch organic food products: a comparison with conventional products. Food Additives & Contaminants 25: 1197–1209.

Huber, B. 2005. Labelling claims, processed products-US NOP 2002. http://organicrules. org/407/.

IFOAM–Organics International. 2016. «The World of Organic Agriculture 2016 FAO». http://www.fao.org/family-farming/detail/fr/c/415723/.

Johnson, R. 2014. Food Fraud and "Economically Motivated Adulteration" of Food and Food Ingredients. Informing the legislative debate since 1914. Congressional research service. USA. 1–39.

Joint FAO/WHO *Codex Alimentarius* Commission, Food and Agriculture Organization of the United Nations, et World Health Organization, éd. 2007. Organically produced foods. 3rd ed. *Codex alimentarius*. Rome: World Health Organization: Food and Agriculture Organization of the United Nations.

Kelly, S.D. and Bateman, A.S. 2010. Comparison of mineral concentrations in commercially grown organic and conventional crops–Tomatoes (*Lycopersicon esculentum*) and lettuces (*Lactuca sativa*). Food Chemistry 119(2): 738–745.

Kuepper, G. 2010. A brief overview of the history and philosophy of organic agriculture. Kerr Center for Sustainable Agriculture, Poteau. http://www.agripress.nl/_STUDIOEMMA_ UPLOADS/downloads/organic-philosophy-report.pdf.

Laursen, K.H., Schjoerring, J.K., Kelly, S.D. and Husted, S. 2014. Authentication of organically grown plants—advantages and limitations of atomic spectroscopy for multi-element and stable isotope analysis. Trends in Analytical Chemistry 59: 73–82.

Le Nguyen, D.D., Hanh, H.N., Dijoux, D., Loiseau, G. and Montet, D. 2008. Determination of fish origin by using 16S rDNA fingerprinting of bacterial communities by PCR-DGGE: an application on *Pangasius* fish from Viet Nam. Food Control 19: 454–460.

Leff, J.W. and Fierer, N. 2013. Bacterial communities associated with the surfaces of fresh fruits and vegetables. PLoS One 8: e59310.

Loi n° 80–502 du 4 juillet 1980 d'orientation agricole, July 4th, 1980, Journal Officiel de la République Française, p. 1670.

Lowdermilk, W.C. 1975. «Conquest of the land through seven thousand years». https://webdisk.ucalgary.ca/~walkerd/public_html/cpescCanada%20ProfessionalPractice/Conquest%20Land%20Lowdermilk%201953.pdf.

Madesis, P., Ganopoulos, I., Sakaridis, I., Argiriou, A. and Tsaftaris, A. 2014. Advances of DNA-based methods for tracing the botanical origin of food products. Food Research International 60: 163–172.

Maffei, D.F., Arruda Silveira, N.F., Penha Longo, M. and Catanozi, M. 2013. Microbiological quality of organic and conventional vegetables sold in Brazil. Food Control 29: 226–230.

Mie, A., Laursen, K.H., Åberg, K.M., Forshed, J., Lindahl, A., Thorup-Kristensen, K., Olsson, M., Knuthsen, P., Larsen, E.H. and Husted, S. 2014. Discrimination of conventional and organic white cabbage from a long-term field trial study using untargeted LC-MS-based metabolomics. Analytical and Bioanalytical Chemistry 406: 2885–2897.

Muilwijk, M., Heenan, S., Koot, A. and Van Ruth, S.M. 2015. Impact of production location, production system, and variety on the volatile organic compounds fingerprints and sensory characteristics of tomatoes. Journal of Chemistry Article ID 981549, 7 pages.

Nawrocka, A. and Lamorska, J. 2013. Determination of food quality by using spectroscopic methods, Advances in Agro-physical Research, Stanisław Grundas (ed.). InTech, doi:10.5772/52722.

Nawrocki, A., Thorup-Kristensen, K. and Jensen, O.N. 2011. Quantitative proteomics by 2DE and MALDI MS/MS uncover the effects of organic and conventional cropping methods on vegetable products. Journal of Proteomics 74: 2810–2825.

Novotná, H., Kmiecik, O., Gałązka, M., Krtková, V., Hurajová, A., Schulzová, V., Hallmann, E., Rembiałkowska, E. and Hajšlová, J. 2012. Metabolomic fingerprinting employing DART-TOFMS for authentication of tomatoes and peppers from organic and conventional farming. Food Additives and Contaminants, Part A Chemical Analytical Control Exposition Risk Assessment 29: 1335–1346.

Organic Food Production Act of 1990, Pub. L. 101–624, title XXI, §2101–2123, Nov. 28, 1990, 104 Stat. 3935–3951.

Oregon Tilth. 2016. Oregon Tilth: Organic Certification & Sustainable Agriculture. Oregon Tilth. https://tilth.org/.

Ottesen, A.R., White, J.R., Skaltsas, D.N., Newell, M.J. and Walsh, C.S. 2009. Impact of organic and conventional management on the phyllosphere microbial ecology of an apple crop. Journal of Food Protection 72: 2321–2325.

Paolini, M., Ziller, L., Laursen, K.H., Husted, S. and Camin, F. 2015. Compound-specific δ15n and δ13c analyses of amino acids for potential discrimination between organically and conventionally grown wheat. Journal of Agricultural and Food Chemistry 63(25): 5841–5850.

Palmieri, L., Bozza, E. and Giongo, L. 2009. Soft fruit traceability in food matrices using real-time PCR. Nutrients 1: 316–328.

Picone, G., Trimigno, A., Tessarin, P., Donnini, S., Rombolà, A.D. and Capozzi, F. 2016. ^{1}H NMR foodomics reveals that the biodynamic and the organic cultivation managements produce different grape berries (*Vitis vinifera* L. cv. Sangiovese). Food Chemistry 213: 187–195.

Reilly, K., Cullen, E., Lola-Luz, T., Stone, D., Valverde, J., Gaffney, M., Brunton, N., Grant, J. and Griffiths, B.S. 2013. Effect of organic, conventional and mixed cultivation practices on soil microbial community structure and nematode abundance in a cultivated onion crop. Journal of the Science of Food and Agriculture 93: 3700–3709.

Scarano, D. and Rao, R. 2014. DNA Markers for Food Products Authentication. Diversity 6: 579–596.

Sturm, M. and Lojen, S. 2011. Nitrogen isotopic signature of vegetables from the Slovenian market and its suitability as an indicator of organic production. Isotopes Environ. Health Studies 47: 214–220.

Shannon, D., Sen, A.M. and Johnson, D.B. 2002. A comparative study of the microbiology of soils managed under organic and conventional regimes. Soil Use and Management 18: 274–283.

Tatsadjieu, N.L., Maiwore, J., Hadjia, M.B., Loiseau, G., Montet, D. and Mbofung, C.M.F. 2010. Study of the microbial diversity of *Oreochromis niloticus* of three lakes of Cameroon by PCR-DGGE: Application to the determination of the geographical origin. Food Control 21: 673–678.

The Organics Institute. 2016. «History of the organic movement and organic farming The organics institute.com». http://theorganicsinstitute.com/organic/history-of-the-organic-movement/.

Tres, A., van der Veer, G., Perez-Marin, M.D., van Ruth, S.M. and Garrido-Var, A. 2012. Authentication of organic feed by near-infrared spectroscopy combined with chemometrics: A feasibility study. Journal of Agriculture Food Chemistry 60: 8129–8133.

USDA National Organic Program, USDA Science and Technology Programs, United States Department of Agriculture (USDA), Agricultural Marketing Service. 2012. USDA 2010–2011 Pilot Study Pesticide Residue Testing of Organic Produce.

Vallverdú-Queralt, A. and Lamuela-Raventós, R.M. 2016. Foodomics: A new tool to differentiate between organic and conventional foods. Electrophoresis 37: 1784–1794.

Van Ruth, S.M., Alewijn, M., Rogers, K., Newton-Smith, E., Tena, N., Bollen, M. and Koot, A. 2011. Authentication of organic and conventional eggs by carotenoid profiling. Food Chemistry 126(3): 1299–1305.

Vetter, W. and Schröder, M. 2010. Concentrations of phytanic acid and pristanic acid are higher in organic than in conventional dairy products from the German market. Food Chemistry 119(2): 746–752.

Wallis, P.D., Haynes, R.J., Hunter, C.H. and Morris, C.D. 2010. Effect of land use and management on soil bacterial biodiversity as measured by PCR-DGGE. Applied Soil Ecology 46: 147–150.

Willer, H. and Kilcher, L. 2012. The World of Organic Agriculture–Statistics and Emerging Trends. IFOAM, Bonn, FiBL, Frick.

Geographical Origin Traceability of Foodstuffs Using a Molecular Technique PCR-DGGE

Didier Montet,[1],* Amenan Clémentine Kouakou,[2]
Yasmine Hamdouche,[1] Corinne Teyssier,[3] Thomas Rychlik,[4]
Léopold N. Tatsadjieu[5] and Edna Froeder Arcuri[6]

1. Introduction

Labeling and traceability of imported food products in European countries became a legal obligation on January 1st, 2005 (EU Regulation 178/2002). The consumers are becoming more and more sensitive to the origin and

[1] Centre de Coopération Internationale en Recherche Agronomique pour le Développement, Cirad, UMR Qualisud, TA 95B/16, 34398 Montpellier Cedex 5, France.
E-mail: yasminehamdouche@gmail.com
[2] Laboratoire de Biotechnologie et Microbiologie (UFR-STA), University Nangui Abrogoua, Abidjan, Côte d'Ivoire; E-mail: kclementine24@yahoo.fr
[3] Université de Montpellier, UMR Qualisud, TA 95B/16, 34398 Montpellier Cedex 5, France.
E-mail: corinne.teyssier@umontpellier.fr
[4] Poznan University of Life Sciences, Institute of Food Technology of Plant Origin, Department of Fermentation and Biosynthesis, 31 Wojska Polskiego street, 60-624 Poznań, Poland.
E-mail: tomrych@up.poznan.pl
[5] Département de Génie Alimentaire et Contrôle Qualité, Institut Universitaire de Technologie, Université de Ngaoundéré. B.P 454 Ngaoundéré, Cameroun.
[6] Embrapa Gado de Leite, Rua Eugenio do Nascimento 610, Bairro Dom Bosco, 36038-330, Juiz de Fora, MG, Brazil; E-mail: edna.arcuri@embrapa.br
* Corresponding author: didier.montet@cirad.fr

the quality of foodstuffs. They think it is important to have food labels that, in addition to the list of ingredients identify the country of origin (Food Standards Agency 2007). Food industry uses commonly simple traceability systems supported by the billing system of bar codes. There is no existing efficient analytical technique which can permit determination of the origin of foods and follow them during processing as well as international trade. In this chapter, we discuss the use of a molecular technique called PCR-DGGE (Polymerase Chain Reaction-Denaturing Gradient Gel Electrophoresis) that permits to link some pertinent microbiological markers to the origin of food. These markers are microorganisms which are isolated from the environment of the food. Analysis of the obtained PCR-DGGE data will assure the link between microbial ecology and geographical origin of the foods.

2. Using Microbial Community to Trace the Origin of Food

The technique, PCR-DGGE was used first to analyze the microbial diversity in *pozol*, a maize-based Mexican fermented food (Ampe et al. 1999). In Cirad France this technique was applied in various scientific studies, i.e., to determinate the origin of food, to discriminate organic from inorganic foods to determinate mycotoxinogenic fungi and to trace hazardous compounds in processed food stuffs. In order to trace geographical origin, the initial idea was to create an analytical "biological bar code" for food (Montet et al. 2004).

This technological development is based on the fact that the skin of fresh foods (fruits, vegetables, cereals, etc.) is not a sterile environment and harbours various groups of microorganisms, microbial fragments, their DNAs and many other molecules of interest. The PCR-DGGE method was based on the assumption that microbial communities present on the surface of the skin of a given food are specific to the geographical area where they grow. The external environment of the food (soil ecology, fungi, insects, treatments, diseases) will impact the presence of various microorganisms. Human activity can also be a source of contaminants when hygienic practices are not followed (Sodeko et al. 1987, Ndiaye et al. 2016). For example, skin of nuts or cereals could be bitten by insects and then contaminated by fungi, chicken farms could be installed on the periphery of a fish pond so as to use the fishes' excreta as chicken feed and thus fishes are contaminated with *Salmonella*. Fishermen could contaminate fish skin with coliforms just by avoiding cleaning of their equipment, boat or hands.

3. Principle of PCR-DGGE Techniques in Food

To detect evolution in the microbial community present on food, it is proposed to combine PCR to DGGE (Ampe et al. 1999, Leesing 2005, Leesing et al. 2011, Le Nguyen 2008, El Sheikha et al. 2009, 2010a). This molecular biology approach permits to analyze in a unique step all microorganisms present on a food—bacteria, yeasts or molds. The advantages of this technique are—its operational cycle speed (24 h), low cost and efficiency. A batch of DGGE gel permits to analyze about 30 samples that could be used to follow a process, to identify microorganisms by sequencing of the bands extracted from the gel or by statistic tools to link microbial ecology to geographical origin or production process (i.e., organic versus traditional).

Universal PCR primers are required to obtain DNA strands of the same size. Different couples of primers are used for bacteria, yeasts and fungi. DGGE is a kind of affinity chromatography which allow to separate DNA strands of the same size while conventional methods of electrophoresis separates the DNA strands by their different sizes (Muyzer et al. 1993, El Sheikha 2010a). For favoring DNA strands migration, acrylamide gel contains a gradient of amides: formamide and urea that creates an affinity between gel and DNA strands.

This technique does not replace the classical analysis of microorganisms by specific medium or by molecular biology approach but permits to establish within 24 h the bacterial, yeast or fungal profiles of around 30 food samples. It is faster than any conventional microbial technique and permits to compare the vertical lines on the gels by image analysis. In addition, sequencing of extracted bands from gels also helps to identify with sufficient precision the microbial species. With the type of primers discussed in this chapter, bacterial DNA strands are composed of 236 base pairs, whereas, yeast and fungal DNA strands are composed of 258 base pairs.

The different steps in this method are described in Fig. 1. The first step consists of washing of all the microorganisms from the food using a buffer. Usually, this extraction is done on the surface of the food and on a crushed sample for big samples. The second step consists in DNA extraction from all microorganisms (bacteria, yeast or mold) in one step. The third step is to put out a PCR amplification performed by the microbial family with a unique couple of universal primer that conducts to DNA strands of the same size. The fourth step consists in separating DNA strands of the same size on DGGE gels, then stain and photograph the spots under UV lamp. The last step is based on the image analysis by statistical tools after alignment of the spots. DGGE profile is thus an image of all the main microorganisms present on a food. Each spot refers to a sequence type or single phylotype, which correspond to species of microorganism. In some particular analysis, spots could recover different DNA that have a similar base composition.

For gel analysis, opposite DNA strands are linked by a clamp composed of around 30 to 40 bases to favor migration and separation on the gel. The more GC has the strand, stronger is the electric link between the strands and thus the shorter is the migration.

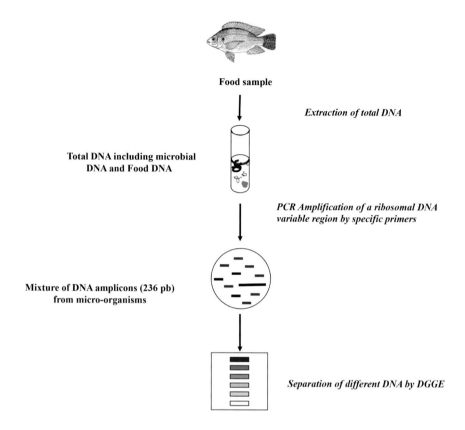

Figure 1. Different Steps of PCR-DGGE Analysis.

To identify strains, sequencing could be done directly from excised bands on the polyacrylamide gels. Sequences are then compared to those in the GeneBank database (http://www.ncbi.nlm.nih.gov/BLAST/) and those of the Ribosomal Database Project (http://rdp.cme.msu.edu/index.jsp) using the blast program. Sequences with a percentage identity of 98% or greater are considered to belong to the same species.

4. Tracing the Origin of Food

The PCR-DGGE method was adapted to different foods (fish, nuts, fruits, cheeses) and different strains (bacteria, yeasts, fungi). Some examples are presented for a couple of food/microorganisms. Three PCR-DGGE methods were developed for bacteria, yeasts and fungi.

4.1 Application to aquaculture fish traceability, the case of Adjuevan in Ivory Coast

In this part, the use of bacteria as geographical origin markers of aquaculture fish is presented. Adjuevan is a traditional Ivorian naturally fermented fish prepared from the Atlantic bumper *Chloroscombrus chrysurus* or the sea fish *Galeoides decadactylus* (Koffi-Nevry et al. 2011, Kouakou et al. 2012a,b). It is produced through a spontaneous and uncontrolled fermentation and by using a high amount of salt. For traditional adjuevan production, the fresh fish is scaled, gutted, washed and left overnight before the seemingly deteriorated fish is treated with salt and allowed to ferment for 3 to 5 days. Two processing methods for adjuevan production have been identified and both seem to lead to different visually end-products (Kouakou et al. 2012c). The final product is highly concentrated in salt with a strong smell and is used as a condiment to season sauces. In order to develop controlled fermentation technology, it was necessary to understand the associated microorganisms and the distribution of the microbial populations in this traditional fermented fish product. Previous studies by classical microbiological methods evaluated the microbial communities in other African fermented fish. The dominant species were found to belong to the genera *Bacillus* and *Staphylococcus* (Sanni et al. 2002, Anihouvi et al. 2007). However, only easily cultivable microorganisms can be detected by these classical microbiological methods based on plate counts.

The PCR-DGGE method permitted to rapidly identify the major microorganisms on this fish. Only a few published works reported the analysis of the bacterial communities of fish samples by PCR-DGGE (Spanggaard et al. 2000, Díez et al. 2001, Huber et al. 2004, Le Nguyen et al. 2008). Determination of fish geographical origin was achieved by comparison of bacterial communities of tilapia fish using PCR-DGGE from different Lakes of Cameroon (Maiwore et al. 2009) and from aquaculture farms in Vietnam (Leesing et al. 2011). Adjuevan was fermented during 5 days. However, no real investigation was focused on microorganism diversity after fermentation of the traditional fermented fish, despite the important place of the adjuevan in Ivorian people's food history.

In this chapter, the application of culture-independent (DGGE analysis) methods to study and compare bacteria and yeast ecology profiles of fermented fish adjuevan is described. This study was done by PCR-DGGE to evaluate the impact of the weather/climate or fish origin on the microbial communities (bacteria and yeasts) directly by analyzing the fermented and salted fish matrix. Tilapia fish samples were collected in a Lake (Bandama) and in a fish farm of Côte d'Ivoire. Tilapia samples were sent to Ivorian laboratory where they were fermented during five days (Kouakou et al. 2012a). Fermented fish adjuevan samples (30), were aseptically transferred to storage bags, then maintained on ice and transported to the Cirad laboratory in France. Bacteria and yeast DNA extraction from the fermented fish adjuevan samples was performed in a unique step according to Kouakou et al. (2012a,b) using phenol-chloroform-isoamyl alcohol extraction and ethanol precipitation.

PCR-DGGE patterns of DNA directly extracted from fermented fish samples collected at the fifth day of fermentation revealed the presence of 10 bands of bacterial DNA in fermented adjuevan from lake and 6 bands of bacterial DNA in the aquaculture river farm (Fig. 2). Some of the bands were common to all the samples. The two profiles were typical for each group of fermented fish. Fermented fish from lake showed an increase in yeast diversity and gave 7 DNA bands while there were 4 DNA bands for the fermented fish from fish farm (Fig. 3). DGGE bacteria profiles were more intensive than of yeasts. There are published works that analyzed the bacterial communities and yeasts in fermented fish samples by PCR-DGGE (Kouakou et al. 2012a,b) according to the process but not according to the origin. The analysis of fermented fish samples from different locations product following the same process, showed some significant differences in the migration patterns on DGGE. Differences in band profiles can be attributed to environmental differences between fresh fish origins and to the differences in the feeding methods between farms and the type of aquaculture system applied. The variations may also be due to the water

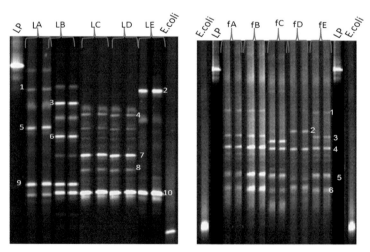

Figure 2. PCR- DGGE Profiles of Bacterial Community from DNA Directly Extracted from *adjuevan* Fermenting; L: Lake, f: fish farm, A, B, C, D, E: fermented fish samples. L pl: *Lactobacillus plantarum*

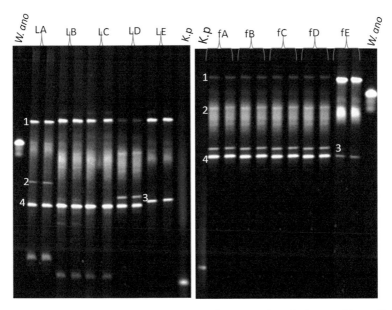

Figure 3. PCR-DGGE profiles of Yeast Community from DNA Directly Extracted from *adjuevan* Fermenting; L: Lake, f: fish farm, A, B, C, D, E: fermented fish samples.

supply which can be affected by the pollution. These results suggested that the DGGE method could be used to discriminate different fermented fish according to fish origin.

4.2 Application of PCR-DGGE to the traceability of fruits

During the European project Innovkar (2007–2011, 6th PCRDT, UE 032037), a Work Package was dedicated to the traceability of shea tree fruits and some specific tools were developed based on the same technique to trace shea tree fruits to their geographical origins.

Shea tree (*Vitellaria paradoxa*) is of the *Sapotaceae* family and grows wild in West Africa within a geographical area called by traders "Shea belt" ranging from Mali to Sudan in the north and Togo and Uganda in the south. In some African countries, shea tree fruits are nicknamed "The Gold of Women", because shea butter is not only used as fat for cooking but also as cosmetic for women, mainly in rural areas which account for 80% of total consumption (El Sheikha 2010a). Nowadays, shea butter is also used for cosmetics in developing countries and to partly replace cocoa butter in chocolate (UE regulation 2000/36/EC).

In this study, mature shea tree fruits were harvested in different districts, Daelan village in the Ségou region and Nafégué village in the Sikasso region (Mali). The fruits were gathered to preserve their initial flora and were collected directly on the tree using gloves and put in sterile bags in July 2008. These bags were kept into a refrigerator then transferred by plane to Cirad laboratory where yeasts and fungi DNA were extracted immediately from the fresh fruits. The origin of the samples was defined by country, site and date of harvest.

On DGGE gel, the observed bands obtained for the samples and the reference DNA of *Wickerhamomyces anomalus* MTF 1103 and *Komagataella pastoris* ATCC 28484 for yeast and *Mucor racemosus* DNA and *Trichoderma harzianum* DNA for fungi permitted to analyze samples of yeasts and molds DNA extracted from shea tree fruits from two different regions of Mali (Fig. 4 and Fig. 5). Vertical line on the gels represents a fruit and each

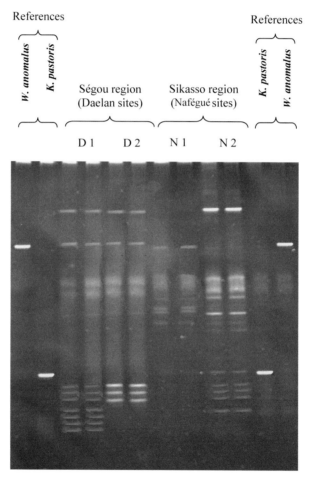

Figure 4. DGGE Profiles of 26S rDNA of Yeast Strains Isolated from Shea Tree Fruits from Two Different Regions of Mali: Ségou Region (D1, D2: Daelan Sites) and Sikasso Region (N1, N2: Nafégué Sites).

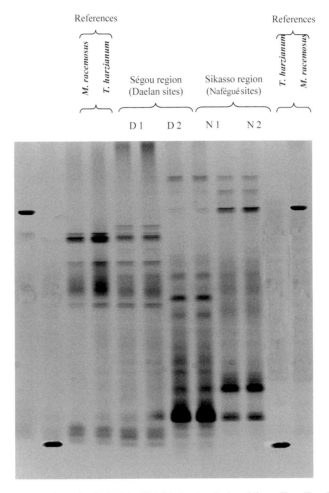

Figure 5. DGGE Profiles of 28S rDNA of Molds Strains Isolated from Shea Tree Fruits from Two Different Regions of Mali: Ségou Region (D1, D2: Daelan Sites) and Sikasso Region (N1, N2: Nafégué Sites).

spot represents a yeast or mold species. PCR-DGGE patterns of shea tree fruits for each location revealed the presence of 6 to 11 DNA bands for each shea tree fruit for yeasts and 8 to 12 bands for molds. Cluster analysis by the single linkage method with Euclidean distance measure was used to determine similarities in yeast and molds communities structures of shea tree fruits. At 83% similarity level, two main clusters were observed for yeasts: the first cluster contained samples from the Sikasso region while the second contained samples from the Ségou region (Fig. 6). At 67% similarity level, two main clusters were observed for fungi samples from the two regions (Fig. 7). Differences in banding patterns can be attributed

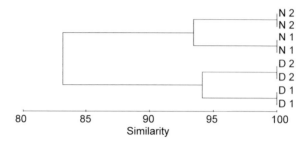

Figure 6. Cluster Analysis of 26S rDNA Profiles of Yeast Strains Isolated from Shea Tree Fruits Two Different Regions of Mali: Ségou Region (D1, D2: Daelan Sites) and Sikasso Region (N1, N2: Nafégué Sites).

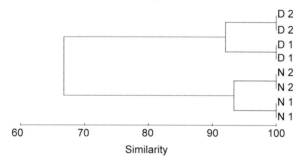

Figure 7. Cluster Analysis of 28S rDNA Profiles of Yeast Strains Isolated from Shea tree Fruits Two Different Regions of Mali: Ségou Region (D1, D2: Daelan Sites) and Sikasso Region (N1, N2: Nafégué Sites).

to environmental differences between the two types of holdings and type of agriculture practice applied. There was a connection between yeast and molds populations and geographical area.

This technique was also applied for the discrimination of geographical origin of some other fruits as Physalis (El Sheikha et al. 2009, 2011b, 2012), coffee (Nganou et al. 2012) and *Hibiscus sabdariffa* (Ndiaye et al. 2016) by the same team.

4.3 Application of PCR-DGGE to the microbial traceability of coffee during post-harvest treatments

This study was conducted with the main objective to trace the microbial community of Cameroonian coffee during its post-harvest processing as well as its impact on the final quality of this product. The effect of types of post-harvest treatment on the structure of microbial communities was studied at different stages of the process (dry cherries, dry parchment, green coffee beans). PCR-DGGE was used to better understand the microbial succession during the process and to identify the microorganisms involved (Hamdouche et al. 2016).

Quality of coffee depends on several parameters such as plant variety, growing conditions, harvest season, geographical origin and of course human expertise (Saltini et al. 2013). Samples were harvested in May 2014 in two regions of Cameroon (Bafoussam, Dschang), in three farms for each region. Two types of post-harvest treatments were applied to Arabica coffee: the wet method (ripe cherries) and dry method (immature or damaged cherries). Five hundred grams of coffee were collected in bags of 50 kg (100 g at 5 different parts in the coffee bag). Samples were introduced into sterile plastic bags and stored at room temperature for two weeks before analysis at Cirad Montpellier, France.

Bacterial DNA was extracted from 10 g of coffee beans then verified on gel and Nanodrop (Shimadzu). DNA was amplified by PCR according to the method described previously by Hamdouche et al. (2016). PCR products were analyzed by DGGE according to the procedure described by Hamdouche et al. (2015, 2016).

Bacterial DGGE profiles extracted from Arabica coffee treated by the wet and dry method were compared. Seven main bacterial DNA were identified as Lactic Acid Bacteria (LAB) during the wet method. In the three farms of Bafoussam district, 4 bands corresponding to *Weissella* sp., *Lactobacillus* sp., *Lactococcus lactis* and *Lactobacillus fermentum* appeared on the green coffee obtained by the wet method. On coffee treated by dry method, strains belonging to Gammaproteobacteria family were detected and seemed to be specific to this method, the majority of these strains were identified as *Pseudomonas* sp.

Images of bacterial DGGE profiles were analyzed statistically by Principal Component Analysis (PCA) to estimate changes in the structure of microbial communities between coffee samples. Two coffee groups were obtained by data analysis separated by the axis F1 (38.2%), corresponding to coffee processed by dry and wet method. Bacterial communities variations are mainly due to the type of treatment used. Canonical Correspondence Analysis (CCA) was then applied in order to explain changes in the structure of bacterial communities of coffee by linking the communities with different parameters such as post-harvest treatment and treatment steps.

Four coffee groups were discriminated by CCA. The first axis explained 76.1% of variance in the bacterial communities of the two treatment methods. There was a strong relationship between coffee treated by dry process and bacteria species identified as *Pseudomonas* sp., *Microbacterium* sp. and *Enterobacter* sp. The two strains *Pseudomonas* and *Microbacterium* have been already isolated from fresh cherries coffees (Vega et al. 2005). These strains have already been detected in the dry process in preliminary studies (Hamdouche, PhD Thesis 2015) and these genii are probably specific communities associated with dry coffee. CCA connected LAB species *Leuconostoc mesenteroides* and *Weissella* sp. to coffee obtained by wet process, these result were repeatable over the two years of harvest. These results show that LAB are good specific microbial markers associated to the wet process given that these bacteria are involved in fermentation steps (Sven and Walter 1990) which are realized during the wet process and not during the dry method. Some other authors also found LAB during coffee fermentations (Avallone et al. 2001, Vilela et al. 2010).

4.4 Application of PCR-DGGE to the microbial traceability of traditional cheeses

4.4.1 Application to traditional cheese from Poland

Protected Geographical Indication (PGI) is a legal European Union system that protects the authenticity of traditional food products all over the world. Fried cheese from Wielkopolska region in Poland is one of the few traditional products labeled as PGI in Poland and thus should not be manufactured in any other part of the country (UE regulation N° 323/2009).

Its original and specific character is due to an old traditional manufacturing process maintained through the ages. The sensory properties of this fried cheese are closely related to the traditional cheese manufacturing technology but also to its associated microflora which plays a key role in creating its typical aroma and taste.

The curd ripened fried cheese was produced on the basis of acid Tvarog obtained from pasteurized cow milk previously treated with lactic acid bacteria (LAB) starter cultures (Cais-Sokolińska and Pikul 2009). During curd ripening process other proteolytic and lipolytic agents such as *Geotrichum candidum* and *Oidium lactis* appeared causing increase of pH from 4.4 to pH over 5.0 (Cais-Sokolińska and Majcher 2010). After ripening, cheese bulk is fried with butter until smooth texture is obtained. Frying process permits to maintain microorganism load less than 10 CFU/g during a long storage period.

It was necessary to create an effective analytical traceability tool for polish cheeses in order to eliminate misleading of consumers and unfair competition by promoting non-genuine products.

The PCR-DGGE was used as an analytical tool that permitted to link microbial ecology to the geographical origin of Wielkopolska cheeses. In this case, PCR-DGGE was used to analyze in a single step all the yeasts present in cheese matrix. For this purpose, 26S rDNA profiles generated by PCR-DGGE were used to assess the variation in the yeast community of traditional fried cheeses from six different producers, five from the Great Poland region and one from the Silesia region (Rychlik et al. 2017).

Ten grams of cheese samples were homogenized with 90 mL of a 2% (w/v) sterile sodium citrate solution at 40°C in a BagMixer 400W stomacher (Interscience, France). Serial of 10-fold dilutions were made in 0.85% (w/v) sodium chloride solution and plated in duplicate on to Yeast-Extract Glucose Chloramphenicol agar (BTL, Poland). Yeast and fungi were enumerated after 3–5 days of incubation at 25°C.

DNA extraction from the cheese samples was performed in a unique step according to Arcuri et al. (2013) using phenol-chloroform-isoamyl alcohol extraction and ethanol precipitation.

The DGGE 26S rDNA band profiles of curd ripened fried cheese from 7 manufacturers was compared by image and statistical analysis. The standardized banding patterns matrix and the Dice similarity coefficient (SD) were used to compare fungi profiles of cheeses.

The cluster analysis was used to organize observed variation in DGGE fingerprint patterns. Cluster analysis is an exploratory data analysis tool which aims at sorting different objects into groups in a way that the association degree between two objects is maximal if they belong to the same group and minimal otherwise. This analysis was used to discover structures in data.

The analysis of the hierarchical tree showed that the fungi profile of sample A is similar to F and G, and these samples formed one cluster (Fig. 8). Sample A, which is a non-registered cheese (PGI), tends to be the most dissimilar but discrimination of this sample with hierarchical cluster analysis is not unequivocal.

Galactomyces geotrichum was present in all analyzed samples from the Wielkopolska region. *Kluyveromyces marxianus* appeared on almost all lanes. It was noted that *Galactomyces geotrichum* and *Pichia kudravzevii* ($p < 0.05$) had sufficient discrimination power. The variable *Kluyveromyces marxianus* properly discriminated cheese samples using LDA results. Principal Component Analysis (PCA) proved to be the most suitable to diversify cheese samples.

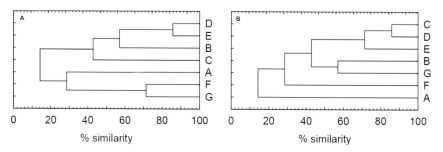

Figure 8. Dendrogram Based on the Fungi DGGE Fingerprints Clustering of Standardized Banding Pattern (A) and the Dice Similarity Coefficient (SD) Matrix (B), of Fried Cheese Samples: A—Sample from Silesia Region; B, C, D, E, F, G—Samples from Wielkopolska Region.

Band profiles of cheeses from different sources were specific for almost each district and in particularly the product from Silesia which is not protected by EU law. This method seems to be an effective traceability tool providing for a biological bar code which allows tracing back the food to their authentic location.

4.4.2 Application to traditional Minas cheese from Brazil

This is the first study that dealt with the use of PCR-DGGE to characterize traditional or "artisanal" Minas cheese on the basis of their dominant bacteria and geographical origin.

Traditional Minas cheese includes some varieties of cheeses that are classified according to their region of origin within the state of Minas Gerais (MG), Brazil. The most important are produced in the regions of Serro, Cerrado, Araxa and Serra da Canastra. Traditional Minas cheese is produced with raw cow's milk employing a natural whey culture as starter, and calf quimosin as coagulant.

Samples of Minas cheese produced in four regions, namely Serro, Cerrado, Araxa and Serra da Canastra, were purchased at the local market. Two traditional cheeses made from raw cow's milk were collected from each region, and two cheeses made of pasteurized milk from the Serro region. The cheeses were refrigerated, and then transported by plane to Cirad, Montpellier in France, where the bacterial DNA was immediately extracted.

Bacterial DNA extraction, PCR-DGGE analysis, and identification of the dominant bacteria were performed as described by Arcuri et al. (2013). DGGE revealed only one band of bacteria in cheese samples made from pasteurized milk that was used as control, and seven to thirteen bands for traditional Minas cheeses (Fig. 9). Only one DNA band was common to all the traditional Minas cheeses from the four regions, however some bands were present in cheeses from more than one region, and some were unique for each region. At 65% similarity, the cluster analysis of the DGGE patterns revealed separation of the traditional Minas cheese samples related to their region of origin. In addition, the band profiles clearly distinguished cheese made from pasteurized and raw milk. Similar results have been reported by Coppola et al. 2001 for Mozzarella cheese produced in Italy. In total, 15 different DGGE bands were encountered and all of them were sequenced (Fig. 9). The results indicated the predominance (eight bands) of species belonging to the genera *Streptococcus* with two bands assigned to uncultured *Streptococcus* sp., followed by two species of *Lactobacillus*, two uncultured bacterium and one *Lactococcus lactis*. Two faint bands were identified because they did not yield good sequencing results, probably due to their low amount. The detection limit of PCR-DGGE depends on the species or perhaps even the strain considered, and it has been indicated for some species, ranging from 10^4 and 10^8 CFU/mL (Ercolini 2004).

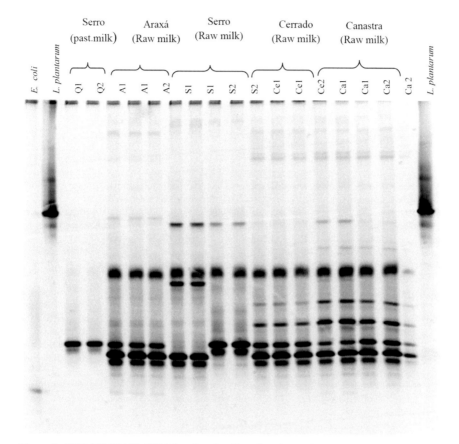

Figure 9. PCR-DGGE 16S rDNA Banding Profiles of Traditional Minas Cheese Bacteria from Four Regions (Araxá, Serro, Cerrado and Canastra), Minas Gerais State, Brazil.

5. Conclusion and Future Perspectives

During these last 15 years, some papers related the link between the geographical origin of food and the analysis by PCR-DGGE of microbial communities (Leesing 2005, Le Nguyen et al. 2008, El Sheikha et al. 2009, 2010a,b, 2011a,b, Arcuri et al. 2013, Dufossé et al. 2013). The DNA band profiles from bacteria, yeasts and molds communities isolated from fish, salts, nuts or fruits obtained by PCR-DGGE are strongly related to the microbial environment of fish and fruit. These biomarkers are statistically specific from places of production and to techniques used by producers and could be applied to differentiate the geographical origins of food.

New sequencing techniques such as NGS (see chapter in this book) will be more effective than PCR-DGGE in the very near future but are based on the identification of DNA strands by data bases. PCR-DDGE permits to visualize microorgnisms by their DNA bands and thus to rapidly interpret the results if there is a problem (pathogenic strains, contamination in fermentations) in a process. These two techniques are thus complementory.

Keywords: Food origin, traceability, microbial communities, biological bar-code, PCR-DGGE

References

Ampe, F., Omar, N.B., Moizan, C., Wacher, C. and Guyot, J.P. 1999. Polyphasic study of the spatial distribution of microorganisms in Mexican pozol, fermented maize dough, demonstrates the need for cultivation-independent methods to investigate traditional fermentations. Applied Environmental Microbiology 65: 5464–5473.

Anihouvi, V.B., Sakyi-Dawson, E., Ayernor, G.S. and Hounhouigan, J.D. 2007. Microbiological changes in naturally fermented cassava fish (*Pseudotolithus* sp.) for lanhouin production. International Journal of Food Microbiology 116: 287–291.

Arcuri, E.F., El Sheikha, A.F., Rychlik, T., Piro-Métayer, I. and Montet, D. 2013. Determination of cheese origin by using 16S rDNA fingerprinting of bacteria communities by PCR-DGGE: Preliminary application to traditional Minas cheese. Food Control 30: 1–6.

Avallone, S., Guyot, B., Brillouet, J.M., Olguin, E. and Guiraud, J.P. 2001. Microbiological and biochimical study of coffee fermentation. Current Microbiology 42: 252–256.

Cais-Solińska, D. and Majcher, M. 2010. Sensory properties and volatile composition of full and non-fat cheese produce from curd–ripened fried acid tvarog. Acta Alimentaria 39: 69–80.

Cais-Sokolińska, D. and Pikul, J. 2009. Cheese meltability as assessed by the tube test and Schreiber test depending on fat contents and storage time, based on curd-ripened fried cheese. Czech Journal Food Sciences 27(5): 301–308.

Coppola, S., Blaiotta, G., Ercolini, D. and Moschetti, G. 2001. Molecular evaluation of microbial diversity occurring in different types of Mozzarella cheese. Journal of Applied Microbiology 90: 414–420.

Directive 2000/36/EC of the European Parliament and of the Council of 23 June 2000 relating to cocoa and chocolate products intended for human consumption.

Díez, B., Pedrós-Alió, C., Marsh, T.L. and Massana, R. 2001. Application of denaturing gradient gel electrophoresis (DGGE) to study the diversity of marine picoeukaryotic assemblage and comparison of DGGE with other molecular techniques. Applied Environmental Microbioly 67: 2942–2951.

Dufossé, L., Donadiao, C., Valla, A., Meile, J.C. and Montet, D. 2013. Determination of specialty food salt origin by using 16S rDNA fingerprinting of bacterial communities by PCR-DGGE: An application on marine salts produced in solar salterns from the French Atlantic Ocean. Food Control 32: 644–649.

El Sheikha, A.F., Condur, A., Métayer, I., Le Nguyen, D.D., Loiseau, G. and Montet, D. 2009. Determination of fruit origin by using 26S rDNA fingerprinting of yeast communities by PCR-DGGE: preliminary application to Physalis fruits from Egypt. Yeast 26(10): 567–573.

El Sheikha, A.F. 2010a. Determination of geographical origin of Shea tree and Physalis fruits by using the genetic fingerprints of the microbial community by PCR/DGGE. Analysis of biological properties of some fruits extract, Ph.D Thesis, University of Montpellier 2, France, p. 220.

El Sheikha, A.F. 2010b. Determination of the geographical origin of fruits by using 26S rDNA fingerprinting of yeast communities by PCR-DGGE: An application to Shea tree fruits. Journal of Life Sciences 4: 9–15.

El Sheikha, A.F., Bouvet, J.M. and Montet, D. 2011a. Biological bar-code for the determination of geographical origin of fruits by using 28S rDNA fingerprinting of fungal communities by PCR-DGGE: An application to Shea tree fruits. Quality Assurance and Safety of Crops & Foods 3: 40–47.

El Sheikha, A.F. and Montet, D. 2011b. Determination of fruit origin by using 28S rDNA fingerprinting of fungal communities by PCR-DGGE: An application to Physalis fruits from Egypt, Uganda and Colombia. Fruits 66: 79–89.

El Sheikha, A.F., Durand, N., Sarter, S., Okullo, J.B. and Montet, D. 2012. Study of the microbial discrimination of fruits by PCR-DGGE: Application to the determination of the geographical origin of Physalis fruits from Colombia, Egypt, Uganda and Madagascar. Food Control 24(1-2): 57–63.

Ercolini, D. 2004. PCR-DGGE fingerprinting: novel strategies for detection of microbes in food. Journal of Microbiological Methods 56: 297–314.

European Commission Regulation No 178/2002 of the European Parliament and of council of 28 January 2002. Laying down the general principles and requirements of food law, establishing the European Food. Safety Authority and laying down procedures in matters of food safety. Official Journal of the European Communities L31, 1.2.2002, p. 3.

Food Standards Agency. 2007. What consumers want-A literature review, Labelling and Packing. Labelling Research, Available from: http://www.food.gov.uk/foodlabelling/researchandreports/litreview [Accessed September 2007].

Hamdouche, Y. 2015. Discrimination des procédés de transformation post-récolte du Cacao et du Café par analyse globale de l'écologie microbienne, phD. Thesis, 234 pages, Supagro, Montpellier, France.

Hamdouche, Y., Guehi, T., Durand, N., Kedjebo, K.B.D., Montet, D. and Meile, J.C. 2015. Dynamics of microbial ecology during cocoa fermentation and drying: Towards the identification of molecular markers. Food Control 48: 117–122.

Hamdouche, Y., Meile, J.C., Nganou, D.N., Durand, N., Teyssier, C. and Montet, D. 2016. Discrimination of post-harvest coffee processing methods by microbial ecology analyses. Food Control 65: 112–120.

Huber, I., Spanggaard, B., Appel, K.F., Rossen, L., Nielsen, T. and Gram, L. 2004. Phylogenetic analysis and *in situ* identification of the intestinal microbial community of rainbow trout (*Oncorhynchus mykiss*, Walbaum). Journal of Applied Microbiology 96: 117–132.

Innovkar European project, 6th PCRDT : Innovative Tools and Techniques for Sustainable Use of the Shea Tree in Sudano-Sahelian zone, Contract no.: 032037, 2007–2011.

Koffi-Nevry, R., Ouina, T.S., Koussemon, M. and Brou, K. 2011. Chemical composition and lactic microflora of *adjuevan*, a traditional Ivorian fermented fish condiment. Pakistan Journal of Nutrition 10: 332–337.

Kouakou, A.C., Nguessan, K.F., Dadier, A.T., Djè, K.M. and Montet, D. 2012a. Application of culture dependent methods and culture-independent methods (DGGE analysis) to study Lactic acid bacteria ecology of Ivorian fermented fish Adjuevan. Challenges of Modern Technology 3(1): 50–56.

Kouakou, A.C., Cissé, M., Kossonou, E., Brou, K.D., Dje, K.M. and Montet, D. 2012b. Identification of yeasts associated with the fermented fish, adjuevan, of Ivory Coast by using the molecular technique of PCR-denaturing gradient gel electrophoresis (DGGE). African Journal of Microbiology Research 6(19): 4138–4145.

Kouakou, A.C., Kouadio, F.N., Dadier, A.T., Montet, D. and Djè, K.M. 2012c. Production et commercialisation de l'adjuevan, poisson fermenté de Côte d'Ivoire. Cahiers Agriculture 22: 559–67.

Le Nguyen, D.D., Ha, N.H., Dijoux, D., Loiseau, G. and Montet, D. 2008. Determination of fish origin by using 16S rDNA fingerprinting of bacterial communities by PCR-DGGE: An application on Pangasius fish from Vietnam. Food Control 19: 454–460.

Leesing, R. 2005. Identification et validation de marqueurs spécifiques pour la traçabilité de poissons d'aquaculture lors de leur import/export. Thesis Montpellier University II, 8 décembre 2005.

Leesing, R., Dijoux, D., Le Nguyen, D.D., Loiseau, G., Ray, R.C. and Montet, D. 2011. Improvement of DNA Extraction and Electrophoresis Conditions for the PCR-DGGE Analysis of Bacterial Communities Associated to Two Aquaculture Fish Species. Dynamic Biochemistry. Process Biotechnology and Molecular Biology 5: 83–87.

Maiworé, J., Tatsadjieu, N.L., Montet, D., Loiseau, G. and Mbofung, C.M. 2009. Comparison of bacterial communities of tilapia fish from Cameroon and Vietnam using PCR-DGGE. African Journal of Biotechnology 8: 7156–7163.

Montet, D., Leesing, R., Gemrot, F. and Loiseau, G. 2004. Development of an efficient method for bacterial diversity analysis: Denaturing Gradient Gel Electrophoresis (DGGE). *In*: Seminar on Food Safety and International Trade, Bangkok, Thailand.

Muyzer, G., De Waal, E.C. and Uitterlinden, A.G. 1993. Profiling of complex microbial populations by denaturing gradient gel electrophoresis analysis of polymerase chain reaction-amplified genes coding for 16S rRNA. Applied Environmental Microbiology 59: 695–700.

Ndiaye, N.A., Hamdouche, Y., Alé, K., Cissé, M., Berthiot, L., Ndeye, C.T.K. and Montet, D. 2016. Application of PCR-DGGE to the study of dynamics and biodiversity of microbial contaminants during the processing of *Hibiscus sabdariffa* drinks and concentrates. Fruits 71(3): 141–149.

Nganou, D.N., Durand, N., Tatsadjieu, N.L., Meile, J.C., El Sheikha, A.F., Montet, D. and Mbufung, C.M. 2012. Determination of coffee origin by using 28S rDNA fingerprinting of fungal communities by PCR-DGGE: Application to the Cameroonian coffee. International Journal of Biosciences 2(5): 18–30.

Rychlik, T., Bednarek, M., Arcuri, E., Montet, D., Mayo, B., Nowak, J. and Czarnecki, K. 2017. Application of the PCR-DGGE technique to the fungal community of traditional Wielkopolska fried ripened curd cheese to determine its PGI authenticity. Food Control 73: Part B, 1074–1081.

Saltini, R., Akkerman, R. and Frosch, S. 2013. Optimizing chocolate production through traceability: A review of the influence of farming practices on cocoa bean quality. Food Control 29: 167–187.

Sanni, A.I., Asieduw, M. and Ayernorw, G.S. 2002. Microflora and chemical composition of momoni, a Ghanaian fermented fish condiment. Journal of Food Composition and Analysis 15: 577–583.

Sodeko, O.O., Izuagbe, Y.S. and Ukhun, M.E. 1987. Effect of different preservative treatment on the microbial population of Nigerian orange juice. Microbios 51: 133–143.

Spanggaard, B., Huber, I., Nielsen, T.J., Nielsen, T., Appel, K. and Gram, L. 2000. The microbiota of rainbow trout intestine: a comparison of traditional and molecular identification. Aquaculture 182: 1–15.

Sven, E.L. and Walter, J.D.1990. Antagonistic activities of lactic acid bacteria in food and feed fermentations. FEMS Microbiology Letters 48: 149–163 .

UE regulation N° 323/2009 of 20 April 2009 entering certain names in the register of protected designations of origin and protected geographical indications.

Vega, F.E., Pava-Ripoll, M., Posada, F. and Buyer, J.S. 2005. Endophytic bacteria in *Coffea arabica* L. Journal of Basic Microbiology 45: 371–380.

Vilela, D.M., De M Pereira, G.V., Silva, C.F., Batista, L.R. and Schwan, R.F. 2010. Molecular ecology and polyphasic characterization of the microbiota associated with semi-dry processed coffee (*Coffea arabica* L.). Food Microbiology 27: 1128–1135.

8

Capillary Electrophoresis-Single Strand Conformation Polymorphism (CE-SSCP) as a Tool for Water Traceability

Jean Jacques Godon

1. Introduction

In the European Union and following the European regulation 178/2002, water is considered as a food since January 2005. The article 2 defines that *"Food (or 'foodstuff') means any substance or product, whether processed, partially processed or unprocessed, intended to be, or reasonably expected to be ingested by humans. Food includes drink, chewing gum and any substance, including water, intentionally incorporated into the food during its* manufacture, preparation or treatment". In the article 18 related to Traceability, it is stated that *"The traceability of food, feed, food-producing animals, and any other substance intended to be, or expected to be, incorporated into a food or feed shall be established at all stages of production, processing and distribution"* and *"Food and feed business operators shall have in place systems and procedures to identify the other businesses to which their products have been supplied. This information shall be made available to the competent authorities on demand"*.

Water and liquids recovered from the natural environment or from human activity are full of microorganisms and in most cases display a wide

INRA, UR 0050 LBE Laboratoire de Biotechnologie de l'Environnement, Narbonne, France.
  Email: Jean-jacques.godon@inra.fr

diversity. These microorganisms can be arrayed into different groups based on set of geographic and physic-chemical parameters, like for example the 16S rRNA sequences. This 'signature' (based on specific 16S rDNA gene sequence) can be perceived easily by molecular fingerprinting techniques such as CE-SSCP (Capillary Electrophoresis Single Strand Conformation Polymorphism) (Zumstein et al. 2000). This technique, in its way similar to a bar code, gives an overview of microbial communities. Such techniques do not propose a rapid identification of microorganisms but, often, the identification of microbial species provides little interest due to the vast number of species along with the lack of available information about them. Many of these species have not been isolated or even detected (Rappe and Giovannoni 2003).

A signature for traceability assumes a priori the stability of microbial communities; however, the dynamics of microbial communities prevent this stability. In fact, two types of dynamics coexist; temporal dynamics whose strengths are in the main usually biotic (competition, predation, etc.); and a spatial dynamics whose strengths are generally abiotic (changes in physicochemical parameters). Because these two dynamics are simultaneous, they often blur information given by a molecular signature. Water lends itself well to such analysis because the weak and slow biotic activity it harbors due to limited resources renders the temporal dynamics negligible compared to the spatial dynamics (Henne et al. 2013). Three examples are presented below to highlight the potential of CE-SSCP for tracking a water network using its microbiology.

2. CE-SSCP Technique

Nowadays, a panel of techniques provides a molecular fingerprint of microbial communities without any prior knowledge of their composition (Rastogi and Sani 2011). The signature of each species is associated with its 16S rDNA sequence. From the total DNA in a microbial community, PCR amplification with universal primers provides a "community" of 16S rDNA. The goal of each technique is to separate and display all 16S rDNA fragments.

Among these techniques, CE-SSCP enables to visualize each of the 16S rDNA of a microbial community as a peak in an electrophoretic profile (Zumstein et al. 2000). The choice of PCR primers used to define the microbial groups of interest is primordial (i.e., Prokaryote, Eukaryote, Bacteria, Achaea, Cyanobacteria, Firmicute, etc.). Each 16S rDNA sequence (i.e., each species) is represented by a peak whose intensity is proportional to the abundance of this sequence in the microbial community. All the peaks together with their heights sign the structure of the microbial community. Separation is based upon the ability of the DNA single strand to adopt a secondary structure (depending on the base composition of its own) thus SSCP analysis permits the separation of DNA fragments of the same size but with different DNA compositions. This principle is associated with the use of a fluorescent primer during the PCR. SSCP was performed with the ABI 3130 genetic analyzer (Applied Biosystems) equipped with four 50-cm capillary tubes filled with conformation analysis polymer (Applied Biosystems) in corresponding buffer and 10% glycerol. The injection of DNA in capillaries required is 5 kV for 3 sec. Electrophoresis is carried out at 15 kV and 32°C for 30 min per sample. Raw SSCP data were exported into a file format easily handle able and statistical analyses were performed with 'StatFingerprints' (Michelland et al. 2009).

3. The Achievement of CE-SSCP Profiles

The CE-SSCP profiles contain reproducible information on both the presence and the abundance of microbial species in a sample. This information is

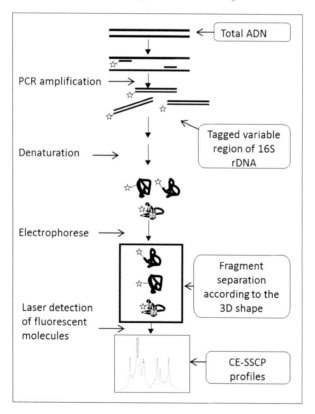

Figure 1. Summarizes How to Obtain a CE-SSCP Profile. A More Complete Description is Available in the Following Reference (Wéry et al. 2008).

considered as semi-quantitative since it is only an image of reality, a picture inevitably distorted by the various steps of the protocol (DNA extraction (Purswani et al. 2011), PCR (Fredriksson et al. 2013), etc).

Which data can be obtained using CE-SSCP profiles (Marzorati et al. 2008)? In terms of diversity markers, richness can not be measured because some rare species are not visible (Haegeman et al. 2013). On the other hand, the Simpson diversity index can be measured (Haegeman et al. 2014). Similarity or difference between large numbers of CE-SSCP profiles can be analyzed using statistical tools such as PCA (Principal Component Analysis) (Fromin et al. 2002). Dedicated free software such as 'StatFingerprints' makes it possible to carry out these analyses (Michelland et al. 2009).

The CE-SSCP technique has the advantage of using capillary DNA sequences. This permits reproducibility and high throughput. Though the equipment is expensive, each analysis is cheap. CE-SSCP has the same drawback as other fingerprint techniques: peak identification is possible but difficult.

4. Examples of Using CE-SSCP on Water Samples

The uses of CE-SSCP in analyzing water samples in specific areas are discussed.

4.1 Tracking an urban water network

Two questions involving water within urban water networks can be addressed by CE-SSCP fingerprinting:

- What differences are there between the water at the consumer's tap and the water leaving the water treatment plant or coming from a natural source? Given that water is defined as drinkable at its point of origin.
- Urban water systems are complex and form an interconnected loop. As a result, it is difficult to know the preferential direction of flow and, thus, to know the distance the water travels and its time of residence in the pipes. CE-SSCP is used to test the spatial dynamics involved and to establish the water's direction of flow.

Figure 2 shows the results obtained for an urban water network (Harmand et al. 2006). The CE-SSCP profiles are different due to the rapid dynamics of microbial communities throughout the water network. According to their similarities, the CE-SSCP profiles have been clustered into four groups: a group of CE-SSCP profiles 1, 2, 3 and 4, corresponding to

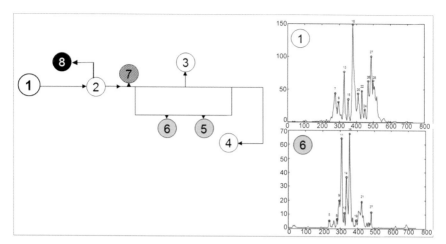

Figure 2. Left Side: Water Network. 1, Corresponds to the Origin of the Water. Arrows Indicate the Known Direction of Flow. Colors Within the Circles Correspond to Sampling Points Clustered Together. Right Side: CE-SSCP Profiles.

coherent dynamics from the origin to the final point of distribution; group of CE-SSCP profiles 5 and 6, which corresponds to two sampling points close to each other isolated within a loop of the network, was found to be very different from all other CE-SSCP profiles; two groups, 7 and 8, that correspond to CE-SSCP profiles far from all other CE-SSCP profiles.

4.2 Spatial dynamics of a network of groundwater

Two aspects are tested in this study: (i) the microbial diversity of karst water; (ii) the impact of storage in a tank. In this context, molecular analysis by the CE-SSCP technique was carried out on five samples from a drinking water network of karst origin.

The storage in the water tank and the flow within the water network induce great changes in microbial communities (Fig. 3). It should be noted that CE-SSCP makes it possible to see changes in the structure of the microbial communities but not their growth.

4.3 Wine

Another example can be the wine. The structure of the wine microbial community includes information about terroir, vintage and grape variety. For classified Grands Crus, this signature will allow to identify the frauds or bad conditions of conservation.

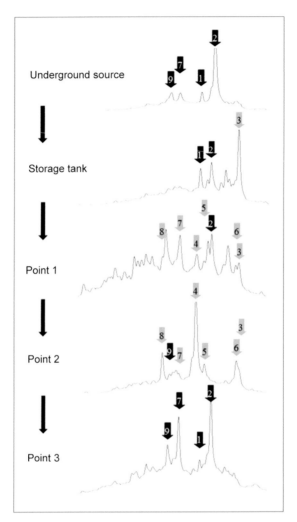

Figure 3. CE-SSCP Profiles Along the Water Network. Black Arrows Correspond to CE-SS-CP Peaks Coming from the Well. Grey Arrows Correspond to CE-SSCP Peaks Coming after Storage.

5. Spatial Analysis of the Diversity of Bacteria in a Cooling Tower Facility

Dispersion of pathogenic *Legionella* spp. is a major microbial problem associated with cooling towers which provide ideal hot habitats for *Legionella* spp. (Sheehan et al. 2005). *Legionella* spp. are facultative intracellular gram-negative bacteria which multiply in protozoan hosts and can also survive within microbial biofilm communities (Fields et al. 2002). Amoebal cysts provide a protective environment for the *Legionella* species, which can then withstand treatments with biocides. Thus, controlling the *Legionella* spp. risk in cooling towers requires a better understanding of the dynamics of the associated microbial communities. Due to the growth requirements of *Legionella* spp., its ability to enter a viable-but-nonculturable state under stress conditions and the association of *Legionella* spp. with protozoa Legionella dynamics must be carried out by cultivation-independent methods such as CE-SSCP (Wullings and van der Kooij 2006).

In order to identify the source of *Legionella* spp. and the role of biocides, bacterial diversity and dynamics were studied using water collected at different locations in a cooling tower facility over a one-year period (Fig. 4). Bacterial diversity varied spatially, i.e., at different locations in the plant. Microbial communities in the input water (before and after treatment) were different from those in the tower area. Bacterial diversity in the water basin and in the biofilm basin was different whereas the diversity in water before dispersion seems to have been a mix of the water and biofilm results (Fig. 5). Tap water was variable over time whereas the other kinds of water were relatively stable despite the use of a biocide.

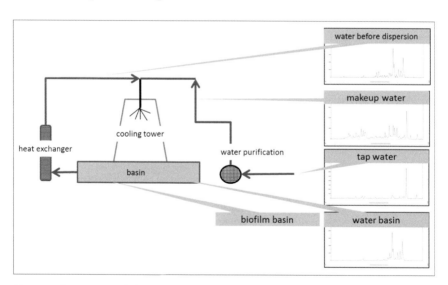

Figure 4. Description of the Cooling Tower Network. On the Right, the CE-SSCP Profiles Correspond to the Sampling Points.

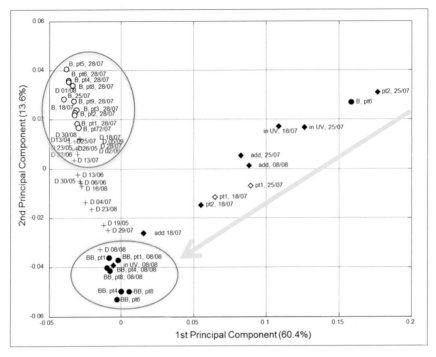

Figure 5. PCA of Spatial and Temporal Dynamics in a Cooling Tower Water Network. Black Diamonds Correspond to Tap Water and Makeup Water; White Circles Correspond to Basin Water; Black Circles Correspond to Basin Biofilm; + Correspond to Water before Dispersion.

6. Conclusion

The CE-SSCP is an excellent tool for highlighting stability or disturbances in a water network, both at the source or throughout the network. An example of possible use of CE-SSCP could be for mineral water springs which are kept under very close surveillance and are surrounded by protected zones. Thus, disturbances must be detected quickly but are not necessarily indicated by the sensors used nowadays. The very sensitive CE-SSCP technique can sign a microbial disturbance and would thus be a valuable indicator.

Keywords: CE-SSCP, 16S rDNA, water, microbial diversity

References

European Commission Regulation No. 178/2002 of the European Parliament and of council of 28 January 2002. Laying down the general principles and requirements of food law, establishing the European Food. Safety Authority and laying down procedures in matters of food safety. Official Journal of the European Communities L31, 1.2.2002, p. 506–526.

Fields, B.S., Benson, R.F. and Besser, R.E. 2002. Legionella and Legionnaires' disease: 25 years of investigation. Clinical Microbiological Review 15: 506+.

Fredriksson, N.J., Hermansson, M. and Wilen, B.M. 2013. The Choice of PCR Primers has great impact on assessments of bacterial community diversity and dynamics in a wastewater treatment plant. PLoS One 8.

Fromin, N., Hamelin, J., Tarnawski, S., Roesti, D., Jourdain-Miserez, K., Forestier, N., Teyssier-Ccvelle, S., Gillet, F., Aragno, M. and Rossi, P. 2002. Statistical analysis of denaturing gel electrophoresis (DGE) fingerprinting patterns. Environmental Microbiology 4: 634–643.

Haegeman, B., Hamelin, J., Moriarty, J., Neal, P., Dushoff, J. and Weitz, J.S. 2013. Robust estimation of microbial diversity in theory and in practice. ISME Journal 7: 1092–1101.

Haegeman, B., Sen, B., Godon, J.J. and Hamelin, J. 2014. Only simpson diversity can be estimated accurately from microbial community fingerprints. Microbial Ecology 68: 169–172.

Harmand, J., Paulou, L., Desmoutiers, J., Garrelly, L., Dabert, P. and Godon, J.J. 2006. The microbial signature of drinking waters: myth or reality? Water Science and Technology 53: 259.

Henne, K., Kahlisch, L., Hoefle, M.G. and Brettar, I. 2013. Seasonal dynamics of bacterial community structure and composition in cold and hot drinking water derived from surface water reservoirs. Water Research 47: 5614–5630.

Marzorati, M., Wittebolle, L., Boon, N., Daffonchio, D. and Verstraete, W. 2008. How to get more out of molecular fingerprints: practical tools for microbial ecology. Environmental Microbiology 10: 1571–1581.

Michelland, R.J., Dejean, S., Combes, S., Fortun-Lamothe, L. and Cauquil, L. 2009. StatFingerprints: a friendly graphical interface program for processing and analysis of microbial fingerprint profiles. Molecular Ecology Resources 9: 1359–1363.

Purswani, J., Manuel Martin-Platero, A., Reboleiro-Rivas, P., Gonzalez-Lopez, J. and Pozo, C. 2011. Comparative analysis of microbial DNA extraction protocols for groundwater samples. Anals of Biochemistry 416: 240–242.

Rappe, M.S. and Giovannoni, S.J. 2003. The uncultured microbial majority. Annual Review of Microbiology 57: 369–394.

Rastogi, G. and Sani, R.K. 2011. Molecular techniques to assess microbial community structure, function, and dynamics in the environment. pp. 29–57. *In*: Ahmad, I. and Ahmad, F. and Pichtel, J. (eds.). Microbes and Microbial Technology: Agricultural and Environmental Applications.

Sheehan, K.B., Henson, J.M. and Ferris, M.J. 2005. Legionella species diversity in an acidic biofilm community in Yellowstone National Park. Applied Environmental Microbiology 71: 507–511.

Wéry, N., Bru-Adan, V., Minervini, C., Delgénes, J.P., Garrelly, L. and Godon, J.J. 2008. Dynamics of *Legionella* spp. and bacterial populations during the proliferation of L. pneumophila in a cooling tower facility. Applied Environmental Microbiology 74: 3030–7.

Wullings, B.A. and van der Kooij, D. 2006. Occurrence and genetic diversity of uncultured *Legionella* spp. in drinking water treated at temperatures below 15 degrees C. Applied Environmental Microbiology 72: 157–166.

Zumstein, E., Moletta, R. and Godon, J.J. 2000. Examination of two years of community dynamics in an anaerobic bioreactor using fluorescence polymerase chain reaction (PCR) single-strand conformation polymorphism analysis. Environmental Microbiology 2: 69–78.

Next Generation Sequencing in Food Authenticity and Safety

Cristina Barbosa, Sofia Nogueira, Ricardo Saraiva and
Sandra Chaves*

1. Introduction

Consumers and the food industry are progressively becoming more apprehensive about what they eat or about the products they are offering. Consumer's choice for food is greatly dependent on socio-economic factors like culture, religion, health concerns, quality, prices, etc. The increase of information and globalization has turned food authenticity and safety into major issues demanding specific authorities and regulations worldwide (Badia-Melis et al. 2015).

Food fraud cases and food borne diseases are examples of problems that enduce consumers to differentiate quality food products and to request a correct and trustworthy labeling along with safe food products. To achieve this, a technical approached to identify false description and mislabeling using genomics or proteomics is essential. A brand, market or even a region can be negatively affected by mislabeling or due to a food borne outbreak since it raises significant consumer-protection issues (Gracia and de-Magistris 2016, Charlebois et al. 2016).

SGS Molecular, Ed. Teclabs, Campus da FCUL, Campo Grande 016, 1749-016 Lisboa, Portugal.
  E-mail: cristina.barbosa@sgs.com; sofia.nogueira@sgs.com; ricardo.saraiva@sgs.com
* Corresponding author: sandra.chaves@sgs.com

Labeling corroboration, verification and identification of food components and food microbial population can be achieved by food testing. These laboratory tests are one of the keys to prevent possible fraudulent practices or disease outbreaks, giving to consumers an accurate information and control of what they eat. Testing can rely on protein, metabolite or DNA-based methodologies. DNA has the advantage of being more stable and less affected by food processing than proteins and metabolites, providing higher sensitivity for testing and being a powerful tool to the agro-food sector (Woolfe and Primrose 2004). From all the methods available, amplification based methods; in particular polymerase chain reaction (PCR) and real-time PCR are the most popular due to their simplicity, sensitivity, specificity, reproducibility, and ability to detect low amounts of the target. These methods rely on the ability to detect specific DNA sequences in the genome of target organisms. This means that in a given food matrix it is possible to identify any specific organism independently if it is an allergen in meat, fish or other, even a microorganism (Cammà et al. 2012, Prado et al. 2015).

A shortcoming of PCR-based methods for food testing is their low versatility, since they are directed to a defined target. PCR-based methods are in disadvantage when used to analyze complex food matrices where multiple species identification is required and when there is no *a priori* indication of which species to investigate. This drawback also exists if it is required to identify closely related species or strains and even to distinguish varieties, breeds or serotypes (Smith and Osborn 2009, Woolfe and Primrose 2004). Sequencing can be one powerful aid to overcome these DNA-based methods drawbacks. Knowledge of the genome sequence turns possible the correct identification of close related species, specific varieties or different taxa.

Next generation sequencing (NGS) is an automated high-throughput sequencing technology that has proven to be potent, reliable and robust, with potential to be implemented in food and feed testing. Presently, there are four major commercially available platforms and several other promising approaches that allow massively parallel sequencing. These platforms are: Roche 454 (Roche), Solid (Thermo Fisher Scientific), Illumina (Illumina, Inc.) and Ion Torrent (Thermo Fisher Scientific). Although technologically different, these platforms share the same workflow: first the DNA is pre-processed into a library with fragments ranging in sizes appropriate for sequencing and then these fragments are immobilized in a solid surface where a single DNA molecule is amplified and sequenced.

2. NGS Technologies

Sequencing technologies include a number of procedures that include template preparation, sequencing, data acquisition and data analysis that determines the characteristics of each platform. The multiple applications of NGS make possible that multiple platforms have their place in the market, with some being more suitable for particular applications than others.

In this section some of the most popular platforms, as well as some of the most recent ones will be reviewed.

2.1 Roche 454 platform

In 2005, Roche released the 454 platform. As mentioned, the first step is the preparation of the DNA fragments to construct a library. This step includes the addition of adapters to both ends of each fragment. This will allow the library fragments to bind to the adapters'complementary sequences, covalently linked to agarose 'nano-beads'. These beads are the solid surface used in this platform to generate sequencing templates. Each fragment is then amplified by a so-called emulsion PCR. The emulsion PCR is a PCR reaction that occurs in an oil-water emulsion that creates small vesicle. Each of these vesicles ideally contains only one bead, a single DNA fragment from the library and the reagents needed for the PCR. This amplification step is required to provide enough signals from the sequencing reaction. After the emulsion PCR, templated beads are loaded onto the Pico Titer Plate device, one bead per well, where parallel sequencing reactions occur. Then the four nucleotides are added sequentially across the plate to enable a sequencing-by-synthesis reaction.

When each nucleotide is incorporated, inorganic pyrophosphate is released and used as substrate for ATP-sulfurylase to produce ATP, which is essential for a luciferase-mediated conversion of luciferin to oxyluciferin. This generates a chemiluminescent light signal that is recorded by the device (Margulies et al. 2005, Buermans and den Dunnen 2014, http://454.com/). Hundreds of thousands of short DNA sequences in a single run are generated.

2.2 Solid platform

In 2007, the Solid platform was introduced. Similarly, to Roche 454, beads are used as the solid surface for the template amplification reaction to occur, but the sequencing process is based on an elaborated sequencing-by-ligation scheme. In this scheme a set of four fluorescently labeled probes, each containing two specific bases in the 5' end and 6 degenerate ones, are added to the templated beads in a physic support (an amino-acid coated glass surface). Each array of two specific bases corresponds to define fluorophore and these probes outcompete for annealing with the template. When the two specific bases are complementary to the template, the ligation reaction to the sequencing primer occurs and the probe is cleaved on a specific site, releasing the fluorophore. Multiple cycles of ligation, detection and cleavage are performed. Following a series of ligation cycles, the extension product is removed and the template is reset with a primer complementary to the n-1 position for a second round of ligation cycles. For each sequence five rounds of primer reset occur, which means that every base is interrogated in two independent ligation reactions by two different primers. Consequently, a 99.99% accuracy of the results is possible using Solid platform (Buermans and den Dunnen 2014, http://www.appliedbiosystems.com/).

2.3 Illumina platform

In the same year Illumina launched a new platform. The major differences to the above mentioned platforms are the solid surface, which is not a bead but a flow cell, and the template amplification strategy. The flow cell contains billions of nanowell substrates across both surfaces which dramatically increases data output. In these nanowells, there are oligonucleotides attached, complementary to the adapters previously added to the template during the library preparation step. There are two different adapters which are complementary to each end of the library fragments. As a result, the fragment binds to the complementary sequence of one of the adapters immobilized in the flow cell and then bend, binding the other end to the second adapter creating a bridge. Due to this conformation the reaction is called bridge amplification. The amplification is isothermal, using cyclic alternations of three specific buffers that mediate the denaturation, annealing and extension steps at 60°C. The final step is to remove one strand of the DNA fragments in the cleavable site of the surface oligo and to block 3′ ends, which could act as sequencing primer sites on adjacent library molecules, with ddNTP. At this point, the library is ready to be sequenced in a sequencing-by-synthesis approach. Four fluorescently-labeled nucleotides are all added at the same time, which allows them to compete for the incorporation, minimizing the bias. During each cycle only one nucleotide is incorporated and the fluorescent dye is imaged to identify the base. The nucleotide label functions as a terminator for polymerization, so it has to be enzymatically cleaved to enable the incorporation of the next nucleotide (Buermans and den Dunnen 2014, http://www.illumina.com/).

2.4 Ion Torrent platform

In 2010, the Ion Torrent platform was presented as a simpler, faster, and more cost-effective technology. On Ion Torrent platform beads are the solid surface and emulsion PCR are used again to immobilize and amplify the library fragments. The method also relies on the sequencing-by-synthesis approach, but each of the four nucleotides is added sequentially across the plate. Detection of base incorporation during the sequencing reaction does not rely on fluorescence, chemiluminescence, or enzyme cascades, but in a pH change detected by a semiconductor technology. Incorporation of the nucleotide generates a pH change due to the formation of a covalent bond and the release of pyrophosphate and a positively charged hydrogen ion. The major advantages of this system is faster and cheaper runs, since it avoids the use of modified nucleotides and optical measurements (Buermans and den Dunnen 2014, https://www.thermofisher.com/).

2.5 Single-Molecule Real-Time (SMRT) sequencing

The above four mentioned technologies are referred to as second-generation sequencing technologies and their major characteristics are summarized in Table 1. A third generation sequencing platform has been developed by Pacific BioSciences (PacBio) and is called Single-Molecule Real-Time (SMRT) sequencing. SMRT sequencing makes use of the natural process of DNA replication and enables real-time observation of DNA synthesis. This technology is also based in sequencing-by-synthesis with the usage of phospholinked nucleotides but eliminates the need for amplification of each library fragment. The SMRT chip surface has wells in which a DNA polymerase/template complex is immobilized. By utilizing zero-mode waveguides each well is illuminated only in the bottom, allowing the detection of the signal generated by the sequencing of a single DNA molecule. The advantages are longest average read lengths, highest consensus accuracy, uniform coverage, simultaneous epigenetic characterization and single-molecule resolution (http://www.pacb.com/, Rhoads and Au 2015).

In the following sections, the application of NGS in the agro-food sector will be discussed, as well the advantages and disadvantages of the usage of this technology to ensure food authenticity and safety.

Table 1. Next-generation Sequencing Technologies—Main Characteristics.

Brand	Platform	Solid surface	Sequence Reaction	Detection	Read length
Roche	Roche 454	Beads	Sequence-by-synthesis	Chemiluminescence	700 bp
Thermo Fisher Scientific	Solid	Beads	Sequence-by-ligation	Fluorescence	35 bp
	Ion Torrent	Beads	Sequence-by-synthesis	Proton release	200–400 bp
Illumina	Illumina	Flow cell	Sequence-by-synthesis	Fluorescence	2 x 150–300 bp

3. Applications of NGS to Food Authenticity

Food authenticity spans from food fraud related to species substitution for low-priced ones, to the presence of non-declared components or to the alteration of certified regional products (Nogueira et al. 2012, Sentandreu and Sentandreu 2014). DNA based methods like PCR and real-time PCR have been used for several years to detect food components (Fig. 1). These techniques have the characteristic of being focused on the detection of a pre-determined species, since they use specific primers to amplify a specific DNA target, requiring a previous knowledge about what to search. Thus, one of its major advantages—specificity—is also a major drawback when the detection of more than one target is needed, requiring a multiplicity of reactions. For such cases a universal method that would allow the detection and identification of several species in a mixture would be highly valuable, especially considering that in food authenticity these are the most frequent situations.

PCR-cloning is a technique that allows the detection of several target DNA sequences in a heterogeneous matrix. This technique comprises a PCR amplification followed by the cloning of the fragments and their sequencing. Then, it is possible to identify the present organisms by comparing the DNA sequence obtained to a reference database (Galimberti et al. 2013, Hellberg and Morrissey 2011). This technique has been applied to different groups which allow to distinguish between species.

PCR-cloning presents several drawbacks since it is more time-consuming and costly than PCR and real-time PCR based methods mainly due to the cloning step. Also, sometimes the use of long DNA fragments is required and they are very difficult to obtain in heavily processed food

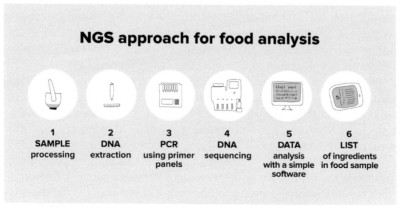

Figure 1. General Workflow of NGS Approach for Food Analysis.

products. Nevertheless, the DNA sequence seems to be of great value to species identification and discrimination. Ultimately, efforts to sequence the whole-genome of organisms will reveal the entire genetic information which includes the DNA markers needed to discriminate species. For these reasons and for new applications next generation sequencing (NGS) seems to be very useful for the agro-food sector (Corrado 2016).

Indeed, the massive data generated by NGS enables the sequencing of entire genomes in a short period of time and represents a cost effective way to audit DNA in heterogeneous samples. Therefore, NGS has potential to be used for the analyses of complex food and feed matrices since in a single instrument run, multiple species from the same sample and also several samples can be simultaneously sequenced. Likewise, it allows for a gain of information from complex or degraded DNA. Nonetheless, NGS application to food industry is still in a nascent stage.

Initially NGS technologies were used mainly for microorganism's whole genome sequencing. Only recently it has started to be applied to food products. Ripp et al. (2014) showed that they were able to identify and quantify several species from sausage using the Illumina HiSeq for sequencing all the present genomes. The sequence reads obtained in this study where mapped to publicly available reference genomes and assigned to the correspondent species, distinguishing mammals, viruses, bacteria and plants. However, the data analysis of this genomic approach (always depending on specific bioinformatics' tools), is time consuming and dependent on public available data that may not exist for all organisms or may be wrongly assemble or annotated. Therefore, strategies to highly simplify and standardize data analysis and output interpretation are essential to implement this approach in routine food laboratories.

Targeted NGS has also been used to identify animal and plant species in mixed samples. Instead of sequencing the whole genome this approach takes advantage of genetic markers like mitochondrial DNA regions, to generate short PCR amplicons with universal PCR primers. These markers have identification potential, allowing the screening of mixed DNA samples and species discrimination. The use of tagged primers—a different tag to amplify each sample—allows the analysis of numerous samples at the same time reducing the cost per sample. Studies using this approach were not on food analysis but reported a successful identification of animal and plant taxa in traditional Chinese medicines and the identification of mammals in forensic cases (Coghlan et al. 2012, Tillmar et al. 2013) showing its potential to identify several species in a complex sample. Even though further studies may be needed to correctly choose the genetic markers to use (considering it should exhibit low intra-species variability but high inter-species variability), the application of NGS in the food industry might allow to assess the food authenticity problematic. For example, meat and fish industries are very prone to food fraud, especially in case of processed products, in which morphological identification is not possible (Flores-Munguia et al. 2000, Cawthorn et al. 2012, Chuah et al., 2016). Indeed, food fraud reached its mediatic peak during the 2013 European horse scandal, when it was discovered that horse meat was being commercialized as beef (Premanandh 2013, http://ec.europa.eu/food/food/horsemeat/). This substitution had the finality of commercial gain since horse meat is low-priced than beef.

Similarly, fraud in fish products is often observed. The substitution of grouper fillets (*Epinephelus marginatus*) for low-priced species is one good example. Also, only *Gadus morhua* and *Gadus macrocephalus* can be commercialized as cod. However, other species with much less economical value are detected in some cod products (Cutarelli et al. 2014). Shrimps and prawns are likewise very susceptible to fraud since there are many species with different commercial values (Ortea et al. 2012). These exemplificative cases and continuous examples of food adulteration call for more food monitoring, traceability and labeling (Ayaz et al. 2006). Thus, species identification either by whole genome sequencing of the species in a food product or by target sequencing is highly valuable to monitor and prevent fraudulent products like those mentioned above.

This same approach can be applied to monitor the adulteration of products with Halal certification. These products are highly important for cultural and religious reasons. The criteria to get this certification include the specification of which components are allowed to be present in food products. For example, the presence of any trace of pork meat in these products is unacceptable to get this certification. Therefore, these products call for a strict regulation and for a strict control of the entire process from farm to fork (Nakyinsige et al. 2012). A DNA-based method to evaluate these products has to be very sensitive and capable to correctly distinguish the species, thus making the use of NGS a valuable tool.

Food fraud is also found in traditional or regional products. These products are commercialized to the customers presenting a number of characteristics that grant them higher quality (e.g., being from a specific variety or from specific geographical area). The use of a specific breed or variety in a food product may increase the commercial value of these products. For example, the use of specific breeds in meat products, like Angus beef, increases the value (Bureš et al. 2015). Also, there are many rice varieties with commercially distinct values (Kamath et al. 2008). Honey is also very prone to fraud. Honey commercial value is dependent on the geographical region, as it correlates with botanic species used by bees to make it. Also, honey's potential for medical and nutritional applications are dependent on the botanical origin (Danezis et al. 2016, Galemberti et al. 2014). In this case, development of methods to ensure the correct variety or breed identification is very challenging and difficult. The entire DNA sequence is sometimes essential for a correct identification. Thus, the evolution of NGS and its application in food industry will undoubtedly help in these cases.

Another application of NGS is in the identification of probiotics. Fermentation is a very ancient process to prepare and preserve food. Nowadays, this process is strictly controlled to ensure the high quality of products. Probiotics are microorganisms present or added to food that benefit the host and have received considerable attention in recent years. Most probiotic products contain *Lactobacillus*, *Bifidobacterium*, *Pediococcus*, *Streptococcus* and *Lactococcus* species as either a mixture or single isolates. Therefore, since different microbes may exert different probiotic activities, identification of the specific microorganisms is of the utmost importance to certify the product (Di Lena et al. 2015, Patro et al. 2016).

In conclusion, the use of NGS as tool for food authenticity is highly interesting and promising as: (i) it is suitable for highly processed and degraded DNA; (ii) there is no need of *a priori* species information; (iii) is cost-effective when processing many samples; (iv) it is possible to detect DNA in very low amounts. The main hindrance to use either whole genome or target sequencing is that they generate a massive data amount that needs specific tools to be analyzed. This is time consuming and requires substantial biological and bioinformatical knowledge. Simple and fast bioinformatics tools must be developed to allow the application of NGS in routine food laboratories. For this reasons, routine application of NGS based methods for food analysis is just now emerging, but probably will be the next big step in food authenticity.

3.1 Applications of NGS to food safety

Food safety is a major issue for consumers and is a global concern. Indeed, it is estimated that 600 million people—almost 1 in 10 persons in the world— fall ill after eating contaminated food and 420000 die every year (WHO 2015). Food borne disease surveillance, food inspection and monitoring, detection and investigation of the origin of outbreaks are becoming progressively more important. This requires a control of the entire food chain, from farm to fork. Consequently, there is an increased awareness about the safety of animal derived food, sanitary status and welfare of farm animals. These issues demand regulation by many organizations and governments, leading to a considerable effort from many countries to improve food control systems (Collins and Wall 2004, Mayo et al. 2014).

Indeed, healthy animals are directly linked to safe food and human health. There are many dangerous pathogens both for animals and humans (e.g., *Salmonella* spp.). However, there are also some pathogens that are innocuous to animals but that may cause serious illness in humans (e.g., the enterohemorrhagic *Escherichia coli* O157). This highlights the need to ensure the health of animals and controlled conditions of farm-derived food making it marketable (Collins and Wall 2004, Oliver et al. 2005, Armstrong et al. 1996). Another potential problem source is transport, storage and handling conditions. Global distribution of food increased stock density and competitive markets might allow a higher dissemination of pathogens causing a potential risk to the consumer. To minimize these issues there are regulations and an international code recommended by the Food and Agriculture Organization of the United Nations (FAO) and the World Health Organization (WHO) that establish hygiene practices from primary production through to the final consumption. For example, *Listeria monocytogenes* is an example of pathogenic bacteria that can contaminate food processing facilities. It is hard to eradicate since it can grow in cold storage conditions (Ryser and Marth 2007, Badia-Melis et al. 2015). Thus, food monitoring before it goes into the market seems to be very valuable both for the consumer and for the supplier, since an outbreak may cause severe consumer's mistrust in the product.

Next generation sequencing (NGS) is a new tool with great potential to provide important improvements in the management of food safety worldwide (FAO 2016). This technology appeared on the market ten years ago and during this past decade tremendous progress has been made in terms of speed, read length, and throughput, along with a sharp reduction in per-base cost (van Dijk et al. 2014). Together, these advances result in a range of new successful applications in food safety, such as chIP-sequencing to identify binding sites of DNA-associated proteins, RNA-sequencing to profile the cellular transcriptome, whole genome sequencing (*de novo* sequencing and re-sequencing) and targeted-sequencing (Solieri et al. 2013).

Within the whole genome sequencing (WGS) there are two main approaches: *de novo* sequencing where no pre-existing information related to the target genome exists and re-sequencing that implies that sequenced genomes from the target species are already available, providing a reference genome for the analysis (Solieri et al. 2013). This last approach is the most used in food safety.

The gold standard technique of microbial subtyping approach is the pulsed-field gel electrophoresis—PFGE (Diaz-Sanchez et al. 2013). When it started to be used, this technique had a tremendous positive impact on food safety. However, this and other subtyping methods have some limitations that can be overcome by NGS technique. With WGS a higher number of samples can be run at once (unlike PFGE where different gels may be needed for different pathogens) and it provides a substantial improvement in subtype discrimination of some highly clonal pathogen populations (Wiedmann 2015). This can be very important to distinguish outbreaks-related strains from other genetically similar ones that are not associated with the same outbreak. In addition, WGS provides additional data to epidemiological tracing, including identification of specific virulence markers that may be strain specific (Diaz-Sanchez et al. 2013).

WGS discriminatory power has already been shown in a number of recent outbreaks. In the summer of 2010, *Vibrio parahaemolyticus* caused an outbreak in Maryland linked to the consumption of oysters. WGS with MLST (Multi-Locus Sequence Typing) allowed finding the real origin of the outbreak strain—Asia (Haendiges et al. 2016). Another example occurred in Germany (2011). A large outbreak of an exceptionally virulent Shiga-like toxins (Stx) producing *Escherichia coli* O104:H4 was responsible for 852 cases of hemolytic uremic syndrome (HUS) and caused 32 HUS-associated deaths. Contaminated raw sprouts from a farm in Germany were the likely source of the outbreak (CDC 2011). WGS was a highly valuable resource that allowed to understand that the cause of the outbreak was a new extremely virulent strain of *E. coli* O104:H4 serotype (Mellmann et al. 2011). Also, a *Salmonella* sp. outbreak related with egg consumption began in 2014 in England. In this outbreak, WGS showed its power by linking the apparently different outbreaks sources to only one egg producer (FAO 2016). Recently a *Listeria monocytogenes* outbreak occurred in the United States affecting 9 states. Epidemiologic and laboratory evidences indicated that packaged salads produced in Springfield, Ohio were likely to be the source of this outbreak. At the same time, in Canada, listeriosis cases also started to increase. WGS analysis of clinical isolates from ill people in Canada showed that isolates from both countries were genetically related with the one founded in the package salads (CDC 2016).

WGS can also be used in surveillance and food monitoring. A good example of its applicability in this area is given by FAO. United States began a real-time surveillance for *Listeria monocytogenes* in food and environmental samples in 2013. One year later, PFGE allowed identifying of two clusters clearly associated with seasonal food products and one cluster associated with both seasonal food and ready-to-eat food. Unlike PFGE, WGS enable the researchers to exclude the ready-to-eat food because the cases linked to this source showed to be caused by isolates without any relationship with those found in seasonal food (FAO 2016). This confirmation, in early stages, that seasonal food was the reservoir of *Listeria monocytogenes* strain that caused infection cases, allowed preventing a large outbreak. This good example of how understanding the true source of an outbreak helps to determine preventive controls and mitigation strategies that can be put in practice for future protection in the food supply chain (FAO 2016).

Targeted-sequencing is another highly used approach of NGS. As described above, by using this strategy only the targeted genes or genomic regions are sequenced. Thus, data generated is much less complex, making data processing much faster. It also enables sequencing more samples per run when comparing with WGS, leading to a lower cost per sample.

Both WGS and targeted-sequencing have the advantage of lacking the need of previous sample enrichments and bacterial cultures. Consequently, it is possible to get faster results, especially using targeted-sequencing, with a higher representativeness of microbial diversity in the sample. Furthermore, bias resulting from the selection and competition among microorganisms during their enrichment/culture growth is avoided.

Target genes must show identification potential (van Dijk et al. 2014, Gardner et al. 2015, Mayo et al. 2014). The presence of conserved gene regions is commonly selected to desing universal primers for amplification (Gardner et al. 2015). The target sequence should contain enough nucleotide heterogeneity to distinguish among different organisms. Hyper-variable regions within the 16S rRNA gene are normally chosen for the assessment of bacterial diversity. Moreover, targeting specific microbial groups may require the amplification of specific hypervariable regions (Mayo et al. 2014). For instance, V1 region can be used to differentiate between *Staphylococcus aureus* and coagulase negative *Staphylococcus* species. Similarly, V6 is able to differentiate between species of *Bacillus*, like *B. anthracis* and *B. cereus* (Chakravorty et al. 2007). Despite their huge potential for microbial community profiling, targeted-sequencing using 16S rRNA gene presents some pitfalls. Universal primers have demonstrated to have a bias in the detection of high GC content bacteria and, in several cases, the 16S rRNA gene may not have enough variability in order to assign microorganisms to genus and species taxonomic levels (Solieri et al. 2013). Enterobacteriaceae is one of the most relevant bacterial families for food safety and a good example of this. It includes harmless species but also many food borne pathogens, such as *Salmonella* spp., *Escherichia coli*, *Yersinia* spp. and *Cronobacter* spp. The different species within this family cannot be distinguished with 16S rRNA gene. The same happens within Vibrionaceae, Listeriaceae, Campylobacteraceae and other families that include food borne pathogens. For these cases a single target approach based on 16S rRNA gene will not lead to conclusive results. However, other approaches based on other marker genes for target groups of food borne pathogens can be used. Instead of a universal target for all microorganisms (16S rRNA genes), specific genomic regions can be very useful. This would give to NGS technology and to targeted-sequencing approach the resolution required to discriminate and identify the different groups of microorganisms in a food matrix. Furthermore, different targets for different groups can be use simultaneously, in a multiple targeted-sequencing approach, allowing an even shorter turnaround time. Indeed, some companies and laboratories are already developing and using these kind of approaches for the study of microorganisms' populations in food matrices which show the huge potential of this technique for food safety.

Although food safety is often related to microorganisms, allergens can also be harmful to human health, being another relevant issue concerning food safety and also requiring a tight control of food components (Fæste et al. 2011). The World Health Organization estimates that around 250 million people worldwide suffer from food allergies. Some allergens are extremely resistant to food processing like denaturation or enzymatic digestion, which means that even in highly processed food the presence of an allergen may induce a strong immune response (Kimber and Dearman 2001, Besler et al. 2001). As mentioned before, DNA amplification-based techniques with only one predetermined and specific target are the gold standard for the detection of undesirable components in food. This is also true for food allergens, even though that DNA based methods are indirect in determining allergenic components, as they detect the "producer" and not the allergen itself. However, with targeted-sequencing approaches it is possible to quickly and simultaneously identify the producers of several allergen groups: (i) cereals containing gluten; (ii) nuts and dried nuts; (iii) spices commonly used in ready-to-eat and canned meals; (iv) fish and other marine animals in non-seafood products. In such cases, a multiple targeted-sequencing approach to identify potential causes of allergies could be applied. Since the presence of allergens, even in trace amounts, may lead to health problems, the high sensitivity of NGS technology along with its different approaches allows to obtain high reliable answers, leading to allergen detection even if present in trace amounts in food.

4. Conclusion and Future Perspectives

NGS technique and its different approaches have already started to show huge impact in the food industry, even though simple and reliable strategies are essential for its routine application in food analysis (Fig. 1). As stated before, it is highly valuable in helping consumers to have access to safer and higher quality food. Furthermore, authorities have a powerful tool for food fraud control and suppliers can comply with and enforce the legal labeling requirements, ensuring authentic and safe food products. In the future it may be envisaged obtaining the composition of a food product, its content in microorganisms and even its content in allergens in a single analytical step.

Keywords: Food authenticity, food safety, next-generation sequencing (NGS), DNA sequencing, food analysis

References

Armstrong, G.L., Hollingsworth, J. and Morris, Jr, J.G. 1996. Emerging food borne pathogens: *Escherichia coli* O157:H7 as a model of entry of a new pathogen into the food supply of the developed world. Epidemiologic Reviews 18(1): 29–51.

Ayaz, Y., Ayaz, N.D. and Erol, I. 2006. Detection of species in meat and meat products using enzyme-linked immunosorbent assay. Journal of Muscle Foods 17(2): 214–220.

Badia-Melis, R., Mishra, P. and Ruiz-García, L. 2015. Food traceability: New trends and recent advances. A review. Food Control 57: 393–401.

Besler, M., Steinhart, H. and Paschke, A. 2001. Stability of food allergens and allergenicity of processed foods. Journal of Chromatography B: Biomedical Sciences and Applications 756: 207–228.

Buermans, H.P.J. and den Dunnen, J.T. 2014. Next generation sequencing technology: Advances and applications. Biochimica and Biophysica Acta (BBA) - Molecular Basis of Disease 1842: 1932–1941.

Bureš, D., Bartoň, L., Kotrba, R. and Hakl, J. 2015. Quality attributes and composition of meat from red deer (*Cervuselaphus*), fallow deer (*Damadama*) and Aberdeen Angus and Holstein cattle (*Bostaurus*). Journal of the Science of Food and Agriculture 95: 2299–2306.

Cammà, C., Di Domenico, M. and Monaco, F. 2012. Development and validation of fast Real-Time PCR assays for species identification in raw and cooked meat mixtures. Food Control 23(2): 400–404.

Cawthorn, D., Steinman, H. and Witthuhn, R. 2012. Evaluation of the 16S and 12S rRNA genes as universal markers for the identification of commercial fish species in South Africa. Gene 491(1): 40–48.

CDC—Centers for Disease Control and Prevention. 2011. Outbreak of Shiga toxin-producing *E. coli* O104 (STEC O104:H4) Infections Associated with Travel to Germany (FINAL UPDATE) (available at http://www.cdc.gov/ecoli/2011/travel-germany-7-8-11.html) Accessed on July 2016.

CDC—Centers for Disease Control and Prevention. 2016. Multistate Outbreak of Listeriosis Linked to Packaged Salads Produced at Springfield, Ohio Dole Processing Facility (available at http://www.cdc.gov/listeria/outbreaks/bagged-salads-01-16/) Accessed on July 2016.

Chakravorty, S., Helb, D., Burday, M., Connell, N. and Alland, D. 2007. A detailed analysis of 16S ribosomal RNA gene segments for the diagnosis of pathogenic bacteria. Journal of Microbiological Methods 69(2): 330–339.

Charlebois, S., Schwab, A., Henn, R. and Huck, C. 2016. Food fraud: an exploratory study for measuring consumer perception towards mislabeled food products and influence on self-authentication intentions. Trends in Food Science & Technology 50: 211–218.

Chuah, L., He, X., Effarizah, M., Syahariza, Z., Shamila-Syuhada, A. and Rusul, G. 2016. Mislabelling of beef and poultry products sold in Malaysia. Food Control 62: 157–164.

Coghlan, M.L., Haile, J., Houston, J., Murray, D.C., White, N.E., Moolhuijzen, P., Bellgard, M. and Bunce, M. 2012. Deep sequencing of plant and animal DNA contained within traditional chinese medicines reveals legality issues and health safety concerns. PLoS Genetics 8(4): e1002657.

Collins, J.D. and Wall, P.G. 2004. Food safety and animal production systems: controlling zoonoses at farm level. Scientific and Technical Review of the Office International des Epizooties 23(2): 685–700.

Corrado, G. 2016. Advances in DNA typing in the agro-food supply chain. Trends in Food Science & Technology 52: 80–89.

Cutarelli, A., Amoroso, M., Roma, A., Girardi, S., Galiero, G., Guarino, A. and Corrado, F. 2014. Italian market fish species identification and commercial frauds revealing by DNA sequencing. Food Control 37: 46–50.

Danezis, G., Tsagkaris, A., Camin, F., Brusic, V. and Georgiou, C. 2016. Food authentication: Techniques, trends & emerging approaches. Trends in Analytical Chemistry (in press).

Di Lena, M., Quero, G.M., Santovito, E., Verran, J., De Angelis, M. and Fusco, V. 2015. A selective medium for isolation and accurate enumeration of *Lactobacillus casei*-group members in probiotic milks and dairy products. International Dairy Journal 47: 27–36.

Diaz-Sanchez, S., Hanning, I., Pendleton, S. and D'Souza, D. 2013. Next-generation sequencing: The future of molecular genetics in poultry production and food safety. Poultry Science 92: 562–572.

FAO—Food and Agriculture Organization of the United Nations in collaboration with the World Health Organization (2016). Applications of Whole Genome Sequencing in Food Safety Management.

Fæste, C.K., Rønning, H.T., Christians, U. and Granum, P.E. 2011. Liquid chromatography and mass spectrometry in food allergen detection. Journal of Food Protection 74(2): 316–345.

Flores-Munguia, M.E., Bermudez-Almada, M.C. and Vazquez-Moreno, L. 2000. A research note: Detection of adulteration in processed traditional meat products. Journal of Muscle Foods 11: 319–325.

Galimberti, A., De Mattia, F., Losa, A., Bruni, I., Federici, S., Casiraghi, M., Martellos, S. and Labra, M. 2013. DNA barcoding as a new tool for food traceability. Food Research International 50: 55–63.

Galimberti, A., De Mattia, F., Bruni, I., Scaccabarozzi, D., Sandionigi, A., Barbuto, M., Casiraghi, M. and Labra, M. 2014. A DNA barcoding approach to characterize pollen collected by honeybees. PLoS ONE 9(10): e109363.

Gardner, S.N., Frey, K.G., Redden, C.L., Thissen, J.B., Allen, J.E., Allred, A.F., Dyer, M.D., Mokashi, V.P. and Slezak, T.R. 2015. Targeted amplification for enhanced detection of biothreat agents by next-generation sequencing. BMC Research Notes 8: 682.

Gracia, A. and de-Magistris, T. 2016. Consumer preferences for food labeling: what ranks first? Food Control 61: 39–46.

Haendiges, J., Jones, J., Myers, R.A., Mitchell, C.S., Butler, E., Toro, M. and Gonzalez-Escalona, N. 2016. A nonautochthonous US Strain of *Vibrio parahaemolyticus* isolated from Chesapeake Bay oysters caused the outbreak in Maryland in 2010. Applied and Environmental Microbiology 82(11): 3208–3216.

Hellberg, R.S.R. and Morrissey, M.T. 2011. Advances in DNA-based techniques for the detection of seafood species substitution on the commercial market. Journal of the Association for Laboratory Automation 16(4): 308–321.

http://454.com/

http://ec.europa.eu/food/food/horsemeat/

http://www.appliedbiosystems.com/

http://www.illumina.com/

http://www.pacb.com/

https://www.thermofisher.com/

Kamath, S., Stephen, J.C., Suresh, S., Barai, B., Sahoo, A., Radhika Reddy, K. and Bhattacharya, K. 2008. Basmati rice: its characteristics and identification. Journal of the Science of Food and Agriculture 88: 1821–1831.

Kimber, I. and Dearman, R.J. 2001. Food allergy: what are the issues? Toxicology Letters 120: 165–170.

Margulies, M., Egholm, M., Altman, W.E., Attiya, S., Bader, J.S., Bemben, L.A., Berka, J., Braverman, M.S., Chen, Y.-J., Chen, Z., Dewell, S.B., Du, L., Fierro, J.M., Gomes, X.V., Godwin, B.C., He, W., Helgesen, S., Ho, C.H., Irzyk, G.P., Jando, S.C., Alenquer, M.L.I., Jarvie, T.P., Jirage, K.B., Kim, J.-B., Knight, J.R., Lanza, J.R., Leamon, J.H., Lefkowitz, S.M., Lei, M., Li, J., Lohman, K.L., Lu, H., Makhijani, V.B., McDade, K.E., McKenna, M.P., Myers, E.W., Nickerson, E., Nobile, J.R., Plant, R., Puc, B.P., Ronan, M.T., Roth, G.T., Sarkis, G.J., Simons, J.F., Simpson, J.W., Srinivasan, M., Tartaro, K.R., Tomasz, A., Vogt, K.A., Volkmer, G.A., Wang, S.H., Wang, Y., Weiner, M.P., Yu, P., Begley, R.F. and Rothberg, J.M. 2005. Genome sequencing in microfabricated high density picolitre reactors. Nature 437: 376–380.

Mayo, B., Rachid, C., Alegría, Á., Leite, A., Peixoto, R. and Delgado, S. 2014. Impact of next generation sequencing techniques in food microbiology. Current Genomics 15(4): 293–309.

Mellmann, A., Harmsen, D., Cummings, C.A., Zentz, E.B., Leopold, S.R., Rico, A., Prior, K., Szczepanowaki, R., Ji, Y., Zhang, W., McLaughlin, S.F., Henkhaus, J.K., Leopold, B., Bielaszewska, M., Prager, R., Brzoska, P.M., Moore, R.L., Guenther, S., Rothberg, J.M. and Karch, H. 2011. Prospective genomic characterization of the German enterohemorrhagic *Escherichia coli* O104:H4 outbreak by rapid next generation sequencing technology. PloS one 6(7): e22751.

Nakyinsige, K., Man, Y.B. and Sazili, A.Q. 2012. Halal authenticity issues in meat and meatproducts. Meat Science 91: 207–214.

Nogueira, C., Iglesias, A. and Estevinho, L.M. 2012. Commercial bee pollen with different geographical origins: a comprehensive approach. International Journal of Molecular Sciences 13: 11173–11187.

Oliver, S.P., Jayarao, B.M. and Almeida, R.A. 2005. Foodborne pathogens in milk and the dairy farm environment: food safety and public health implications. Foodborne Pathogens & Disease 2(2): 115–129.

Ortea, I., Pascoal, A., Cañas, B., Gallardo, J.M., Barros-Velázquez, J. and Calo-Mata, P. 2012. Food authentication of commercially-relevant shrimp and prawn species: from classical methods to Foodomics. Electrophoresis 33: 2201–2211.

Patro, J.N., Ramachandran, P., Barnaba, T., Mammel, M.K., Lewis, J.L. and Elkins, C.A. 2016. Culture independent metagenomic surveillance of commercially available probiotics with high throughput next-generation sequencing. mSphere 1(2): e00057–16.

Prado, M., Ortea, I., Vial, S., Rivas, J., Calo-Mata, P. and Barros-Velázquez, J. 2015. Advanced DNA- and protein-based methods for the detection and investigation of food allergens. Critical Reviews in Food Science and Nutrition. http://dx.doi.org/10.1080/10408398.2 013.873767, pp. 2511–2542.

Premanandh, J. 2013. Horse meat scandal—A wake-up call for regulatory authorities. Food Control 34: 568–569.

Ripp, F., Krombholz, C.F., Liu, Y., Weber, M., Schäfer, A., Schmidt, B., Köppel, R. and Hankeln, T. 2014. All-Food-Seq (AFS): a quantifiable screen for species in biological samples by deep DNA sequencing. BMC Genomics 15: 639.

Rhoads, A. and Au, K.F. 2015. PacBio sequencing and its applications. Genomics, Proteomics & Bioinformatics 13(5): 278–289.

Ryser, E.T. and Marth, E.H. (eds.). 2007. *Listeria*, Listeriosis, and Food Safety. CRC Press.

Sentandreu, M. and Sentandreu, E. 2014. Authenticity of meat products: tools against fraud. Food Research International 60: 19–29.

Smith, C. and Osborn, A. 2009. Advantages and limitations of quantitative PCR (Q-PCR)-based approaches in microbial ecology. FEMS Microbiology Ecology 67: 6–20.

Solieri, L., Dakal, T.C. and Giudici, P. 2013. Next-generation sequencing and its potential impact on food microbial genomics. Annals of Microbiology 63: 21–37.

Tillmar, A.O., Dell'Amico, B., Welander, J. and Holmlund, G. 2013. A universal method for species identification of mammals utilizing next generation sequencing for the analysis of DNA mixtures. PLoS ONE 8(12): e83761.

van Dijk, E.L., Auger, H., Jaszczyszyn, Y. and Thermes, C. 2014. Ten years of next-generation sequencing technology. Trends in Genetics 30: 418–426.

WHO—World Health Organization (2015). WHO estimates of the global burden of food borne diseases: food borne disease burden epidemiology reference group 2007–2015. Geneva, Switzerland. (available at http://apps.who.int/iris/bitstream/10665/199350/1/9789241565165_eng.pdf). Accessed on July 2016.

Wiedmann, M. 2015. Use of whole-genome sequencing in food safety. Food Safety Magazine June/July.

Woolfe, M. and Primrose, S. 2004. Food forensics: using DNA technology to combat misdescription and fraud. Trends in Biotechnology 22(5): 222–226.

10

Detection of Biomolecules Using Surface Plasmon Resonance (SPR) Technology for Food Quality and Traceability

Christian Jay-Allemand,[1,]* *Nathalie Rugani,*[1] *Ruba Nasri*[1] and *Luc P.R. Bidel*[2]

1. Introduction

Surface plasmon resonance (SPR) is an electronic evanescent wave that can be triggered in a thin metallic layer as gold when special optical conditions such as incident light angle and refraction indexes in glass and liquid medium are combined (Lahmani et al. 2007, Schasfoort and Tudos 2008). The source of energy used is a polarized infrared light, and the decrease of reflected light intensity at a special reflection angle indicates an increase of molecular masses fixed to the coated gold film. So, this physical phenomenon, that is independent to solution turbidity, allows to access in a dynamic manner to the capacity of macromolecules as proteins to interact specifically or not with relatively smaller molecules such as DNA fragments,

[1] UMR IATE (UM/INRA/SupAgro) CC024, Université de Montpellier, place Eugène Bataillon, 34095 Montpellier cedex 05, France.
[2] UMR AGAP (INRA/SupAgro/CIRAD), INRA, Place Viala, 34000 Montpellier, France.
E-mail: nathalie.rugani@umontpellier.fr; luc.bidel@inra.fr; ruba.nasri@etu-umontpellier.fr
* Corresponding author: christian.jay-allemand@umontpellier.fr

RNA, peptides, oligosaccharides, terpenes and phenolics (Malmqvist 1999). The technology was developed since more than three decades and named as BIAcore (Malmqvist 1999) and is based on the use of biosensors according to the types of molecules to be studied (Rich and Myszka 2000). Interestingly, it was shown that it is more efficient to graft relatively small molecules as phenolics, oligosaccharides, oligonucleotides, and peptides to dextran coating the gold layer to optimize the signals when macromolecules such as proteins move in a microfluidic system connected to the biosensor itself. More the variations of molecular masses at the biosensor level that are important (> 180 Da), the detection of formed complexes will be better (Douat-Casassus et al. 2009). Furthermore, this technology can be also used to detect microorganisms by direct or indirect assays through specific interactions using antibodies that react with specific antigenic structures at species, sub-species or strain level (Bergwerff and van Knapen 2003). In all cases, not only we can know which molecules interact specifically with other ones at high sensitivity (less than picomolar concentration), but also get new information on their putative functions based on such interactions. So this technology coupled to mass spectrometry (Williams and Addona 2000) would offer large applications including food quality and traceability.

2. Physical and Technical Aspects

This method is based on the SPR phenomenon occurring in a metallic layer (gold, titan, silver) coated by a linker layer with dextran from which ligands are bounded for identifying putative specific targeted soluble molecules that move in a connected microfluidic system. The SPR phenomenon is an electronic evanescent wave (free electrons = plasmons) occurring in the gold film that itself absorbs energy from photons provided by a polarized infrared light source (often used at 850 nm) applied at a specific angle going through two successive media defined respectively by a high (glass) and a low (water) refractive indexes (Fig. 1). When soluble molecules bound to the ligands, the molecular masses change at biosensor level. That modifies the physical conditions of the biosensor (refractive index), inducing a shift of the angle (corresponding to resonance unit = RU) for which the intensity of the reflected light will be the lowest. So, for standard biosensors as carboxymethyl-dextran sensor CM5, when this angle varies by 0.1 degree (1000 RU) that means 1 ng of proteins/mm² was fixed to ligand.

The key practical point is to perform a standardized surface of the biosensor chip that determines its quality. This operation consists in the fixation of the ligand to carboxymethyl-dextran layer using the best procedure for its immobilization (Fig. 2) that may often represent a very difficult step to be achieved (Schasfoort and Tudos 2008). Then, the

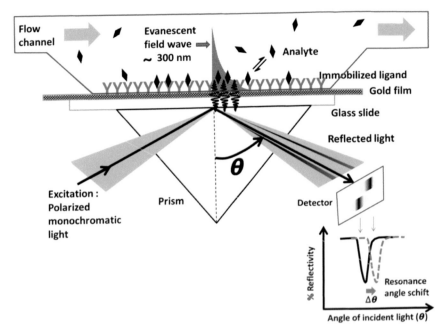

Figure 1. BIAcore's SPR technology is composed by a prism focusing a polarized light (incident light) on the sensor chip connected to a flow channel in which macromolecules (analytes) are moved (5–100 μL/min) within 100–200 seconds for the shorter recordings. The sensor chip is made of a gold film coated by dextran on which ligands are bounded. The measure of resonance signal (% reflectivity) results in an increase of the reflected light angle when free macromolecules in solution are bounded to ligands for a maximum decreasing of reflected light intensity showed by the sensorgram (below).

sensor chip quality must be tested through the maximum capacity of the surface determined by Rmax = (MW_A/MW_L) x R_L x Surface (L, ligand; A, free macromolecule; MW, molecular weight). Nevertheless, non-specific interactions between ligands and macromolecules as proteins can occur, requiring a special attention to the SPR signal analysis and appropriated molecules as negative control.

BIAcore's SPR technology enables to detect and record in real time the molecular events occurring at the biosensor chip during a period 100 to 500 seconds. It provides quantitative data on the:

- Specificity in the interactions between 2 molecules through the evaluation of affinity constant between 2 molecular partners (K_D = 100 μM to 200 pM),
- Concentration on a given soluble molecule that is possible to identify by coupling this technology to mass spectrometry analyses (1 mM to 10 pM),

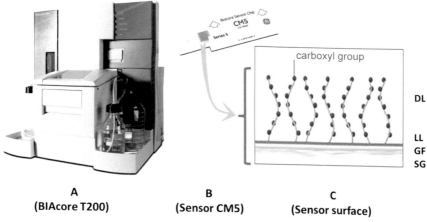

Figure 2. BIAcore T200 pumps (A) is completed by an optical system, a microfluidic system and a biosensor chip (B) as CM5, itself composed (C, down to top) by strip glass (SG), gold film (GF), linker layer (LL) and dextran layer (DL, 100 nm thick) on which molecules considered as ligands are grafted (small discs) using covalent coupling *via*-NH2, -SH, -COH, -COOH groups.

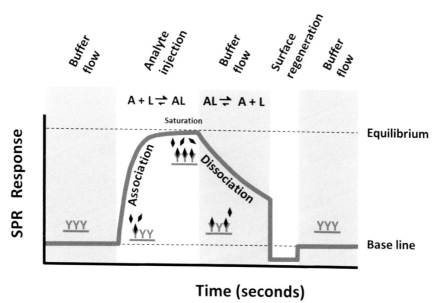

Figure 3. Sensorgram consists in the association phase when the analyte (A) is injected over the immobilized ligand (L) on the biosensor surface (Y). With increasing interaction between analyte and ligand, an increasing response is detected (SPR response). When a maximal number of molecules (analytes) bounded to ligands is reached (saturation/equilibrium), the injection of analyte is stopped by switching the system back to buffer (dissociation phase). Then, an injection with an appropriate regeneration solution is performed. Adapted from Hahnefeld et al. (2004).

- Association and dissociation kinetics for understanding the allosteric or cooperative effects between enzymes and their modulators ($K_a = 10^3$ to 10^7 M^{-1} s^{-1} and $K_d = 10^{-1}$ to 10^{-5} s^{-1}),
- Putative activity of ligands through the nature of macromolecules (DNA fragment, enzyme, transcription factor, nuclear receptor, …) interacting with.

So, the SPR technology can be considered as a biosensor combining a biological receptor (ligand) and a transductor (optical system) that is totally independent to liquid medium turbidity in the microfluidic channel in which targeted molecules are circulating.

3. Applications from SPR Technology Related to Food Quality and Traceability

Although this detection method has large application in the human health both for physiology and therapy, other applications concern the field of agronomy including food quality and traceability. Indeed, the food produced, stored, transformed, commercialized and consumed by human and animals undergoes different controls to determine its nutritional quality and to limit risks due to microbiological and/or chemical contaminants. In addition, food must be included in a traceability procedure from the farmer to the consumer in order to identify its origin in respect to the production chain and storage steps. In this context, this chapter will present the use of SPR technology through different cases previously mentioned and already published in scientific journals.

3.1 SPR technology for evaluating nutritional quality

3.1.1 Study of specificity in the interaction between phenolics and proteins (Quideau et al. 2005, Douat-Casassus et al. 2009)

The objective was to develop SPR technology in order to show whether phenolic structures can interact specifically with an enzyme involved in DNA replication during mitotic processes. This study pinpoints the roles of tannins on the alpha-topoisomerase II activity in order to discover new cellular functions.

The ligand is vescalin, a phenolic compound belonging to the class of tannins formed by one gallic acid bounded to 2 molecules of glucose (Fig. 4). The chemical compound of vescalin was transformed in sulfhydryl thioether deoxyvescalin then grafted to the sensor surface in two steps through sulfur bounds as shown in Fig. 4.

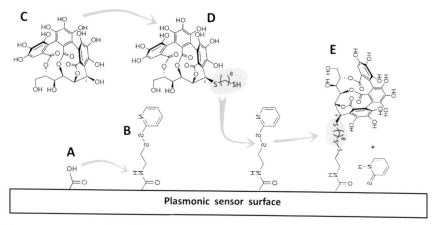

Figure 4. Process used for fixing the ligand on the plasmonic sensor surface. From dextran carboxylated groups (A) a disulfite linker (N-succinyl-carboxylated surface) is synthesized (B). Then the chemical structure of vescalin (C) is transformed in sulfhydryl thioether deoxyvescalin (D) which is grafted (E, ligand) to dextran coated the sensor surface in 2 steps including disulfite bound exchange (E). Adapted from Douat-Casassus et al. (2009).

Alpha-topoisomerase II (Top2α = 170 kDa) was produced in a recombinant system (*Escherichia coli*) then isolated and lyophilized. Different concentrations of enzyme (6 to 100 nM) were injected into the flow channel connected to a standardized sensor chip with vescalin. Controls with BSA and streptavidin were also tested for specificity evaluation. Interestingly, the reverse system where proteins are immobilized on the sensor chip was tested after injection of different concentrations of vescalin (0.5 to 20 µM) using the same controls as ligands. The results demonstrated clearly that the SPR technology sensitivity and specificity were strongly higher when vescalin is the ligand as expected. While BSA and streptavidin were not detected by vescalin, Top2α at 6.25 nM interacted with vescalin showing a relative low SPR signal reaching to 14 RU within 500 s (Fig. 5). At 100 nM the SPR signal increased to 35 RU that could indicate 35 pg/mm² of proteins have been fixed to the vescalin ligand. No signal with BSA and streptavidin was detected with higher concentrations, indicating the high specificity of the SPR detection when phenolics are used as ligands.

In conclusion, this method for detecting relative small molecules as phenolics is very sensitive giving relevant data on their putative roles through the identification of proteins which are able to establish specific binding (hydrogen bridge, hydrophobic interaction, van der Vaals bind, …) with phenolics (Quideau et al. 2011).

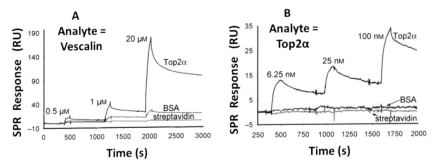

Figure 5. Sensorgrams for binding of (A) vescalin (analyte) to a Top2α coated sensor surface (vescalin is injected at 3 concentrations in µM) or (B) Top2α (analyte) to deoxyvescalin-coated sensor surface. The 2 sensorgrams (A and B) were recorded for 3 different proteins (Top2α, Topo-isomerase2 alpha; BSA, bovine serum albumin; streptavidin). The sensorgram B was recorded for immobilized deoxyvescalin at 3 concentrations in nM of each protein tested. Adapted from Douat-Casassus et al. (2009).

3.1.2 *Determination of vitamins in infant formula and milk proteins*
(Indyk and Mcwhirter 2001, Indyk 2006, Kalman et al. 2003)

The water-soluble vitamins such as folic acid, biotin, riboflavin, B12 are constitutive components or added to many different foods in respect to current legislation. Thus, their accurate concentration must be estimated during food processing to ensure that vitamin amounts are relevant with the expected commercialized products. Finally, the control of food quality devoted to humans and animals needs a rapid and reproducible analytical procedure. The second objective is to detect minor bioactive proteins as immunoglobulin G (IgG), folate binding protein (FBP), and lactoferrin (Lf) in milk with physiological properties targeted for being of benefit to early neonatal health.

SPR-immunoassays determine vitamin concentrations through BIAcore technology 3000. The assays are kit-based with standards, antibody/antigen, sensor chip and running buffers included. When starch is present in food matrix as cereal foods, more sophistical procedures are necessary for analyte release prior to SPR based-immunoassays. Sensor surfaces were prepared with affinity-purified goat anti-bovin IgG or Lf, a folic acid derivative on a carboxymethyl-dextran coated gold sensor chip (CM5). Samples were prepared in a Hepes buffer and analyzed after different dilutions between 3 to 8 min. Binding responses required 30 sec. Surfaces were regenerated by injecting 25 mM phosphoric acid (IgG), 75 mM sodium hydroxide (FBP) and 10 mM glycine-HCl pH 1.75 (Lf).

It has been demonstrated that SPR-immunoassays are accurate, rapid, sensitive, and highly suited to a routine program. It was shown that BIAcore provides data statistically equivalent to the references microbiological assays or HPLC analyses. For quantification of vitamin B2 (riboflavin), the limit of detection by BIAcore instrument was established at 12 ng/mL in various food matrices using a sensor chip CM5 coated with a layer of carboxymethyl-dextran on which riboflavin as ligand was immobilized. The assay was based on an inhibition format where the association between a riboflavin-specific binding protein (or antibodies) to riboflavin at sensor chip surface level is inhibited by the presence of a free riboflavin in a given sample. Extraction buffer was spiked to give final concentrations between 150 to 400 ng/mL of vitamin B2 with a suitable coefficient variation determined at around 5%. In baby food composite and whole milk powder, vitamin B2 measurements using BIAcore are 3.1 and 11.2 ng/mg of food samples.

As conclusion and perspectives, this current work is proceeding to optimize such strategies, incorporating various enzyme treatments, protein precipitation, and acid or alkaline hydrolysis for more reliable estimates of food folate and other vitamins. BIAcore has great potential in routine laboratories where a variety of vitamins must be quantified each day (Indyk et al. 2000).

3.2 SPR technology for detecting chemical contaminants in food (Rich et al. 2002)

The objective was to identify and quantify toxic compounds in various biological matrices that have direct effects on human health. Kinetic analyses of oestrogen receptor/ligand interactions are studied through the SPR technology.

Comparative SPR analyses of specific interactions between one ligand corresponding to the protein isoform of the human oestrogen receptor (ERα) and 12 analytes with MW 200 to 500 Da was the method used (Fig. 6). Sensorgrams from 9 molecules including bisphenol A and prosterone are presented in Fig. 7 and give the cycles of fixation between antibody/receptor/analytes with their kinetic parameters.

During the first 30 sec, the response is very variable according to the nature of molecules tested. The association constants with the oestrogen receptor are high for typical agonists as estriol, estradiol and estrone but also for diethylstilbestrol considered as a non-oestrogen agonist with even a higher response thus showing the strongest affinity found during this experiment. By contrast, bisphenol A and prasterone have lower interaction capacities to typical agonists. However, bisphenol A which is considered as a toxic molecule released from plastic polymers which are used for human feeding can interact with oestrogenic receptors at a lower level that is estimated to two folds less that estriol only. These results indicate that SPR technology can enable both to quantify small residual molecules

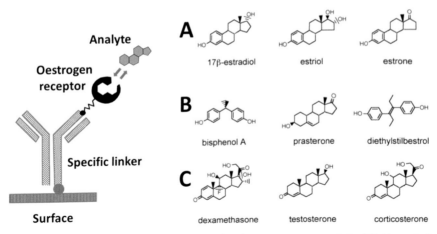

Figure 6. Interactions between 9 analytes (A, B, C) and oestrogen receptor alpha (ERα) bounded to biosensor surface with a specific linker (left) studied through SPR analyses. (A) Oestrogen agonists; (B) non-oestrogen agonists; (C) controls. Adapted from Rich et al. (2002).

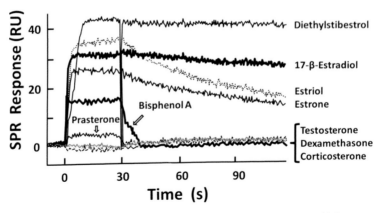

Figure 7. SPR analyses expressed in Resonance Unit (RU) during 100 sec of 9 drugs screened for binding to oestrogen receptor (ERα). Responses were generated from an injection of 1 μM each of non-oestrogen agonists (Bisphenol A, Prasterone, Diethylstibestrol), oestrogen agonist (17-ß-Estradiol, Estriol, Estrone), and control compounds (Testoterone, Dexamethasone, Corticosterone). Adapted from Rich et al. (2002).

contained in food samples and to evaluate the relative affinity with a specific protein responsible for physiological regulation in organisms. For example, testosterone and corticosterone (negative controls) have shown no response with the oestrogen receptor tested, thus giving much interest to this technology in term of its ability to make comparison of the specificity between compounds tested. Such classification based on the SPR response levels represents a good approach for toxin detection in food, evaluating directly the risk (danger level) of food contaminants according to their concentration. After 30 sec of analysis, it is possible to evaluate the dissociation levels of tested compounds giving more useful data on the function of chemical contaminants. The di-ethylstilbestrol seems to be a strong competitive agonist to natural oestrogens.

These results showed that BIAcore can be used for evaluating the level of toxicity of molecules from our environment with high sensitivity through targeted physiological processes playing very important functions in human.

3.3 SPR technology for detecting pathogenic microorganisms and aflatoxins B1 in foods and beverages (Bergwerff and van Knapen 2003, Jongerius-Gortemaker et al. 2002, Daly et al. 2000)

The objective was to develop new relevant methodologies in order to reduce the risks for population health when contaminated foods or beverages are consumed. This may include contaminants in food/water used by bioterrorists that can inflict dramatic impacts in the society. Detection of dangerous bacteries (*Salmonella, Listeria, Toxoplasma, E. coli, Cryptosporium, Hepatitis, Staphylococcus,* …) by SPR analyses has been in progress during the last two decades (Swanenburg et al. 2005).

BIAcore's SPR technology was used for detection of antibodies against *Salmonella*-specific flagellar antigens in chicken sera. Salmonella could be detected using 4×10^6 CFU/mL (Medina 1997). Furthermore, Bokken et al. (2003) have shown that optimum detection required a monoclonal antibody as the first immobilized protein determining specificity, then a polyclonal antibody as the second immunoglobulin determining the sensitivity. Other special proteins can be used for SPR detection. The 27 KDa N-terminal domain of fibronectin can be bound specifically by particular species as *Staphylococcus aureus* with the highest specificity. Indirectly, the detection and quantification of biotoxins as aflatoxin B1 from *Aspergillus* (Daly et al. 2000) can reveal the current or previous presence of pathogenic microorganisms in food and beverages.

Using the BIAcore Q apparatus, BIAcore's tests were compared to ELISA tests to detect porcine *Salmonella* antibodies directed to sero-group B and D. For that, lipopolysaccharides originating from two species of *Salmonella* were extracted then immobilized onto carboxymethylated dextran sensor (CM5) and used as ligands to antibodies against *Salmonella* in serum generated by pigs. By comparison to ELISA tests, the SPR assays showed very similar results based on the number of positive sera per test and farm: 52/211 (24,6%) for SPR and 54/211 (25,6%) for ELISA. The coefficient variations never exceed 3%.

As conclusion and perspectives, the developed SPR assays to detect antibodies against bacteria in liquid samples or extracted tissues were relevant with the objective fixed and can be used for many applications including direct or indirect detection of pathogenic microorganisms either in food and beverages or in animal or human sera for their protection. However, many strategies can be developed in order to maximize the detection in terms of sensibility and specificity that should be challenged through the use of the most reliable sensor chip. What kind of ligands must be used? A large variety of molecules are available and can be bounded to dextran by chemists as proteins, oligosaccharides, phenolics, alkaloids, terpenes, and DNA sequences, with the possibility to combine two ligands for increasing the specificity of detection, but in any case it should be very useful to immobilize relatively small molecules that will be recognized by bigger analytes or microorganisms or cells for increasing the sensitivity of response (SPR signal). The field of applications is very open in the agro-environment as well as in medicine (interactions between oncoproteins from human papillomavirus and ubiquitine ligase E6AP for example).

3.4 SPR technology for food traceability

Traceability is widely used by developed countries covering different fields as human foods, but remains to be developed in emerging countries at world level. The question to be addressed is what are the strategies and methods to be developed in order to limit the risks when consumers buy contaminated foods without high nutritive value? In other words, how to give the best evidences that each food is controlled from the farmers to the consumers to ensure that its quality is optimal. Moreover, the consumers make choices concerning the geographical origin of their food, including the company that is in charge of their conditioning and transport. But the main problem concerns the safety of consumers when chemical and microbiological contaminants are detectable in relatively high concentrations. So, in all cases food traceability is needed for protection of populations by using efficient and suitable technologies.

Food traceability is managed more and more by electronic bar codes (RFID chips), even genetic bar codes (identification of species), when we are faced to food and beverage adulteration or animal/plant protection or water/food contamination. But in all cases, we need relevant biomarkers (with high social impacts) that should be identified by sophisticated technologies as the BIAcore system in a first step (Lackman 2001), then used more and more in traceability procedures to be set up by public organisations in charge of security and safety.

4. Conclusions

All molecular and biochemical biological processes can provide an infinite of putative markers devoted to food traceability. But on the base of European programs (REACH) directed by recent legislation and on the dramatic impact on human health that begins to be well evaluated (Fréry et al. 2012), the security and quality of foods in the world should be improved by the development of a new (bio)traceability using relevant biomarkers at three different levels: (1) Exposition, (2) Effects and (3) Susceptibility. We can consider that these markers could form the lacking chain elements between the environment and diseases induced by contaminants assimilated by organisms mainly through water, air, food and beverages. Therefore, it seems very urgent to develop a whole chain of "bio-traceability" that should be based on biomarkers that have strong social impacts and high relevant functions for humans, but also for animals and plants. Whereas the usual methods always play an important roles for environmental, food quality and medicinal diagnostics (Myszka and Rich 2000, Swanenburg et al. 2005), we need today both sophisticated and expensive technologies but also very cheap biosensors to be made available for each citizen for vastly improving the "conditions" in which we live, whichever the country in the world, because all people are now concerned. So, one of the next challenge to be attained is to demonstrate that the effects of contaminants result in the combination of different chemicals or pathogens (antagonist and synergic effects) in order to adjust the list of chemical polluants prescribed by the international Stockolm Convention (2001) refered by Sandholtz and Whytock (2017): (i) the first classification of contaminants to be eradicated, (ii) the maximal concentrations authorized for each chemical product and published by the EU and states, and (iii) the new reliable evidences that each contaminant or combination between them are responsible for serious diseases in young and old populations. That is probably one of the more crucial points to be solved, on which as yet only a few scientific groups are working on. The methods developed to reach such objectives are:

i) Those based on genetic transformation (plants or animals) with a vector containing specific promoters fused to reporter genes (luciferin, GFP). These promoters have cis-elements that are regulated by nuclear receptors, themselves able to interact specifically with xenobiotics that triggered expression of the reporter genes in tissues or in whole organisms as mouse or zebra fish (Balaguer et al. 2001). This approach gives new insights on both the cellular and organ localization of target molecules and their cellular impacts.

ii) Those using up to date technologies that allow to determine directly the association or dissociation constants between at least two chemical partners. Several approaches can be developed by means of fluorescence, enzymology, docking and SPR detection as previously shown. This approach should give very informative data.

Thus SPR analyses can be considered today as one of the more suitable technologies for routine analyses in the field of "interactomic" giving necessary functional information (Verneret 2014) and for identifying new toxic components that it would make it possible to classify according to their risk level for organism integrity (mutagenic, carcinogenic and reprotoxic effects). Concerning the last point, by coupling HPLC system and/or mass spectrometry analyses to BIAcore apparatus, we have the capacity to identify new biochemical partners that could also disturb the cellular processes. However, new technologies are emerging for discovering both new drugs and toxins through electromagnetic detection when proteins are submitted to targeted molecules (toxins), themselves grafted with nanoparticles containing iron oxide (Brajalal et al. 2013).

Perspectives

In future, this technology can be used for reaching one of the best societal challenges that is to give suitable evidence on natural or artificial compounds from our near environment which trigger specific physiological responses (Slepak 2000) through nuclear receptors, transporters or enzymes, that are involved in well-known diseases (cancer, diabetes, Alzheimer). The capacity of such technology to quantify the specific interactions between molecular partners and to identify the proteins or DNA sequences targeted by putative toxins will be the major way to impact politicians in their decision making and citizens at European level in order to improve the quality of food and environment with efficiency. SPR analyses will be not only be used to confirm the presence of a binding partner but also to identify the unknown chemical structures just after the dissociation phase *via* coupling high performance liquid chromatography—tandem mass spectrometry (Zhang et al. 2013). Non-specific interactions and «bulk refractive index» variations are probably the most difficult problems to be solved for reliable kinetic and concentration applications. This technology based on SPR signals may be still improved or better adapted to routine analyses that we are progressively more in need of (Verneret 2014). It would be interesting to increase the sensitivity and specificity using the SPR phenomenon triggered by optic fiber through nano-gold-particles bounded to bioactive molecules when proteins are added in the buffer solution in order to follow the kinetic of association/dissociation at very low concentration (Rousseau personal communication). Indeed, our research group aims to study the specific interactions between recombinant Lipases and phenolics through this new methodology developed at University Montpellier. It seems even possible to develop simple sensors and to detect the complex formed in a unique cell or in mitochondria (Lasne et al. 2007). Such methodology has revealed a very high sensitivity since its resolution could reach the detection of a unique molecule.

For food traceability, we have shown that the potentialities of BIAcore technology are considerable in a global food diagnostics market estimated to grow to 1.4 billion \$US, knowing that the application of biosensors is always currently under-exploited (Indyk 2006), probably due to an inherent conservatism by governments and industries. The number one problem is its cost, but as we have seen, this technology offers real low cost potential through the generation of new biosensors with high sensitivity and also new alternatives for their use. Nevertheless, even if SPR analyses coupled with mass spectrometry may give very high-throughput screening and identification of drug targets and pathogens of important diseases from complex mixtures, it should be necessary to used complementary biosensors combining different strategies (electrochemical aptamer sensors, plastic antibodies, methacrylate probes, double antibodies) for detecting, with the highest specificity as possible, the high diversity of chemical structures we have to face.

Keywords: Surface plasmon resonance, detection, biomolecule, interaction, specificity, traceability, food

References

Balaguer, P., Bouissioux, A.M., Demirpence, E. and Nicolas, J.C. 2001. Reporter cell lines are useful tools for monitoring biological activity of nuclear receptor ligands. Luminescence 16: 153–158.

Bergwerff, A.A. and van Knapen, F. 2003. Sensing pathogens: screening strategies in food and environmental safety. Biacore Journal 2: 10–15.

Brajalal, S., Hung, T.Q., Sri, R.T. and Terki, F. 2013. Planar Hall resistance ring sensor based on NiFe/Cu/IrMn trilayer structure. Journal of Applied Physics 113: 063903.

Bokken, G.C.A.M., Corbee, R.J., Van Knapen, F. and Bergwerff, A.A. 2003. Immunochemical detection of *Salmonella* group B, D, E using an optical surface plasmonic resonance biosensor. FEMS Microbiological Letters 222: 75–82.

Daly, S.J., Keating, G.J., Dillon, P.P., Manning, B.M., O'Kennedy, R., Lee, H.A. and Morgan, M.R. 2000. Development of surface plasmon resonance-based immunoassay for aflatoxin B1. Journal of Agricultural and Food Chemistry 48: 5097–50107.

Douat-Casassus, C., Chassaing, S., Di Primo, C. and Quideau, S. 2009. Specific or nonspecific protein-polyphenol interactions? Discrimination in real time by surface plasmon resonance. ChemBioChem 10: 2321–2324.

Fréry, N., Saoudi, A., Guldner, L., Garnier, R., Zeghnoun, A., Falq, G., Bidondo, M.L., Bérat, B., Maître, A., Olichon, D., Cirimèle, V., Leblanc, A., Goën, T. and Castetbon, K. 2012. Exposition de la population française aux substances chimiques de l'environnement.

Volet environnemental d'ENNS. 50ème congrès de la STC – 29 Novembre 2012, INVS. Oral communication.

Hahnefeld, C., Drewianka, S. and Herberg, F.W. 2004. Determination of kinetic data using surface plasmon resonance biosensors. Methods in molecular medicine, Molecular diagnosis of infectious diseases, 2/e, Edited by J. Decker, U. Reischl, Humana Press Inc., Totowa, NJ 94: 299–319.

Indyk, H.E., Evans, E.A., Bostrom Caselunghe, M.C., Persson, B., Finglas, P.M., Woollard, D.C. and Filonzi, E.L. 2000. Determination of biotin and folate in infant formula and milk by optical biosensor-based immunoassay. Journal of AOAC International 83: 1141–1148.

Indyk, H.E. and Mcwhirter, A. 2001. Determination of vitamin concentrations in food samples by BIAcore's SPR technology. Biacore Journal 2: 4–7.

Indyk, H.E. 2006. Optical Biosensors: Making Sense of Interactions. Chemistry in New Zealand, July 42–46.

Jongerius-Gortemaker, B.G., Goverde, R.L., van Knapen, F. and Bergwerff, A.A. 2002. Surface plasmon resonance (BIAcore) detection of serum antibodies against *Salmonella enteritidis* and *Salmonella typhimurium*. Journal of Immunological Methods 266: 33–44.

Quideau, S., Jourdes, M., Lefeuvre, D., Montaudon, D., Saucier, C., Glories, Y., Pardon, P. and Pourquier, P. 2005. The chemistry of wine polyphenolic C-glycosidic ellagitannins targeting human topoisomerase II. Chemistry European Journal 11: 6503–6513.

Quideau, S., Douat-Casassus, C., Delanoy Lopez, D.M., Di Primo, C., Chassaing, S., Jacquet, R., Saltel, F. and Genot, E. 2011. Binding filamentous actin and winding into fibrillar aggregates by the polyphenolic C-glucosidic ellagitannin vescalagin. Angewandte Chemie International Edition 50: 5099–5104.

Kalman, A., O'Kane, A., Caelen, I., Trisconi, M.J. and Wahiström, L. 2003. BIAcore's vitamin B2 kit for the measurement of riboflavin concentration in food products. Biacore Journal 2: 18–21.

Lackmann, M. 2001. Isolation and characterization of "orphan-RTK" ligands using an integrated biosensor approach. Methods in Molecular Biology 124: 335–59.

Lahmani, M., Boisseau, P. and Houdy, P. 2007. Les nanosciences. 3. Nanobiotechnologies et nanobiologie. Editions Belin 1150 p. ISBN: 978-2-7011-4470-2.

Lasne, D. 2007. Label-free optical imaging of mitochondria in live cells. Optic press 15: 14184–14193.

Malmqvist, M. 1999. BIAcore: an affinity biosensor system for characterization of biomolecular interactions. Biochemical Society Transactions 27: 335–340.

Medina, M.B. 1997. SPR biosensor: food science application. Food Test Analysis 3: 14–16.

Myszka, D.G. and Rich, R.L. 2000. Implementing surface plasmon resonance biosensors in drug discovery. Pharmaceutical Science and Technology Today 3: 310–317.

Rich, R.L. and Myszka, D.G. 2000. Advances in surface plasmon resonance biosensor analysis. Current Opinion Biotechnology 11: 54–61.

Rich, R.L., Hoth, L.R., Geoghegan, K.F., Braown, T.A., Lemotte, P.K., Simon, S.P., Hensley, P. and Myszka, D.G. 2002. Kinetic analysis of estrogen receptor/ligand interactions. PNAS 99: 8562–8567.

Sandholtz W. and Whytock C.A. 2017. Research Handbook on the Politics of International Law. EE Edward Elgar publishing, Cheltenham UK/Northampton USA. p 608. ISBN: 9781783473977.

Schasfoort, R.B.M. and Tudos, A.J. 2008. Handbook of surface plasmon resonance. Edited by The Royal Society of Chemistry, RSC Publishing. 395 p. ISBN: 978-0-85404-267-8.

Slepak, V.Z. 2000. Application of surface plasmon resonance for analysis of protein-protein interactions in the G protein-mediated signal transduction pathway. Journal of Molecular Recognition 13: 20–26.

Swanenburg, M., Bloemraad, R., Achterberg, R. and Maassen, K. 2005. Validation of a surface plasmon resonance based assay to detect Salmonella antibodies in serum of pigs. Congress on Safe Pork 100–102.

Verneret, M. 2014. Surface Plasmon resonance: molecular interactions and ligand binding analysis. Life Science, Technical Bulletin SGS, February 1–4.

Williams, C. and Addona, T.A. 2000. The integration of SPR biosensors with mass spectrometry: possible applications for proteome analysis. Tibtech 18: 45–48.

Zhang, Y., Shi, S., Guo, J., You, Q. and Feng, D. 2013. On-line surface plasmon resonance-high performance liquid chromatography-tandem mass spectrometry for analysis of human serum albumin binders from radix Astragali. Journal of Chromatography A 1293: 92–99.

11

Advances in Authentication Methods for Seafood Species Identification in Food Products

Véronique Verrez-Bagnis

1. Introduction

The consumption of fish and seafood products has considerably increased in the world from an average of 9.9 kg per capita in the 1960s to 19.2 kg in 2012 (FAO 2014). Fish and fisheries products are among the most traded food commodities worldwide. In 2012 they represented about 10% of total agricultural exports and 1% of world merchandise trade in value (FAO 2014). This increase in global trade in seafood products is accompanied by greater complexity in commodity flows, with some products crossing multiple national borders during the supply chain, including movements into territories without stringent traceability requirements (D'Amico et al. 2014). Owing to the high perishability of seafood products, 50% of the traded catches are processed (i.e., fillets, portions, elaborated products) and so, they lose their morphological characteristics to be identified in their whole forms.

Moreover, many fisheries worldwide are in decline because of their overexploitation and required rebuilding (FAO 2014) and global trade effectively masks the successive depletion of stocks (Srinivasan et al. 2012). Mislabelling has been identified by numerous studies for different species and types of seafood, and it is thought to contribute to fish decline because

Ifremer, rue de l'Île d'Yeu, B.P. 21105, F-44311 Nantes 03, France.
  E-mail: Veronique.Verrez@ifremer.fr

it frequently hides illegal, unreported and unregulated (IUU) fishing posing a major threat to sustainable fisheries, and consisting approximately one-fifth of the global catch (Agnew et al. 2009).

As more than 32,500 species of finfishes exist worldwide, there is a huge diversity of species and products available on the global markets, i.e., U.S. Food and Drug Administration includes approximately 1850 fish and shellfish in its seafood list (FDA 2016).

Today, in most of the countries, the use of scientific names is not legally prescribed. However, regarding the European Union, a recent regulation (EU No. 1379/2013) requires the labelling of the commercial designation of the fishery or aquaculture products with their scientific name for fish products sold to the final consumers. But, throughout the developed countries where most of the market of fishery products has evolved, regulations are in general different. Different states can legally:

- designate official commercial fish species names which can be different from one country to another,
- give a common commercial name for a specific genus, and/or
- group some fish species for commercial purposes [e.g., European regulation stipulates that preserved products labelled as tuna must be prepared either from *Thunnus* fish species or from skipjack tuna (*Katsuwonus pelamis*)].

Global trade operations require a stringent line of certifications with regards to fish labels and other related aspects (Maralit et al. 2013). It generates an increased need of international harmonization of species-naming convention, of seafood species identification tools and of easy identification of fishing or farmed area to provide reliable information to customs officers, fishery inspectors, fish industries and consumers and to prevent fraud (FAO 2012).

This chapter, after a brief part on the methods commonly used for fish identification, reviews the different approaches for development of quick DNA-based methods and new identification procedures, and of specific methods for identification of fish groups. Finally, with regard to the requirements of the European legislation on seafood product traceability and labelling, a review on the development of methods to identify fishing areas and farmed fish was done.

2. Commonly Used Methods for Fish Species Identification

Most of traditional and official methods used in fish species identification are based on the biochemical analysis of specific proteins, e.g., isoelectric focusing (IEF), high performance chromatography or immunoassays (Martinez et al. 2005, Rasmussen and Morrissey 2008, Teletchea 2009, Lavilla et al. 2013, Lago et al. 2014). These approaches have many disadvantages, and the most significant is that they cannot be applied on highly processed samples (i.e., cooked or canned products) because proteins become denatured upon the heating. In comparison, DNA is a more thermostable molecule, and even if DNA may be degraded during processing, short fragments are recoverable and can be used for identification analysis in processed foods (Mackie et al. 1999, Quintero et al. 1998). In recent decades, the rapid and widespread use of genetic identification methods has transformed food authenticity analysis, and particularly fish species identification. Detailed surveys of the scientific literature on seafood authenticity reveal that over 150 peer-reviewed papers on the topic were published in the period (1995–2008) (Rasmussen and Morrissey 2009, Teletchea 2009), and more than 160 peer-reviewed papers were published in the period (2009–2016) (personal analysis). As reported by Lavilla et al. (2013), numerous studies related to traceability, fish populations or fishery technologies have been recently completed or are being carried out covering the use of modern molecular technologies and the construction of databases. In spite of all this, access to clear conclusions is difficult and these fragmented and often isolated projects have led to some dispersal of generated data. Information transfer for application of the methods in analytical laboratories must be made by collection of data and databases with standardized formats and easy accessibility. Alongside the exponential development of methods and studies, there are very few guidance on what techniques should be applied. A survey on the current methods frequently used in control laboratories for seafood authenticity testing in Europe reveals that DNA sequencing appears to be the most commonly applied approach, although IEF on sarcoplasmic proteins is still commonly utilized (Griffiths et al. 2014). This last study shows also that there is a variety of DNA-based method approaches utilized by seafood identification testing facilities. And this underlines the need for a more rigorous standard operation procedure to be applied, giving the example of the successful development of standard protocols for seafood identification at a national scale, both within the UK (Wolfe et al. 2013) and the U.S. (Handy et al. 2011). Moreover, as most of the techniques are sequencing-based approaches, the need of DNA public databases compiling standardized and validated data is crucial; all specimens from which sequences are derived have to be authenticated by trained taxonomic experts, and if possible, specimen have to be also vouchered in permanent and curated collections (Table 1: example of seafood dedicated reference databases). As explained by Clark (2015), policymakers cannot ignore how the expanding adoption of DNA barcoding technology is changing the way how cases of intentional mislabelling and fraud are detected and prosecuted. Till now, there is no recognition of an "official" method of DNA-based analysis for fish species identification, neither by AOAC,

Table 1. Examples of Online Resources Dedicated to DNA-based Fish and Seafood Species Identification (from Rasmussen and Morissey 2008 and updated on July 2016).

Online resource	Description	Target DNA	Organization	URL
Fish Barcode of Life (FISH-BOL)	Part of the barcoding of Life project; focused on fish species identification based on DNA sequences.	mt COI gene	Consortium for the Barcoding of Life (CBOL)	http://www.fishbol.org/
FDA Reference Standard Sequence Library for Seafood Identification (Barcode Library)	Reference Standard Sequence Library (RSSL) for seafood identification contains DNA reference barcode sequences (FASTA format) and other data for 1046 specimens.	mt COI gene	U.S. Food and Drug Administration	http://www.accessdata.fda.gov/scripts/fdcc/?set=seafood_barcode_data
FishTrace Database	Provides species identification tools for fish species common to Europe.	mtcyt *b* and rhodopsin genes	FishTrace Consortium	https://fishtrace.jrc.ec.europa.eu/dataviz

nor Codex Alimentarius Commission. The successful adoption of DNA barcoding is also dependent upon its acceptance by leading international organizations like the FAO/WHO, Codex Alimentarius, FDA, WTO as well the European Council as a methodological standard. Multi-level policy-makers can contribute to fight the problems of mislabelling and fraud with proper cooperative policing strategies. Thus the study of Mariani et al. (2015) using DNA barcoding on seafood products shows an apparent sudden reduction of seafood mislabelling in Europe and emphasizes the repercussions of improved legislation, continued surveillance and adoption of forensic genetic tools.

3. Emergent Fish Species Identification Methods and Development of Quick DNA-Based Methods

Alongside these current techniques to identify fish species, other methods, mainly based on DNA, but also on peptides, are currently in development. A large number of these actual studies are focusing on quick or automated and easy-to-handle assays to be applied by companies involved in fish trade as well as authorities concerned with fisheries control and consumer protection.

3.1 Identifying fishes through DNA microarrays

As reported by Rasmussen and Morrissey (2009), DNA microarray (or DNA chip) technology is well-suited for automated and high-throughput operations. It may contain thousands of different oligonucleotide probes that are immobilized on the surface of a glass side or microscopic beads. Generally, to identify species with such device, a targeted fragment is labelled with a fluorophore during PCR amplification before being applied to a microarray where it hybridizes with any probes exhibiting a complementary nucleotide sequence.

Kochzius et al. (2010) have investigated mitochondrial markers to develop DNA microarray containing 64 functional oligonucleotide probes for the identification of 30 European fish species. Their study showed that 16S rRNA sequences can be recommended for designing oligonucleotide probes, and in order to allow the discrimination of closely related species, additional markers such as *cytochrome b* gene would be helpful. The authors noted that unfortunately, *cytochrome oxidase subunit I gene* (COI) was not suitable for the design of oligonucleotide probes for the target species, discouraging the utilisation of the huge number of COI barcode sequences in the Barcoding of Life Database (BOLD) (Ratnasingham and Hebert 2007) as a data source for the development of DNA microarrays for the identification of fish species.

Nevertheless, Kim et al. (2011) have succeeded to develop a DNA microarray based on species-specific oligonucleotide probes targeted COI gene to quickly identify five different filefish (Monacanthidae). Likewise, Park et al. (2014) have successfully developed a DNA microarray based on specific probes based on COI sequences for identifying origins and species of 8 eels. In a previous study, Park et al. (2012) were able to identify 39 fish species in fish cake samples using both next generation sequencing (NGS) techniques and microarray.

In the same idea to differentiate closed fish, Handy et al. (2014), in a proof-of-concept study have developed DNA microarrays coupled with a mid-infrared imaging detection method for identification of seven species of catfish. Species-specific DNA probes targeting three regions per species of the COI gene were developed. Briefly, according to the authors, deoxyribonucleic acid targets labelled with biotin were hybridized to their complementary probes using a strategy that allowed the selective formation of a silver layer on hybridized spots needed for mid-infrared imaging detection.

Using *de novo* transcriptome analysis of *Solea solea* and *S. senegalensis*, two economically important species, the study of Benzekri et al. (2014) have provide new tools to the scientific community, and particularly a new microarray with 43,403 features in Senegalese sole which could be further adapted as forensic tool.

3.2 Next generation sequencing (NGS)

NGS methods use advanced techniques to generate large volumes of sequence data simultaneously and relatively inexpensively. As reported by Rasmussen and Morrissey (2009), some of the applications of NGS include metagenomics, gene expression, and ancient genome sequencing. NGS has also proven to be very powerful for the discovery of SNPs that can be used to develop microarrays. Recently, pyrosequencing technology have been be applied to seafood species identification, especially when using short sequences known to exhibit diagnostic variation, such as minibarcodes to perform metabarcoding. That is the technology used by de Battisti et al. (2014) for the rapid identification of processed products from two groups of fish (Clupeidae and Pleuronectidae). They have developed a process (length of duration of about 3.5 h) based on three PCR reactions to amplify 16S rRNA, *NADH dehydrogenase subunit II gene* and *cytochrome b* fragment genes followed by pyrosequencing. With this method, it was possible to identify the specimens at the species level in most of the analyzed samples; however, in some cases the identification was limited to the genus level (e.g., *Arnoglossus* sp., *Coilia* sp., and *Alosa* sp.) which could be still considered sufficient for commercial purposes depending on the national legislations.

That is also the DNA metabarcoding approach used by Galal-Khallaf et al. (2016) for fish species detection in aquafeed samples using pyrosequencing of a short fragment of the 5′ terminal region of the COI gene (124 bp excluding the Uni-MinibarF1/Uni-MinibarR1 primers from Meusnier et al. (2008). After BLAST comparisons, a total of 13 fish species were found inside fish feed samples belonging to four orders. This DNA metabarcoding approach has permitted to show that approximately 46% of all fish species detected are either overexploited or suffering strong decline. Since these last years, NGS techniques are in continuous improvement and, without doubt more studies would be done in the near future on the identification of fish species, and particularly those fish groups that are still difficult to identify like tuna species of which some of them (i.e., *Thunnus albacares* and *Thunnus obesus*) are genetically closed species. Another challenge with these high throughput techniques is, not only to identify these tuna species, but also to quantify them in canned products (quantification is required because of the European regulation—see introduction to this chapter) when qPCR-based techniques are, at the present time, the methods of first choice for fish species quantification.

3.3 Isothermal amplification of DNA or RNA

According to Clark (2015), thanks to the existence of globally accessible, expert-authenticated databases of genetic information of fish species, and beside metabarcoding using NGS technology, development of 'hand-held' devices with DNA identification capabilities could democratize the authentication process and reduce the time necessary for species identification. Up to now few 'hand-held' devices based on isothermal amplifications have been developed.

Loop mediated isothermal amplification (LAMP), firstly developed, is a single tube technique for the amplification of DNA carried out at a constant temperature, and so a thermal cycler is not required. This method, invented by Notomi et al. (2000) utilizes a DNA polymerase with high strand displacement activity and a set of four specifically designed primers that recognize a total of six distinct sequences on the target DNA. Because LAMP recognizes the target by six distinct sequences initially and by four distinct sequences afterwards, it is expected to amplify the target sequence with high selectivity. The amount of DNA produced in LAMP is considerably higher than standard PCR-based amplification.

The first report of the application of a loop mediated isothermal amplification (LAMP) for the determination of fish species was recently done by Saull et al. (2016). They have designed a LAMP assay for Atlantic cod (*Gadus morhua*) *cytochrome b* gene to clearly identify this species and to distinguish this commercialy important white fish species from the two other species legally identifiable as cod in the UK [Pacific cod (*Gadus macrocephalus*) and Greenland cod (*Gadus ogac*)]. Reactions were carried out in a heating block at 63°C for 60 min and reactions were terminated by heating to 80°C for 1 min in order to denature the *Bst* polymerase. Their developed method was capable of detecting 0.1% w/w *Gadus morhua* in a homogenised raw fish mix. The LAMP assay showed a higher tolerance regarding amplification inhibitors than PCR. Nevertheless, before this LAMP assay being adopted as a routine tool for the future a more comprehensive validation will be required involving greater numbers of samples (in particular processed products) and fish species.

Another isothermal nucleic acid amplification technique, the recombinase polymerase amplification (RPA) has been firstly developed to simplify the laboratory instrumentation by Piepenburg et al. (2006). Their novel approach operated at constant low temperature coupled isothermal recombinase-driven primer targeting of template material with strand-displacement DNA synthesis. Exponential amplification are carried out with no need for pretreatment of sample DNA. RPA is described as a sensitive, specific and rapid methodology. This method has proved to be sensitive to fewer than ten copies of genomic DNA and products can be detected in a simple sandwich assay, thereby establishing an instrument-free DNA testing system. This is the technology used by TwistDx Ltd. (biotechnology company based in Cambridge, UK) for the identification of red snapper (*Lutjanus campechanus*), a highly prized food fish caught commercially along the Atlantic coast of North America and Mexico, in 20 min (TwistFlow®

Red Snapper). As it is a commercial detection kit, the functional set of reaction components of the TwistAmp kit is available only from the TwistDx supplier in comparison to reaction mixtures for PCR or LAMP for which the composition of the reaction mixture is freely published.

Compton developed another isothermal amplification technique, nucleic acid sequence based amplification (NASBA) (Compton 1991). The technique is inspired from the transcriptional process observed in viruses and cells, and it is carried out isothermally at 41°C. Firstly, a target RNA is copied into its complementary DNA using reverse transcriptase. Then RNase H eliminates RNA and the cDNA is transcribed into RNA using a RNA polymerase. This process results in a few cycles to an amplification of about 100 times the amount of RNA initially present in the sample. This method can be made quantitative by introducing internal controls in the sample. Immediately after the invention of NASBA it was used for the rapid diagnosis and quantification of HIV-1 in patient sera (Kievits et al. 1991), and it has been further used to develop rapid diagnostic tests for several pathogenic viruses with single-stranded RNA genomes. An extension of conventional NASBA by the incorporation of fluorescently-labeled molecular beacons (Tyagi and Kramer 1996) allowing real-time detection of NASBA amplicons (RT-NASBA) was first described by Leone et al. (1998). However, this technique has been used for the first time in the field of fish authenticity only in 2013. Indeed, Ulrich et al. (2013) have developed a 90 min multiplex RT-NASBA assay targeting a portion of the mitochondrial 16S ribosomal RNA (rRNA) gene for the accurate identification of most grouper species listed on the FDA Seafood List with no cross-reactivity with non-target species. Following this previous study, Ulrich et al. (2015) have developed a hand-held genetic sensor that employs RT-NASBA method for the analysis of fish tissue in the field. They have demonstrated that the field sensor (80 min for assay completion and can be performed outside of the lab in its entirety) is only slightly less sensitive than the benchtop instrument, and could discern 49 of the 61 FDA allowable species (80.3% of groupers—no target sequence available for three species). As mentioned by the authors, the format of the handheld heated fluorometer allows potential implementation of alternate RT-NASBA assays utilizing oligonucleotide sets to target other commercially important finfish groups such as snappers and tunas, which had the highest mislabel ingrates according to the Oceana investigation (Warner et al. 2013).

3.4 Multiplex PCR-ELISA

Briefly, a first PCR-ELISA strategy for fish identification was recently developed by Santaclara et al. (2015) focusing on the identification of tuna species at genus level and to discriminate the *Thunnus* species and skipjack tuna (*Katsuwonus pelamis*) (all of these species could be labelled as tuna species on food product) from other scombroids. An important step of the PCR–ELISA methodology is the incorporation of a triphosphate nucleotide DIG labelled (*digoxigenin-11-dUTP*) during the PCR process, resulting in the labelling of amplicons which renders them suitable for subsequent ELISA detection. As described by Santaclara et al. (2015), after PCR, ELISA methodology comprises the following steps: (1) hybridisation of the DIG labelled PCR product and the biotinylated probe; (2) binding of the hybridisation product to streptavidin coated microplate; (3) washing; (4) binding of antibody anti-DIG linked to enzyme labelled; (5) washing; (6) addition of specific substrate to the revealing enzyme; and (7) evaluation of results. The method developed by Santaclara et al. (2015) is composed by four systems based on cytochrome b (cytb) and cytochrome oxidase subunit I (COI) fragment sequence-based primers that can be used in a hierarchical way allowing the identification of several scombroids species; or each individual system independently. A first step is performed to assign one sample to the *Thunnus* genus. Next, if the result is positive, several tests can be applied to assign the sample to some particular species of the *Thunnus* genus. In the case that the result is negative (absence of *Thunnus* species), the presence of *Katsuwonus pelamis* in the sample is tested. The method even allows the detection of mixtures of these species in relatively low amounts (up to 1%). The authors have applied the method to eleven commercial samples to verify the labelling status of tuna products in the market, and they found that 18% were mislabelling.

4. Development of Methods to Identify Fishing or Farmed Areas

According to countries, specific information must accompany fishery and aquaculture products sold to the consumers. For the European Union, the Common Organisation of the Markets has established new rules applicable on 13 December 2014. Fish, molluscs, crustaceans and algaes, products sold to consumers or mass caterers must bear the following information:

- the commercial and scientific name of the species
- whether the product was caught at sea or in freshwater, or farmed
- the catch or production area and the type of fishing gear used to catch the product
- whether the product has been defrosted and the date of minimum durability (also known as the 'best before' or 'use by' date), in line with general food labelling rules.

Moreover during the past 20 years, eco-label schemes have emerged on the EU market, a development linked to increased public awareness on the need to ensure sustainable exploitation of marine resources. Indeed, according to the definition provided by the United Nations' Food and Agricultural Organisation (FAO), eco-labels entitle a product to bear a logo or statement certifying compliance of its production with conservation and sustainability standards. While in fisheries the focus is mainly on stock conservation, in aquaculture the emphasis is mostly on potential negative externalities generated by production such as disruption of natural ecosystems or water pollution (European Commission 2016). Besides the labelling of mandatory information and the eco-labels, many fishery products have additional labels and marks related to quality or reputation linked to origin (Martinez et al. 2005). Geographical indications (GIs), defined in the Agreement on Trade-Related Aspects of Intellectual Property Rights, value this particular quality. Depending on the country, different systems of protecting property rights for GIs are used, for example, the EU-wide GI or the certification mark of the United States (Lavilla et al. 2013). Usually, GIs consist of names or words associated with a place and serve as marketing tools for adding economic value to products, for example, 'Scottish farmed salmon' and 'Scottish wild salmon'.

In order to ensure that the fishing area or the country of farmed fish products, the eco-labelled products and the products labelled with GIs come from certified sources and geographical origin, measures shall be put in place to ensure traceability of both the products and their certification. Traceability of certified products could be maintained using relatively straight forward handling and record-keeping procedures (FAO 2014). By contrast, competent authorities still lack cost-effective ready-to-use methods that enable the thorough implementation of validation procedures to address seafood certification of origin (Leal et al. 2015). Nonetheless, practical methodologies to be applied for tracing the origin of specific fishing/production areas based on analysis of trace elements, isotope ratio analysis, analysis of fatty acids and DNA-based analysis have been recently developed (for a review, see Leal et al. 2015).

4.1 Analysis of trace elements

Geochemical tools, particularly trace element fingerprinting (TEF), allow the distinction of populations or stocks using the elemental profile of mineral structures, such as mollusk shells (Ricardo et al. 2015), invertebrate statoliths (Manríquez et al. 2012) and fish otoliths (Chang and Geffen 2013). As reported by Leal et al. (2015), given that these mineral structures grow throughout the year and their compositions are affected by local environmental conditions, TEF appears to be a reliable and accurate method to distinguish specimens from geographically close populations. Furthermore, TEF is also a relatively fast and low-cost method compared with biochemical and molecular tools. The main limitations of TEF is the need for hard-mineralized structures to be present in the products that are to be screened and a data base. This limits the scope of geochemical methods to particular groups of organism and, most importantly, impairs their application to a range of processed products, such as canned food and fish fillets, that have been previously cleaned from shells or fish bones (Leal et al. 2015). It should be noted that the use of microchemical analysis (chemical analysis of elemental and isotopic components) may be critical because of the heterogeneity of the samples and the low concentrations of many of the constituents of mineralized structure. Geffen et al. (2013), in an intercalibration study on microchemical analysis of fish otoliths reported that because the measurement of otolith elements may vary with methodology, it is imperative to consider the sensitivity, accuracy and precision for each application as the suite of elements of interest may differ. Geffen et al. (2013) pointed also that the development of reference materials has been the first step in standardization of the procedures, but there is a need for wider ranging calibration and comparison studies.

Beside the use of TEF in mineral structure, Ibanez (2014) has applied geometric morphometric techniques to the fish scale shapes with discriminant analysis. And, he has shown that no significant differences in scale shape were detected by year, though it is possible to make a distinction on the geographical level. Ibanez pointed out that fish scales may be available in markets and could possibly be useful to trace fish, having reference of scale shape by zone.

Most of the trace elements analysis for a geographical determination are on mineral structures. However, Chaguri et al. (2015) in their study on the assessment of geographic origin and seasonality of croaker (*Micropogonias furnieri*) using macro and trace elements, fatty acid profile and stable isotopes analysis on fish muscle found that significant differences were detected between seasons for Cl, Ca, Fe, Sr and S, whereas differences between geographic origins were only observed with K. Nevertheless, the authors pointed out that further investigation is still required with larger samples of croaker before the implementation of fatty acid profile, elements or stable isotope analysis as traceability tools by food control agencies.

4.2 Stable isotope analysis

Few researches have focused on determining the origin of seafood by stable isotope analysis on muscle tissue, and most of the studies have focused on the discrimination on fish production systems (wild, farmed and organically farmed) (reviewed by Vinci et al. 2013, Moretti et al. 2003). Using isotope ratio mass spectrometry (IRMS), Molkentin et al. (2007) have investigated the ratios of carbon (δ13C) and nitrogen (δ15N) stable isotopes in raw fillets of differently grown Atlantic salmon

(*Salmo salar*) in order to develop a method for the identification of organically farmed salmon. IRMS allowed to distinguish organically farmed salmon from wild salmon, with δ15N-values being higher in organically farmed salmon, but not from conventionally farmed salmon. The authors of this study emphasized that the general applicability in the context of consumer protection should be checked with further samples, particularly regarding the variability of feed composition and possible changes in smoked salmon. Whereas Bell et al. (2007) have investigated whether farmed and wild European sea bass (*Dicentrarchus labrax*) could be discriminated by flesh fatty acid compositions, δ13C of individual fatty acids, δ13C and δ18O of total flesh oil, and δ15N of the glycerol choline fraction of flesh phospholipids. They pointed out the interesting possibilities of isotopic fingerprinting of individual fatty acids for future analytical methodologies to provide basis for discrimination between wild and farmed fish over a range of species. Turchini et al. (2009) have also demonstrated the utility of this analysis to be able to discriminate farmed fish (Australia Murray cod) between farms. Whereas Sant'Ana et al. (2010) found that the combined measurement of δ15N and δ13C provided traceability of the Brazilian freshwater fish cachara (*Pseudoplatystoma fasciatum*) comparing wild and farmed fish in dry and rainy seasons. Finally, Molkentin et al. (2015) have successfully combined stable isotope analysis of δ15N and δ13C analysis in defatted dry matter to differentiate of organically farmed from conventionally farmed salmon and brown trout, whether raw, smoked or graved. For the additional distinction of organic and wild salmon, they demonstrated that a second analysis of δ13C in fish lipids was required.

Regarding the geographical origin of wild fish, Tanaka et al. (2010) have investigated the seasonal migration pattern of anchovy, *Engraulis japonicus*, using a stable isotope approach in the Tachibana bay. More recently, Kim et al. (2015) used a carbon and nitrogen stable isotope ratios (δ13C and δ15N) analysis to investigate this tool to authenticate and establish geographical origin of three commercial fishes, mackerel, yellow croaker and pollock, originating from various countries. And they showed that apart from the species-dependent variation in the isotopic values, marked differences in the δ13C and δ15N ratios were also observed with respect to the country of origin. Rolli et al. (2014) also reported that currently the traceability of European perch (*Perca fluviatilis*) to distinguish between perch caught in Swiss lakes from the majority of perch originating from Estonia and Russia can be assessed through chemical isotopic analysis. Nevertheless, the authors reported that even if the 180/160 isotopic abundance ratio could be used as geographical traceability marker, several aspects affect the accuracy of the method and makes the analysis more complicated:

- the distinct geographical area ratio that differs only very slightly with overlapping standard deviation,

- the need for a large amount of fish material requires the mix of many fillets,
- the impossibility of analyzing processed matrix, and
- the comparison of the ratio with the ratio of a sample of the presumed originating water makes the analyses more complicated.

Nevertheless, Ortea and Gallardo (2015) showed that stable isotope ratio (SIR) analysis of carbon and nitrogen and/or multi-element composition (Pb, Cd, As, P, S) using multivariate statistics methods allowed the correct classification of 100% of the 45 shrimp individuals according to their origin and method of production, and 93.5% according to the species (seven different shrimp species tested). In spite of the promising results obtained, Ortea and Gallardo (2015) insisted on the need of further research with a larger number of samples in each group and including several geographic origins for all the species studied to confirm the validity of these approaches. Whereas Oleivera et al. (2011) showed that nitrogen ($\delta15N$) in tissues with lower turnover rates (bone and skin) and in tissues with greater turnover rates (muscle) can be used to authenticate the species of salted fish samples when distinguishing between Atlantic cod and saithe. Similarly, Chaguri et al. (2015) in their study on the assessment of geographic origin and seasonality of croaker (*Micropogonias furnieri*) using macro and trace elements, fatty acid profile and stable isotopes analysis on fish muscle found that $\delta13C$ and $\delta15N$ were statistically different between geographic origins, whereas differences between seasons were only detected in $\delta15N$ ratio of croaker from only one of the two locations studied.

4.3 Analysis of fatty acids

As reported by Leal et al. (2015) fatty acids (FA) are a diverse group of lipids and their composition is affected by several intrinsic (e.g., age, sex, reproductive cycle, and phylogeny) and extrinsic (e.g., diet, temperature, depth, and salinity) factors. Extrinsic factors are particularly useful for geographical traceability as the diet available for aquatic organisms varies with habitat and ecosystem. Since there is a strong relationship between FAs in fish, the diet and environmental factors, the FA profile is an important tool for distinguishing wild-caught and aquacultured fish species and for establishing seafood origin. Nevertheless, food availability is generally seasonal in most aquatic environments (Ezgeta-Balić 2012) and to minimize the effect of season, it is important to use tissues, such as the adductor muscle of bivalves or the cardiac muscle of fish that are rich in polar lipids. The contents of total fat and different higher chain FAs, including saturated (SFAs), mono-unsaturated (MUFAs) and poly-unsaturated fatty acids (PUFAs), in extracted oil are mostly used for authentication purposes (Lavilla et al. 2013). About 20 FAs are present in most of the fish species in concentrations higher than 1% (m/m), so that it can be determined in an easy way (reported by Lavilla et al. 2013).

Chemometric analysis, in particular, pattern recognition techniques, is essential for the classification of fish as wild or farmed, and for the establishment of their geographical origin. Among them, principal component analysis (PCA), linear discriminant analysis (LDA) and artificial neural networks (ANNs) are the most common. Boukouvala et al. (2012) also used discriminant function analysis on muscle fatty acid composition to discriminate the flatfishes *Solea aegyptiaca* and *Solea solea*, two morphological similar species sharing the same habitat.

Fatty acid profiles in fish fillet have been successively used to distinguish between farmed and wild Australian Murray cod by Turchini et al. (2009), but FAs are less informative in discriminating fish originating from different farms and appear to be more reflective of diet than different farming systems (intensive, semi-intensive, or extensive). Thus, Molkentin et al. (2015) pointed out that while a special feed composition is required in organic aquaculture, the percentages of animal and vegetable components of conventional aquaculture feed have changed considerably within the last decade. The authors have showed also that these diet differences clearly result in distinctive features in terms of stable isotope or fatty acid composition that are utilisable for the authentication of organic salmonid products (raw, smoked or graved products). However, Molkentin et al. (2015) in this study noted that differences between organic and wild salmon were less distinct than between organic and conventional. Bell et al. (2007) have demonstrated in their study on discrimination of wild European sea bass (*Dicentrarchus labrax*) from southern England and sea bass farmed in Scotland or in Greece that the dietary fat has a greater influence on flesh fat levels than the size and age of the animal, and that the intensive culture of sea bass using diets with a lipid content of ~ 20% results in increased lipid deposition in farmed bass compared to wild fish.

4.4 DNA-based techniques

While identification of species by DNA-based methodologies is already common practice in food control laboratories, validation of the authentication of origin by application of population genetic structure markers is possible only in a few cases (Behrman et al. 2015). The use of genetically based methods will be of great interest to trace the origin of processed fish products as the methods based on rare elements, stable element ratio and fatty acid profiles are mostly limited to raw or slightly processed fish. Currently, population genetic investigations of marine fishes are mostly been undertaken to answer biological questions or to manage fisheries areas. The determination of fish biological population limits is a necessary first step to the development of tools to identify the geographic origin of samples (Ogden 2008). As reported also by Ogden, genetic variation among populations reflects levels of gene exchange among spawning stocks and does not necessarily relate to feeding aggregations, and consequently molecular markers may be used to trace an individual salmon to the specific river drainage where it spawned, but possibly not to where it was caught. Furthermore, as underlined by Ogden (2008) non-alignment of political or management regions (e.g., ICES zones, transnational rivers…) and biological distributions is common in fisheries and represents one of the principal limitations to the use of genetic population markers for geographic origin authentification.

For the development of specific markers for fish species used for human consumption, mainly two types of genetic population markers can be used, polymorphic microsatellites and single nucleotide polymorphisms SNPs. Their development for authentification purpose is very laborious and requires a large number of data and samples to be collected as well as for their validation Behrmann et al. (2015). For a sufficient database a set of at least five microsatellites is required (Reiss et al. 2009), and for studies based on SNPs more single-nucleotide polymorphisms are necessary (Martinsohn 2011) (as reported by Behrmann et al. 2015). Till date, very few studies have successfully found markers to trace consumed fish species. For example, Nielsen et al. (2012) in the frame of the EU FishPopTrace project have applied SNP assays to the following four commercial marine fish: Atlantic cod, herring, common sole and hake. They finally found 93 to 100% of individuals to be correctly assigned to origin using a minimum of 8, 16, 50 and 13 SNP outliers to clearly trace respectively, Atlantic cod, herring, common sole and hake in European waters. Yue et al. (2012) have also tested the efficiency of microsatellites to assign wild and farmed individuals to four Asian seabass populations consisting of 354 spawners. Therefore, they pointed out that for routine genetic traceability, it is essential to conduct further feasibility studies.

Before concluding this section, it is also worth noting the study of Le Nguyen et al. (2008) which tested the possibility to trace fish, not with fish components as do the methods described above, but by analyzing the bacterial communities of fish (gills, skin and intestine ecosystem study). They found that, using 16S rDNA PCR-DGGE (Polymerase chain reaction—denaturing gradient gel electrophoresis) fingerprinting to detect variation in bacterial communities, the band profiles of farmed Pangasius fish from South Viet Nam are different according to the origin, but also according to the season (dry or rainy). Nevertheless, the authors pointed out that some common bands throughout the seasons could be used as a unique bar code to trace back the Pangasius fish to their original locations. The findings of this study could be updated with the high-throughput sequencing technologies which are currently developed for microbiome profiling.

In conclusion to this section on the methods to identify geographical fishing or farming areas, it could be highlighted that all these methodologies are, or could be in the near future methods of choice for seafood authentication, but the important work on the acquisition of reference sample and new data is still needed to build dedicated databases (SIR, TEF, SNPs, FA profiles and PCR-DGGE profiles/whole bacterial community profiling) regarding seafood species, fishing areas, and farms and farm country. With the exception of DNA-based methodologies, all of the methods described here are not suitable for fish processed product authentication. The combination of different methods using both elemental and isotopic markers and fatty acid profiles, associated to multifactorial data analysis, is recommended.

5. Conclusion

Of the wide range of analytical methods available for fish species identification, it is more likely the DNA-based techniques which are the favorite approaches due to their ease to use, their possibility to be adapted as rapid and hand-held analysis devices, and as a result of the scientific community efforts to facilitate species identification through standtardised reference sequence libraries. Nevertheless, the plethora of DNA-based identification methods available underlines the need of efficient, validated and standardised procedures that may help promote a safe and transparent food supply chain.

Keywords: Traceability, identification, authentication, seafood, fish, DNA-based methods

References

Agnew, D.J., Pearce, J., Pramod, G., Peatman, T., Watson, R., Beddington, J.R. and Pitcher, T.J. 2009. Estimating the worldwide extent of illegal fishing. PLoS One 4(2): e4570.

Behrmann, K., Rehbein, H., Appen, A.V. and Fischer, M. 2015. Applying population genetics for authentication of marine fish: the case of saithe (*Pollachius virens*). Journal of Agricultural and Food Chemistry 63(3): 802–809.

Benzekri, H., Armesto, P., Cousin, X., Rovira, M., Crespo, D., Merlo, M.A., Mazurais, D., Bautista, R., Guerrero-Fernández, D., Fernandez-Pozo, N., Ponce, M., Infante, C., Zambonino, J.L., Nidelet, S., Gut, M., Rebordinos, L., Planas, J.V., Bégout, M.L., Gonzalo Claros, M. and Ponce, M. 2014. *De novo* assembly, characterization and functional annotation of Senegalese sole (*Solea senegalensis*) and common sole (*Solea solea*) transcriptomes: integration in a database and design of a microarray. BMC Genomics 15: 952.

Bell, J.G., Preston, T., Henderson, R.J., Strachan, F., Bron, J.E., Cooper, K. and Morrison, D.J. 2007. Discrimination of wild and cultured European sea bass (*Dicentrarchus labrax*) using chemical and isotopic analyses. Journal of Agricultural and Food Chemistry 55: 5934–5941.

Boukouvala, V., Cariani, A., Maes, G., Sevilla, R., Verrez-Bagnis, V., Jérôme, M., Guarniero, I., Monios, G., Tinti, F., Volckaert, F., Bautista, J.M. and Krey, G. 2012. Restriction fragment length analysis of the *cytochrome b* gene and muscle fatty acid composition differentiate the cryptic flatfish species *Solea solea* and *Solea aegyptiaca*. Journal of Agricultural and Food Chemistry 60(32): 7941–7948.

Chaguri, M.P., Maulvault, A.L., Nunes, M.L., Santiago, D.A., Denadai, J.C., Fogaça, F.H., Silvia Sant'Ana, L., Ducatti, C., Bandarra, N., Carvalho, M.L. and Marques, A. 2015. Different tools to trace geographic origin and seasonality of croaker (*Micropogonias furnieri*). LWT-Food Science and Technology 61(1): 194–200.

Chang, M.Y. and Geffen, A.J. 2013. Taxonomic and geographic influences on fish otolith microchemistry. Fish and Fisheries 14: 458–492.

Clark, L.F. 2015. The current status of DNA barcoding technology for species identification in fish value chains. Food Policy 54: 85–94.

Compton, J. 1991. Nucleic acid sequence-based amplification. Nature 350(6313): 91–92.

D'Amico, P., Armani, A., Castigliego, L., Sheng, G., Gianfaldoni, D. and Guidi, A. 2014. Seafood traceability issues in Chinese food business activities in the light of the European provisions. Food Control 35(1): 7–13.

De Battisti, C., Marciano, S., Magnabosco, C., Busato, S., Arcangeli, G. and Cattoli, G. 2014. Pyrosequencing as a tool for rapid fish species identification and commercial fraud detection. Journal of Agricultural and Food Chemistry 62(1): 198–205.

European Commission. 2016. Report from the commission to the European parliament and the councilon options for an EU eco-label scheme for fishery and aquaculture products. http://eur-lex.europa.eu/legal-content/EN/TXT/?uri=COM:2016:263:FIN (Accessed 25.07.2016).

Ezgeta-Balić, D., Najdek, M., Peharda, M. and Blažina, M. 2012. Seasonal fatty acid profile analysis to trace origin of food sources of four commercially important bivalves. Aquaculture 334: 89–100.

FAO (Food and Agriculture Organization of the United Nations). 2012. State of world fisheries and aquaculture. Rome, Italy, 230 pp.

FAO (Food and Agriculture Organization of the United Nations). 2014. State of world fisheries and aquaculture. Rome, Italy, 243 pp.

FDA (Food and Drug Administration). 2016. The Seafood List. http://www.accessdata.fda.gov/scripts/fdcc/?set=seafoodlist (Accessed 29.06.2016).

Galal-Khallaf, A., Osman, A.G.M., Carleos, C.E., Garcia-Vazquez, E. and Borrell, Y.J. 2016. A case study for assessing fish traceability in Egyptian aquafeed formulations using pyrosequencing and metabarcoding. Fisheries Research 174: 143–150.

Geffen, A.J., Morales-Nin, B., Pérez-Mayol, S., Cantarero-Roldán, A.M., Skadal, J. and Tovar-Sánchez, A. 2013. Chemical analysis of otoliths: cross validation between techniques and laboratories. Fisheries Research 143: 67–80.

Griffiths, A.M., Sotelo, C.G., Mendes, R., Perez Martin, R.I., Schröder, U., Shorten, M., Silva, H.A., Verrez-Bagnis, V. and Mariani, S. 2014. Current methods for seafood authenticity testing in Europe: is there a need for harmonization? Food Control 45: 95–100.

Handy, S.M., Deeds, J.R., Ivanova, N.V., Hebert, P.D., Hanner, R.H., Ormos, A., Weigt, L.A., Moore, M.M. and Yancy, H.F. 2011. A single-laboratory validated method for the generation of DNA barcodes for the identification of fish for regulatory compliance. Journal of AOAC International 94(1): 201–210.

Handy, S.M., Chizhikov, V., Yakes, B.J., Paul, S.Z., Deeds, J.R. and Mossoba, M.M. 2014. Microarray chip development using infrared imaging for the identification of catfish species. Applied Spectroscopy 68(12): 1365–1373.

Ibáñez, A.L. 2014. Fish scale shape variation by year and by geographic location, could scales be useful to trace fish? A case study on the Gulf of Mexico. Fisheries Research 156: 34–38.

Kievits, T., Van Gemen, B., Van Strijp, D., Schukkink, R., Dircks, M., Adriaanse, H., Malek, L., Sooknanan, R. and Lens, P. 1991. NASBA isothermal enzymatic *in vitro* nucleic acid amplification optimized for the diagnosis of HIV-1 infection. Journal of Virological Methods 35(3): 273–86.

Kim, J.H., Park, J.Y., Jung, J.W., Kim, M.J., Lee, W.S., An, C.M., Kang, J.H. and Hwang, S.Y. 2011. Species identification of filefishes (Monacanthidae) using DNA microarray in Korean marketplace. Biochip Journal 5(3): 229–235.

Kim, H., Kumar, K.S. and Shin, K.H. 2015. Applicability of stable C and N isotope analysis in inferring the geographical origin and authentication of commercial fish (mackerel, yellow croaker and pollock). Food Chemistry 172: 523–527.

Kochzius, M., Seidel, C., Antoniou, A., Botla, S.K., Campo, D., Cariani, A., Vazquez, E., Hauschild, J., Hervet, C., Hjörleifsdottir, S., Hreggvidsson, G., Kappel, K., Landi, M., Magoulas, A., Marteinsson, V., Nölte, M., Planes, S., Tinti, F., Turan, C., Venugopal, M., Weber, H. and Blohm, D. 2010. Identifying fishes through DNA barcodes and microarrays. PLoS ONE 5(9): 12620.

Lago, F.C., Alonso, M., Vieites, J.M. and Espiñeira, M. 2014. Fish and seafood authenticity– species identification. pp. 419–452. In: Boziaris, I.S. (ed.). Seafood Processing: Technology, Quality and Safety. Wiley Blackwell, Chichester.

Lavilla, I., Costas-Rodríguez, M. and Bendicho, C. 2013. Authentication of fishery products. Comprehensive Analytical Chemistry 60: 657–717.

Leal, M.C., Pimentel, T., Ricardo, F., Rosa, R. and Calado, R. 2015. Seafood traceability: current needs, available tools, and biotechnological challenges for origin certification. Trends in Biotechnology 33(6): 331–336.

Le Nguyen, D.D., Ngoc, H.H., Dijoux, D., Loiseau, G. and Montet, D. 2008. Determination of fish origin by using 16S rDNA fingerprinting of bacterial communities by PCR-DGGE: An application on Pangasius fish from Viet Nam. Food Control 19(5): 454–460.

Leone, G., van Schijndel, H., van Gemen, B., Kramer, F.R. and Schoen, C.D. 1998. Molecular beacon probes combined with amplification by NASBA enable homogeneous, real-time detection of RNA. Nucleic Acids Research 26(9): 2150–2155.

Martinez, I., James, D. and Loréal, H. 2005. Application of modern analytical techniques to ensure seafood safety and authenticity. FAO Fisheries Technical Paper 455, Rome.

Martinsohn, J.T. 2011. Deterring illegal activities in the fisheries sector: genetics, genomics, chemistry and forensics to fight IUU fishing and in support of fish product traceability. European Commission - Joint Research Centre, Luxembourg.

Mackie, I.M., Pryde, S.E., Gonzales-Sotelo, C., Medina, I., Pérez-Martın, R.I., Quinteiro, J., Rey-Mendez, M. and Rehbein, H. 1999. Challenges in the identification of species of canned fish. Trends in Food Science & Technology 10(1): 9–14.

Manríquez, P.H., Galaz, S.P., Opitz, T., Hamilton, S., Paradis, G., Warner, R.R., Castilla, J.C., Labra, F.A. and Lagos, N.A. 2012. Geographic variation in trace-element signatures in the statoliths of near-hatch larvae and recruits of Concholepas concholepas (loco). Marine Ecology Progress Series 448: 105–118.

Maralit, B.A., Aguila, R.D., Ventolero, M.F.H., Perez, S.K.L., Willette, D.A. and Santos, M.D. 2013. Detection of mislabeled commercial fishery by-products in the Philippines using DNA barcodes and its implications to food traceability and safety. Food Control 33(1): 119–125.

Mariani, S., Griffiths, S.M., Velasco, A., Kappel, K., Jérôme, M., Perez-Martin, R.I., Schröder, U., Verrez-Bagnis, V., Silva, H., Vandamme, S.G., Boufana, B., Mendes, R., Shorten, M., Smith, C., Hankard, E., Hook, S.A., Weymer, S.A., Gunning, D. and Sotelo, C.G. 2015. Europe gets to grips with seafood labels. Frontiers in Ecology and the Environment 13(10): 536–540.

Meusnier, I., Singer, G.A., Landry, J.F., Hickey, D.A., Hebert, P.D. and Hajibabaei, M. 2008. A universal DNA mini-barcode for biodiversity analysis. BMC Genomics 9: 214.

Molkentin, J., Meisel, H., Lehmann, I. and Rehbein, H. 2007. Identification of organically farmed Atlantic salmon by analysis of stable isotopes and fatty acids. European Food Research and Technology 224(5): 535–543.

Molkentin, J., Lehmann, I., Ostermeyer, U. and Rehbein, H. 2015. Traceability of organic fish–authenticating the production origin of salmonids by chemical and isotopic analyses. Food Control 53: 55–66.

Moretti, V.M., Turchini, G.M., Bellagamba, F. and Caprino, F. 2003. Traceability issues in fishery and aquaculture products. Veterinary Research Communications 27: 497–505.

Nielsen, E.E., Cariani, A., Mac Aoidh, E., Maes, G.E., Milano, I., Ogden, R., Taylor, M., Hemmer-Hansen, J., Babbucci, M., Bargelloni, L., Bekkevold, D., Diopere, E., Grenfell, L., Helyar, S., Limborg, M.T., Martinsohn, Y.T., McEwing, R., Panitz, F., Patarnello, T., Tinti, F., Van Houdt, J.K.J., Volckaert, F.A.M., Waples, R.S., Albin, J.A.J., Vieites, VietesBaptista, J.M., Barmintsev, V., Bautista, J.M., Bendixen, C., Bergé, J.P., Blohm, D., Cardazzo, B., Diez, A., Espiñeira, M., Geffen, A.J., Gonzalez, E., González-Lavín, N., Guarniero, I., Jeráme, M., Kochzius, M., Krey, G., Mouchel, O., Negrisolo, E., Piccinetti, C., Puyet, A., Rastorguev, S., Smith, J.P., Trentini, M., Verrez-Bagnis, V., Volkov, A., Zanzi, A. and Carvalho, G.R. 2012. Gene-associated markers provide tools for tackling illegal fishing and false eco-certification. Nature Communications 01/2012 3: 851.

Notomi, T., Okayama, H., Masubuchi, H., Yonekawa, T., Watanabe, K., Amino, N. and Hase, T. 2000. Loop-mediated isothermal amplification of DNA. Nucleic Acids Research 28(12): e63–e63.

Ogden, R. 2008. Fisheries forensics: the use of DNA tools for improving compliance, traceability and enforcement in the fishing industry. Fish and Fisheries 9(4): 462–472.

Oliveira, E.J.V.M., Sant'Ana, L.S., Ducatti, C., Denadai, J.C. and de Souza Kruliski, C.R. 2011. The use of stable isotopes for authentication of gadoid fish species. European Food Research and Technology 232(1): 97–101.

Ortea, I. and Gallardo, J.M. 2015. Investigation of production method, geographical origin and species authentication in commercially relevant shrimps using stable isotope ratio and/or multi-element analyses combined with chemometrics: an exploratory analysis. Food Chemistry 170: 145–153.

Park, J.Y., Lee, S.Y., An, C.M., Kang, J.H., Kim, J.H., Chai, J.C., Chen, J.Y., Kang, J.S., Ahn, J.J., Lee, Y.S. and Hwang, S.Y. 2012. Comparative study between next generation sequencing technique and identification of microarray for species identification within blended food products. Biochip Journal 6(4): 354–361.

Park, J.Y., Cho, H., Kang, J.H., Kim, E.M., An, C.M., Kim, J.H., Lee, W.S. and Hwang, S.Y. 2014. Development of DNA microarray for species identification of eels (*Anguilliformes* and *Myxiniformes*) in Korean fisheries markets. BioChip Journal 8(4): 310–316.

Piepenburg, O., Williams, C.H., Stemple, D.L. and Armes, N.A. 2006. DNA detection using recombination proteins. PLoS Biol 4(7): e204.

Quinteiro, L., Sotelo, C.G., Rehbein, H., Pryde, S.E., Medina, I., Pérez-Martın, R.I., Rey-Mendez, M. and Mackie, I.M. 1998. Use of mt-DNA polymerase chain reaction (PCR) sequencing and PCR-restriction fragment length polymorphism methodologies in species identification of canned tuna. Journal of Agricultural and Food Chemistry 46: 1662–1669.

Rasmussen, R.S. and Morrissey, M.T. 2008. DNA-based methods for the identification of commercial fish and seafood species. Comprehensive Reviews in Food Science and Food Safety 7(3): 280–295.

Rasmussen, R.S. and Morrissey, M.T. 2009. Application of DNA-based methods to identify fish and seafood substitution on the commercial market. Comprehensive Reviews in Food Science and Food Safety 8(2): 118–154.

Ratnasingham, S. and Hebert, P.D. 2007. BOLD: The Barcode of Life Data System (http://www. barcodinglife. org). Molecular Ecology Notes 7(3): 355–364.

Reiss, H., Hoarau, G., Dickey-Collas, M. and Wolff, W.J. 2009. Genetic population structure of marine fish: mismatch between biological and fisheries management units. Fish and Fisheries 10(4): 361–395.

Ricardo, F., Génio, L., Leal, M.C., Albuquerque, R., Queiroga, H., Rosa, R. and Calado, R. 2015. Trace element fingerprinting of cockle (*Cerastoderma edule*) shells can reveal harvesting location in adjacent areas. Scientific Reports 5: 11932.

Rolli, J., Girardet, S., Monachon, C. and Richard, C. 2014. Microsatellite analysis of perch (*Perca fluviatilis*) and its genetic authentication of geographical localization. CHIMIA International Journal for Chemistry 68(10): 726–731.

Santaclara, F.J., Velasco, A., Pérez-Martín, R.I., Quintero, J., Rey-Méndez, M., Pardo, M.A., Jimenez, E. and Sotelo, C.G. 2015. Development of a multiplex PCR–ELISA method for the genetic authentication of *Thunnus* species and *Katsuwonus pelamis* in food products. Food Chemistry 180: 9–16.

Sant'Ana, L.S., Ducatti, C. and Ramires, D.G. 2010. Seasonal variations in chemical composition and stable isotopes of farmed and wild Brazilian freshwater fish. Food Chemistry 122(1): 74–77.

Saull, J., Duggan, C., Hobbs, G. and Edwards, T. 2016. The detection of Atlantic cod (*Gadus morhua*) using loop mediated isothermal amplification in conjunction with a simplified DNA extraction process. Food Control 59: 306–313.

Srinivasan, U.T., Watson, R. and Sumaila, U.R. 2012. Global fisheries losses at the exclusive economic zone level, 1950 to present. Marine Policy 36(2): 544–549.

Tanaka, H., Ohshimo, S., Takagi, N. and Ichimaru, T. 2010. Investigation of the geographical origin and migration of anchovy *Engraulis japonicus* in Tachibana Bay, Japan: A stable isotope approach. Fisheries Research 102(1): 217–220.

Teletchea, F. 2009. Molecular identification methods of fish species: reassessment and possible applications. Reviews in Fish Biology and Fisheries 19(3): 265–293.

Turchini, G.M., Quinn, G.P., Jones, P.L., Palmeri, G. and Gooley, G. 2009. Traceability and discrimination among differently farmed fish: a case study on Australian Murray cod. Journal of Agricultural and Food Chemistry 57: 274–281.

Tyagi, S. and Kramer, F.R. 1996. Molecular beacons: probes that fluoresce upon hybridization. Nature Biotechnology 14(3): 303–308.

Ulrich, R.M., John, D.E., Barton, G.W., Hendrick, G.S., Fries, D.P. and Paul, J.H. 2013. Ensuring seafood identity: Grouper identification by real-time nucleic acid sequence-based amplification (RT-NASBA). Food Control 31(2): 337–344.

Ulrich, R.M., John, D.E., Barton, G.W., Hendrick, G.S., Fries, D.P. and Paul, J.H. 2015. A handheld sensor assay for the identification of grouper as a safeguard against seafood mislabeling fraud. Food Control 53: 81–90.

Vinci, G., Preti, R., Tieri, A. and Vieri, S. 2013. Authenticity and quality of animal origin food investigated by stable-isotope ratio analysis. Journal of the Science of Food and Agriculture 93(3): 439–448.

Warner, K., Timme, W., Lowell, B. and Hirshfield, M. 2013. Oceana study reveals seafood fraud nationwide. Oceana 11: 1–69.

Woolfe, M., Gurung, T. and Walker, M.J. 2013. Can analytical chemists do molecular biology? A survey of the up-skilling of the UK official food control system in DNA food authenticity techniques. Food Control 33: 385–392.

Yue, G.H., Xia, J.H., Liu, P., Liu. F., Sun, F. and Lin, G. 2012. Tracing Asian seabass individuals to single fish farms using microsatellites. PLoS One 7(12): e52721.

12

Food Traceability and Authenticity Based on Volatile Compound Analysis

Dufossé Laurent

1. Introduction

Between December 2014 and January 2015, Operation Opson IV, an Interpol and Europol coordinated operation, seized more than 2,500 tons of counterfeit, contaminated and adulterated food, in raids across 47 countries. The seizures included mozzarella cheese, strawberries, eggs, cooking oil and dried fruits. Previously Opson III, in the summer of 2014, similarly resulted in the seizure of 1,200 tons of fake or substandard foods and nearly 430,000 liters of counterfeit drinks (oil, vinegar, biscuits, spices and condiments, cereals, dairy products and honey) (Masters 2015).

Lest anyone be in any doubt, the adulteration, falsification, counterfeiting and mis-selling/labelling of food and drinks, of ingredients, is a huge problem for the food and drink industry. Finding technical, and practical solutions to thwart the wrongdoers, is also a major challenge. It should be clear that there is no single test or "black-box" able to fix the analysis or verification of an authentic food or drink (Masters 2015). As explained in this book, many techniques are in use or are emerging and this chapter focuses on volatile compound analysis, through gas chromatography, mass spectrometry, and even (bio)electronic noses.

Laboratoire de Chimie des Substances Naturelles et des Sciences des Aliments, ESIROI Agroalimentaire, Université de La Réunion, F-97490 Sainte-Clotilde, Ile de La Réunion, France; E-mail: laurent.dufosse@univ-reunion.fr

2. Some Words about Volatile Compounds, Gas Chromatography, Mass Spectrometry and other Techniques

The whole set of foods produced in the world contains thousands of volatile compounds. As an example the chemical identities and CAS/FEMA-GRAS/ EC numbers (if known) of more than 7,100 flavor chemicals, the qualitative and the quantitative occurrence of these flavor chemicals in about 450 food products, linked to more than 4,500 references (including reviews) related to these products are described in databases such as Volatile Compounds in Food 16.2 from Triskelion™.

Studying food traceability and authenticity is not only a question of analytical chemistry as data handling using statistical tools and soft wares is at least as important as chemistry to obtain useful conclusions. Up to now the best analytical combination is based on gas chromatography (GC) and mass spectrometry (MS) applied on the headspace (HS) of the food product, avoiding the tedious and artifact-introducing extraction step. Developments in the solid phase micro extraction (SPME) also brought huge improvements in the laboratories.

Such techniques are however very costly (huge initial investment and maintenance) and perhaps users will see in the next decade applications of bioelectronic noses that are a mix of nanotechnologies and biological components, as previous e-noses did not perform very well, with their quite primitive sensors.

3. Food Traceability and Authenticity Based On Volatile Compound Analysis: Applications in Various Foods and Drinks

3.1 Honey

Assessment of the botanical origin of honey is of great concern in food analysis, since authenticity (i) guarantees the quality of the product, (ii) prevents overpayment and (iii) helps to identify frauds. In addition, European Community legislation concerning honey, i.e., Directive EC/74/409 amended by the Proposal COM/95/0722 (1997) requires labelling of the floral origin. Usually the determination of the botanical origin of honey is carried out by melissopalynological analysis, which is based on the identification of pollen by microscopic examination (Bianchi et al. 2005). This analysis is time-consuming, requires a very skillful analyst for data interpretation and sometimes results do not allow a reliable identification of the botanical origin, especially in the case of honey characterized by low amounts of pollen. Consequently, due to difficulties in analysis, great attention has been paid in order to find specific chemical markers for proving the botanical source of honeys. One of the most typical features of honey is its aroma profile, which could be used to characterize volatile marker compounds specific to a given botanical origin. In particular, the gas chromatography pattern can be considered as a chemical "fingerprint", since the nature and relative amount of the compounds present in the volatile fraction are a distinctive feature of different floral sources (Bianchi et al. 2005). Volatile compounds in honey may arise, among others, from the nectar or honeydew source, from the transformations of these raw materials carried out by honey bees, from honey processing and storage or from microbial and environmental contamination.

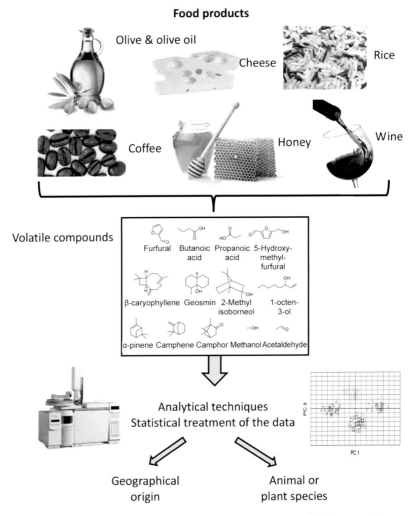

Figure 1. Description of the process, allowing determination of food traceability and authenticity through volatile compound analysis.

Gas chromatography coupled to mass spectrometry (GC–MS) combines the high sensitivity and efficacy required for the analysis of the very complex mixtures of volatiles present in honey at low concentrations and provides structural information (mass spectrum) for their qualitative analysis. Although GC–MS has become the technique of choice for characterization of the volatile fraction of honey, its application requires a previous fractionation step in which volatiles are isolated from the major components of the honey matrix (sugars and water) and pre-concentrated (Soria et al. 2008). Several studies have been published on the application of different fractionation techniques to the study of honey volatiles: solvent extraction (SE), simultaneous steam distillation–extraction (SDE), mixed procedures-based on SE followed by SDE, solid phase extraction, static headspace, solid phase dynamic extraction (SPDE), solid phase micro-extraction (SPME) and dynamic headspace (Purge and Trap system, P&T).

Up to over 100 to 110 volatiles with different functionality have been identified in honey samples. Norisoprenoid compounds such as α-isophorone, β-isophorone and 4-oxoisophorone, were recognized as specific floral origin markers of the strawberry-tree (*Arbutus unedo* L.) honey. In heather (*Calluna vulgaris*) the presence of isophorone and 2-methyl butyric acid and the absence of linalool were quite characteristic (Seisonen et al. 2015). The occurrence of lilac aldehyde [2-(5-methyl-5-vinyltetrahydro-2-furanyl) propanal] and 2-aminoacetophenone were proposed as indicators of rhododendron honey (Senyuva et al. 2009). Excess amount of volatiles such as thymol (trace) and carvacrol (0.66%) that originate from thyme plants indicated adulteration of thyme honey by thyme essential oil in some commercial samples (Mannaş and Altuğ 2007). In another study on the volatiles of coriander, lime and sunflower honeys, 3-hexen-1-ol, isoeugenol, cinnamaldehyde, cinnamyl alcohol, pyridine, quinoline, salicylic aldehyde, undecanol and octanol were identified as the major compounds. The characteristic aroma components of buckwheat honey were shown to be 3-methylbutanal, sotolon and β-damascenone, whereas thistle, tulip, sourwood and mountain laurel honeys were found to involve volatiles such as benzaldehyde, furfural, isovaleraldehyde and phenylacetaldehyde (Mannaş and Altuğ 2007).

As a conclusion often given by the authors in this research field, multivariate statistical analysis has to be applied to GC–MS relative data, with promising results being obtained for the characterization of honey source.

3.2 Coffee

Coffee is the world most popular beverage after water, considering that over 400 billion cups are consumed annually. Consequentially, coffee is one of the most important agricultural products in the international trade.

More than 800 volatile compounds have been identified to be present in roasted and ground coffee. The most common classes of compounds reported in the headspace include acids, aldehydes, alcohols, sulphur compounds, phenolic compounds, pyrazines, pyridines, thiophenes, pyrroles, ketones, esters, benzenic compounds, thiazoles, oxazoles, lactones, alkanes, alkenes, and furans (Yang et al. 2016). These components are present in variable concentrations and each of them contributes uniquely to the final aroma quality (see the excellent Master of Science in Chemistry thesis from Risticevic 2008).

Research dedicated to the recognition of coffee adulteration practices has been conducted for many years. Regarding the adulterate addition of roasted grains to coffee, the following complete list of ways to perform fraudulent practices as pertaining to coffee is: (i) adulteration with coffee substitutes (such as roasted soybean and wheat added in ground roasted); (ii) mixing two coffee bean varieties and (iii) mixing of high commercial value coffee beans from one region with lower commercial value ones originating from other regions (Risticevic 2008).

HS-SPME–GC–TOFMS methodology was developed and optimized for the purposes of verifying its capability in terms of tracing back the coffee samples to their production areas (Zambonin et al. 2005). Among the tested commercially available fiber coatings, the mixed-phase DVB/CAR/PDMS fiber demonstrated best performance characteristics for a wide range of analytes having different physico-chemical characteristics and hence this coating was used in super elastic metal fiber assembly form for the completion of overall sequence of coffee samples. The SPME method optimization was completed by the utilization of multivariate experimental design and accordingly the optimum set of conditions for the two identified influential parameters was 55°C for 12 min for extraction time and temperature, respectively (Risticevic et al. 2008). The utilization of high-speed data acquisition rate option offered by the TOFMS instrument ensured the completion of one GC–MS run of a complex coffee sample in 7.9 min. The complete list of benefits provided by TOF software, including the fully automated background subtraction, baseline correction, peak finds and mass spectral deconvolution algorithms were exploited during the data evaluation procedure. Finally, the acquired data set was submitted to principal component analysis and the corresponding geographical origin discriminations of coffees originating from South and Central America, Africa and Asia were successfully established (Risticevic et al. 2008).

At this point it is important to emphasize that in addition to successful geographical discrimination of: (i) authentic sample collections from Brazil and Colombia and (ii) non-authentic sample collections from South America, Central America, Africa and Asia, this classification study was also successful in detecting potential compositional changes that coffee undergoes due to the limited shelf-life stability over extensive storage conditions. This finding is crucial in the realization that imported food commodities are quite likely to undergo authentic aroma losses before they are distributed to consumer populations. Finally, the conducted geographical origin verification of collected samples proved that this rapid analytical methodology demonstrates great potential for the assessment of quality and detection of adulterations in worldwide coffee industry (Risticevic et al. 2008).

3.3 Wine

Wine aroma is formed by a complex mixture of many natural and processing variables, such as varietal origin, degree of maturity at harvest, percent of solid present in the fermented grape juice, wooded/unwooded process and bacterial/yeast strains (Cozzolino et al. 2005). A great number of papers have been published about the classification of wines according to geographical origin, based on their volatile compounds.

As an example the work of Berna et al. (2009) deals with Sauvignon Blanc wines. Analysis of 34 Sauvignon Blanc wine samples from three different countries and six regions were performed by gas chromatography–mass spectrometry (GC-MS). Linear discriminant analysis (LDA) showed that there were three distinct clusters or classes of wines with different aroma profiles. Wines from the Loire Region in France and Australian wines from Tasmania and Western Australia were found to have similar aroma patterns. New Zealand wines from the Marlborough region as well as the Australian ones from Victoria were grouped together based on the volatile composition. Wines from South Australia region formed one discrete class (Berna et al. 2009). There are similarities qualitatively and quantitatively in the volatile profile of wines from different regions. Seven analytes, most of them esters (ethyl ethanoate, nerol oxide, 1-hexanol, trans-2-hexenyl butanoate, pentyl ethanoate, hex-4-enoic acid ethyl ester and propyl decanoate), were found to be the relevant chemical compounds that characterized the classes (Berna et al. 2009). However, this technique was not strong enough to sort all the six wines. Isotope mass spectrometry (IRMS) seems to be the most suitable complementary technique.

3.4 Spirits

The production of fermentation ethanol and spirits is an important branch of the agrofood industry in the world. The quality and the price of these products are determined by the variety, purity and origin of the raw material (Bauer-Christoph et al. 1997). National food and customs regulations have to be respected when checking the quality of these products for marketing and sales and also for official control. There have always been attempts to adulterate spirits and other food products, for instance by blending high-quality distillates with ethanol made from a cheaper raw material, by adding synthetic, volatile components to neutral alcohol or by misleading labelling of the variety and origin of the raw material used (Bauer-Christoph et al. 1997).

Determination of the botanical origin of raw spirit used for alcoholic beverage production is of great importance for rectifying units, control laboratories, and proper product labeling. Raw spirit samples produced from rye, corn, and potato were analyzed using a solid phase microextraction-mass spectrometry (SPME-MS) method, which involved volatiles pre-concentration by SPME with subsequent volatile fraction characterization by MS without particular compounds separation by GC (Jeleń et al. 2010). The main groups of volatile fermentation by-products that raw spirits contain are carbonyl compounds, higher alcohols (fusel alcohols or oils), esters, and acetals. Some of them, such as fusel alcohols and ethyl acetate, are present in relatively large amounts (often several hundred milligrams per liter) and can be determined by direct injection of the sample without a pre-concentration step, whereas others, such as carbonyl compounds (except acetaldehyde) and fatty acid ethyl esters, occur at microgram per liter concentrations and require specific isolation or detection procedures. Obtained data were treated using principal component analysis and linear discriminant analysis (LDA) to test the possibility of sample classification. SPME sampling conditions allowed rapid extraction in 2 min at 50°C using a carboxen/divinylbenzene/polydimethylsiloxane fiber, followed by rapid MS analysis. Use of LDA made possible the classification of raw spirits based on the material they were produced from. The classification ability of the developed SPME-MS method was 100%, whereas its predictive ability was 96% (Jeleń et al. 2010).

3.5 Vinegar

Sherry vinegar and traditional balsamic vinegar (TBV) are high-quality vinegars that are famous all over the world. Sherry wine vinegar is produced in a series of wooden oak casks in which acetification and aging take place simultaneously. TBV is obtained by a traditional method that ferments the cooked must and ages the product in a set of wooden barrels (Cerezo et al. 2009). During this period, the finished product develops the desired organoleptic properties that make it highly appreciated.

The EU-funded WINEGAR project (ref. COOP-CT-2005-017269) tested four different kinds of woods (acacia, chestnut, cherry, and oak) for wine vinegar production. The quality and chemical composition were checked throughout the process and can be used for a variety of purposes, including authentication and product classification according to quality criteria. For example, the analysis of the vinegars aged in acacia (*Robinia pseudoacacia*) revealed a compound that was not present in vinegars aged in the other woods. This compound named (+)-dihydrorobinetin was likely to serve as a chemical marker of acacia wood for authenticity purposes (Cerezo et al. 2009).

3.6 Rice

Basmati rice is taken here as an example. There is increasing concern regarding the authenticity of basmati rice throughout the world. For years, traders have been passing off a lesser quality rice as the world's finest long-grained, aromatic rice, Basmati, in key markets like the US, Canada, and the EU. A DNA rice authenticity verification service in India has concluded that more than 30% of the basmati rice sold in the retail markets of the US and Canada is adulterated with inferior quality grains. In Britain, the Food Standard Agency found in 2005 that about half of all basmati rice sold was adulterated with other strains of long-grain rice (Cleland et al. 2014). Waters™ Company developed a complete set of methods to assess the authenticity of basmati rice using off the shelf supermarket samples with the latest advancements in high resolution GC-MS hardware and informatics. Based on 2907 statistically significant ions from the 3885 originally generated, the authors conclude they are able to discriminate basmati rice, jasmine rice, long grain rice and any combination of these (Cleland et al. 2014).

3.7 Olive oil

Olive oil represents an important ingredient in the popular Mediterranean diet and is worldwide appreciated both for nutritional and sensory properties, often related to geographical origin and cultivar of olive fruits employed. Fraudsters trying to seek financial gain can adulterate the product causing economic repercussions and, sometimes sanitary risks. The "protected designation of origin" (PDO) European label insures a relative protection of both consumers and honest producers, since it prescribes production techniques and specific geographical origin, but one of the main problems is to set down objective tools to control these specifications (Janin et al. 2014). The sensorial attributes of olive oils are mainly generated by volatile constituents, including carbonyl compounds (aldehydes, ketones, etc.), alcohols, esters and hydrocarbons. These attributes may have a high influence on the flavor of olive oil. Thereby, the potential discrimination of geographical origin of olive oil can be based on their volatile profile characterization (Cerrato Oliveros et al. 2005). In some cases, up to 65 volatile compounds (aldehydes, alcohols, furans, hydrocarbons, acids, ketones, esters) have been listed and compared to show that virgin olive oil authentication is closely related with the olive variety from which oil is obtained by strictly physical means (Aparicio et al. 1997). One variety could show the maximum concentration of esters, responsible for the green (grass) perception, while another one has the maximum concentration of furans, which are responsible for sweet (ripe fruit) perception. In the frame of oil characterization, the techniques analyzing volatiles have been applied with success for the determination of adulteration such as mixing of olive oil with hazelnut oil, sunflower oil and/or olive–pomace oil, for the detection of contaminants (such as benzene, toluene, ethylbenzene, xylene isomers and styrene) (Janin et al. 2014).

3.8 Cheese

The origins of volatile compounds found in cheese are diverse and can be classified into two groups: the first one contains native volatile compounds already present in milk, which are not transformed during cheese manufacturing while the second group includes components produced in the cheese itself during manufacture or maturation. Forage is an important factor influencing the composition of the volatiles of the first group. The volatile compounds of the second group are formed during manufacture and ripening of cheese by microbial, enzymatic and (bio) chemical transformations (Pillonel et al. 2003). As a first example, the volatile compounds of Emmental cheese samples from different European regions were investigated by GC-MS. Each region could be separated from the others using compounds which were more or less specific to one or two regions. For instance, the concentrations of butan-2-one, 3-hydroxybutanone, butan-2-ol and octane made it possible to separate "Switzerland" origin from the other cheeses. These investigations showed the potential of volatile compounds to discriminate cheese samples with different origins (Pillonel et al. 2003).

Analysis of the volatile fraction of Provola dei Nebrodi, a typical Sicilian *pasta filata* cheese, was performed using Solid Phase Micro-extraction High-Resolution Gas Chromatography/Mass Spectrometry. The cheeses were sampled and analyzed at four different stages of ripening (0, 7, 30 and 90 days). A total of 60 components were identified: fatty acids (11), esters (15), lactones (2), methyl ketones (8), aldehydes (9), alcohols (4), hydrocarbons (3), terpenes (1), chlorinated compounds (1), and aromatic compounds (6). The main components were found to be hexanoic and octanoic acids and ethyl hexanoate. The 60 components were present in all the samples, regardless of the ripening stage, although their ratio showed statistically significant variations with aging. Fatty acids and esters increased, whereas the aldehyde content drastically decreased after 30 days. The profile of volatile components identified appeared to be different from that of other *pasta filata* cheeses (Ziino et al. 2005).

In another case, the limited number of volatile compounds measured in the headspace of the Montasio cheese, a typical protected designation of origin (PDO) Italian semi-hard and semi-cooked cheese produced in northeast Italy from raw or thermized cow's milk, did not allow clear determination of origin. However the information could be useful in preserving this cheese in the market and provides an additional means to evaluate the effects of new technologies or changes in the traditional production techniques described in the PDO regulations for this cheese (Innocent et al. 2013).

3.9 Meat

Besides sensory tests of meat flavor, the instrumental determination of volatile compounds and their profile is an alternative. In order to determine the geographic origin of meat, especially of processed meat, studying the volatile compounds occurring in the headspace and in meat itself could be interesting. Gas chromatography (GC) combined with mass spectroscopy (MS) and olfactometry was used to determine and identify volatile compounds in Italian-type dry-cured meat products, but could not describe single-flavor compounds eliciting the typical salami or Parma ham flavor. There have also been several studies using different methods investigating volatile compounds in raw and cooked chicken meat and during spoilage (Senter et al. 2000), or monitoring the stage of ripening from volatile components in salami (Franke et al. 2005). Particularly promising is the analysis of volatile compounds for the determination of the geographic origin in the case of processed meat. It has to be emphasized that, in this case, these compounds code for the site where the processing is done and not for the origin of the raw meat, as these sites are not necessarily identical. Processing would add flavors, e.g., from bacteria, smoke or air, which do not only characterize the specific product, but also may be specific for the geographic origin. This approach, therefore, seems unsuitable for unprocessed meat where the concentration of volatile compounds is very low (Franke et al. 2005).

4. Electronic Nose (E-Nose), Bioelectronic Nose

Since the first applications of solid state gas sensors in arrays, some thirty three years ago, "electronic noses" (e-noses) have undergone a great deal of development. Around 4800 articles on this subject have been published on all aspects and among them about 850 specifically for food testing.

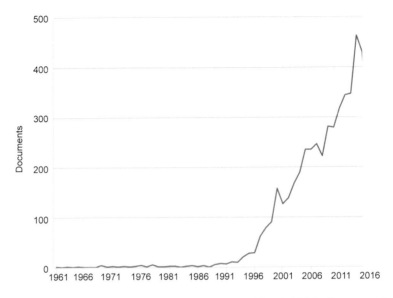

'Electronic' AND 'nose' in Scopus database   Total 4838 documents

with 'food' term added
842 documents

database prospecting
done on Sept 8, 2016

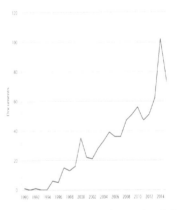

Figure 2. Bibliometrics about e-noses (Scopus database query done on September 8, 2016). Future will tell us whether bioelectronic nose will surpass sensor array technology.

The name "electronic nose" comes from a certain parallel of the measurement concept of the instrument and that of the mammalian olfactory system (Ampuero and Bosset 2003), even if some authors reject such a name, and prefer sensor array technology. Electronic noses base the analysis of the cross-reactivity of an array of semi-selective sensors. Hence, products with similar aroma generally result in similar sensor response patterns (similar "fingerprints") whereas products with different aroma show differences in their patterns (different "fingerprints"). The interaction of volatiles with the array of sensors provokes a series of signals which are then processed by the computer via a pattern recognition program. Just like the human olfactory system, electronic noses do not need to be specially designed to detect a particular volatile. In fact, they can learn new patterns and associate them with new odors via training and data storage functions as humans do (Ampuero and Bosset 2003).

However, the electronic nose has many limitations to overcome (low sensitivity, low selectivity) and appear now to be something from the past. Recently, the bioelectronic nose (Lim and Park 2014), using biological components, has been developed. The bioelectronic nose seems to have a bright prospect as a powerful and effective biosensing system, capable of detecting and discriminating a huge variety of odorant molecules (Lim et al. 2013). The most meaningful characteristics of the bioelectronic nose are that it mimics the human olfactory system. The bioelectronic nose is expected to replace the sensory evaluation method. It can be used for standardization of smell, development of code for each smell, and visualization of smell (Ko et al. 2014). Consequently, the development of the bioelectronic nose is expected to open up many new possibilities to improve the procedures for assessment of authenticity and traceability of food, when problems such as cost and limited service life will be solved.

5. Conclusion and Future Perspectives

The past two decades have seen the advent of high throughput technologies in biology, making it possible to sequence genomes cheaply and quickly, to measure gene expression for thousands of genes in parallel, and to test large numbers of potential regulatory interactions between genes in a single experiment. The large amounts of data created by these technologies have given rise to entire new research areas in biology, such as computational biology and systems biology. The latter, which attempts to understand biological processes at a 'systems' level, is particularly indicative of the potential advantage that large datasets and their analysis can offer to biology (Ahnert 2013), and to other fields of research such as food chemistry. Thanks to big data in the cloud, making big data technology more accessible regardless of expertise or budget, we can expect to see more application of data analytics throughout the food industry in the future (Thusoo 2014). Based on thousands of food flavor tests done every day, millions and billions food flavor investigations dataset, it will increase our capacity to determine food traceability and authenticity based on volatile compound analysis.

Abbreviations

CAS : CAS Registry Number, also referred to as CASRN or CAS Number, is a unique numerical identifier assigned by the Chemical Abstracts Service (CAS) to every chemical substance described in the open scientific literature.

EC : European Community, European Community number for chemicals within EU regulatory schemes.

FEMA : Flavor and Extract Manufacturers Association. Expert Panel of FEMA is the primary body for the safety evaluation of food flavoring for the flavor industry and the public through its assessment of flavoring substances.

GRAS : "GRAS" is an acronym for the phrase Generally Recognized As Safe. Under sections 201(s) and 409 of the USA Federal Food, Drug, and Cosmetic Act (the Act), any substance that is intentionally added to food is a food additive, that is subject to premarket review and approval by FDA, unless the substance is generally recognized, among qualified experts, as having been adequately shown to be safe under the conditions of its intended use, or unless the use of the substance is otherwise excepted from the definition of a food additive.

HS : Headspace. The air or empty space left above the contents in a sealed container, where flavor compounds could be extracted.

LDA : Linear discriminant analysis (LDA) is a generalization of Fisher's linear discriminant, a method used in statistics to find a linear combination of features that characterizes or separates two or more classes of objects or events.

MS : Mass spectrometry. An instrumental method for identifying the chemical constitution of a substance by means of the separation of gaseous ions according to their differing mass and charge—also called mass spectroscopy.

PCA : Principal Component Analysis. Principal component analysis (PCA) is a statistical procedure that uses an orthogonal transformation to convert a set of observations of possible correlated variables into a set of values of linear uncorrelated variables called principal components.

PDO : Protected Designation of Origin. Among the three European Union schemes of geographical indications and traditional specialties, known as protected designation of origin (PDO), protected geographical indication (PGI), and traditional specialties guaranteed (TSG) that promote and protect names of quality agricultural products and foodstuffs.

SPME   : Solid Phase Micro Extraction. Solid-phase micro extraction, or SPME, is a solid phase extraction sampling technique that involves the use of a fiber coated with an extracting phase, that can be a liquid (polymer) or a solid (sorbent), which extracts different kinds of analytes (including both volatile and non-volatile) from different kinds of media, that can be in liquid or gas phase. The quantity of analyte extracted by the fiber is proportional to its concentration in the sample as long as equilibrium is reached or, in case of short time pre-equilibrium, with the help of convection or agitation.

TOF   : Time-Of-Flight. Time-of-flight mass spectrometry (TOFMS) is a method of mass spectrometry in which an ion's mass-to-charge ratio is determined via a time measurement. Ions are accelerated by an electric field of known strength. http://www.kore.co.uk/tutorial.htm

Keywords: Authenticity, food traceability, volatile compound, gas chromatography, mass spectrometry

References

Ahnert, S.E. 2013. Network analysis and data mining in food science: the emergence of computational gastronomy. Flavour 2: 4.

Ampuero, S. and Bosset, J.O. 2003. The electronic nose applied to dairy products: a review. Sensors and Actuators B: Chemical 94(1): 1–12.

Aparicio, R., Morales, M.T. and Alonso, V. 1997. Authentication of European virgin olive oils by their chemical compounds, sensory attributes, and consumers' attitudes. Journal of Agricultural and Food Chemistry 45(4): 1076–1083.

Bauer-Christoph, C., Wachter, H., Christoph, N., Roßmann, A. and Adam, L. 1997. Assignment of raw material and authentication of spirits by gas chromatography, hydrogen- and carbon-isotope ratio measurements I. Analytical methods and results of a study of commercial products. Zeitschriftfür Lebensmittel-Untersuchung und -Forschung A 204(6): 445–452.

Berna, A.Z., Trowell, S., Clifford, D., Cynkar, W. and Cozzolino, D. 2009. The geographical origin of Sauvignon Blanc wines predicted by mass spectrometry and metal oxide based electronic nose. Analytica Chimica Acta 648(2): 146–152.

Bianchi, F., Careri, M. and Musci, M. 2005. Volatile norisoprenoids as markers of the botanical origin of Sardinian strawberry-tree (*Arbutus unedo* L.) honey: Characterisation of aroma compounds by dynamic headspace extraction and gas chromatography–mass spectrometry. Food Chemistry 89: 527–532.

Cerezo, A.B., Espartero, J.L., Winterhalter, P., Garcia-Parrilla, M.C. and Troncoso, A.M. 2009. (+)-Dihydrorobinetin: a marker of vinegar aging in acacia (*Robinia pseudoacacia*) wood. Journal of Agricultural and Food Chemistry 57(20): 9551–9554.

Cerrato Oliveros, C., Boggia, R., Casale, M., Armanino, C. and Forina, M. 2005. Optimization of a new headspace mass spectrometry instrument. Discrimination of different geographical origin olive oils. Journal of Chromatography A 1076(1-2): 7–15.

Cleland, G., Ladak, A., Lai, S. and Burgess, J. 2014. The use of HRMS and statistical analysis in the investigation of Basmati rice authenticity and potential food fraud. Waters™ application note 10 pages.

Cozzolino, D., Smyth, H.E., Cynkar, W., Dambergs, R.G. and Gishen, M. 2005. Usefulness of Chemometrics and mass spectrometry-based electronic nose to classify Australian white wines by their varietal origin. Talanta 68(2): 382–387.

European community. 1974. Council Directive 74/409/EEC on 22 July 1974 on the harmonization of the laws of the Member States relating to honey.

European community. 1997. Report on the Commission proposal concerning the simplification of the vertical directives on food. Proposals for Council directives relating to: (i) certain sugars intended for human consumption (COM(95)0722 - C4-0402/96 - 96/0113 (CNS), (ii) honey (COM(95)0722 - C4-0403/96 - 96/0114 (CNS)), (iii) fruit juices and certain similar products intended for human consumption (COM(95)0722 - C4-0404/96-96/0115 (CNS)), (iv) certain partly or wholly dehydrated preserved milk for human consumption (COM(95)0722 - C4-0405/96 - 96/0116(CNS)), (v) fruit jams, jellies and marmalades and chestnut purée intended for human consumption (COM(95)0722 - C4-0406/96 - 96/0118(CNS)). Committee on the Environment, Public Health and Consumer Protection.

Franke, B.M., Gremaud, G., Hadorn, R. and Kreuzer, M. 2005. Geographic origin of meat-elements of an analytical approach to its authentication. European Food Research and Technology 221(3-4): 493–503.

Innocente, N., Munari, M. and Biasutti, M. 2013. Characterization by solid-phase microextraction-gas chromatography of the volatile profile of protected designation of origin Montasio cheese during ripening. Journal of Dairy Science 96(1): 26–32.

Janin, M., Medini, S. and Técher, I. 2014. Methods for PDO olive oils traceability: state of art and discussion about the possible contribution of strontium isotopic tool. European Food Research and Technology 239(5): 745–754.

Jeleń, H.H., Ziółkowska, A. and Kaczmarek, A. 2010. Identification of the botanical origin of raw spirits produced from rye, potato, and corn based on volatile compounds analysis using a SPME-MS method. Journal of Agricultural and Food Chemistry 58(24): 12585–12591.

Ko, H.J., Lim, J.H., Oh, E.H. and Park, T.H. 2014. Applications and perspectives of bioelectronic nose. pp. 263–283. *In*: Bioelectronic Nose: Integration of Biotechnology and Nanotechnology. ISBN 9789401786133.

Lim, J.H., Park, J., Ahn, J.H., Jin, H.J., Hong, S. and Park, T.H. 2013. A peptide receptor-based bioelectronic nose for the real-time determination of seafood quality. Biosensors and Bioelectronics 39: 244–249.

Lim, J.H. and Park, T.H. 2014. Concept of bioelectronic nose. pp. 1–22. *In*: Bioelectronic Nose: Integration of Biotechnology and Nanotechnology. ISBN 9789401786133.

Mannaş, D. and Altuğ, T. 2007. SPME/GC/MS and sensory flavor profile analysis for estimation of authenticity of thyme honey. International Journal of Food Science and Technology 42: 133–138.

Masters, K. 2015. How technology can weed out the fraudsters. The World of Food Ingredients 4: 55–56.

Pillonel, L., Ampuero, S., Tabacchi, R. and Bosset, J.O. 2003. Analytical methods for the determination of the geographic origin of Emmental cheese: Volatile compounds by GC/MS-FID and electronic nose. European Food Research and Technology 216(2): 179–183.

Risticevic, S. 2008. HS-SPME-GC-TOFMS methodology for verification of geographical origin and authenticity attributes of coffee samples. Master of Science in Chemistry thesis, 101 pages, Waterloo, Ontario, Canada.

Risticevic, S., Carasek, E. and Pawliszyn, J. 2008. Headspace solid-phase microextraction-gas chromatographic-time-of-flight mass spectrometric methodology for geographical origin verification of coffee. Analytica Chimica Acta 617: 72–84.

Seisonen, S., Kivima, E. and Vene, K. 2015. Characterization of the aroma profiles of different honeys and corresponding flowers using solid-phase microextraction and gas chromatography-mass spectrometry/olfactometry. Food Chemistry 169: 34–40.

Senter, S.D., Arnold, J.W. and Chew, V. 2000. APC values and volatile compounds formed in commercially processed, raw chicken parts during storage at 4 and 13°C and under simulated temperature abuse conditions. Journal of the Science of Food and Agriculture 80(10): 1559–1564.

Senyuva, H.Z., Gilbert, J., Silici, S., Charlton, A., Dal, C., Gürel, N. and Cimen, D. 2009. Profiling Turkish honeys to determine authenticity using physical and chemical characteristics. Journal of Agricultural and Food Chemistry 57(9): 3911–3919.

Soria, A.C., Martínez-Castro, I. and Sanz, J. 2008. Some aspects of dynamic headspace analysis of volatile components in honey. Food Research International 41: 838–848.

Thusoo, A. 2014. How big data is revolutionizing the food industry. Wired https://www.wired.com/insights/2014/02/big-data-revolutionizing-food-industry/.

Yang, N., Liu, C., Liu, X., Kreuzfeldt Degn, T., Munchow, M. and Fisk, I. 2016. Determination of volatile marker compounds of common coffee roast defects. Food Chemistry 211: 206–214.

Zambonin, C.G., Balest, L., De Benedetto, G.E. and Palmisano, F. 2005. Solid-phase microextraction-gas chromatography mass spectrometry and multivariate analysis for the characterization of roasted coffees. Talanta 66: 261–265.

Ziino, M., Condurso, C., Romeo, V., Giuffrida, D. and Verzera, A. 2005. Characterization of "ProvoladeiNebrodi", a typical Sicilian cheese, by volatiles analysis using SPME-GC/MS. International Dairy Journal 15(6-9): 585–593.

13

Analytical Tools in Authenticity and Traceability of Olive Oil

Noelia Tena,[1,]* *Ramón Aparicio-Ruiz,*[1] *Anastasios Koidis*[2] and *Diego L. García-González*[1]

1. Introduction

The criteria that define the genuineness of a food product depend on the commodity but the following definition fits to a general viewpoint of what it is in terms of authenticity: "A product is authentic as long as it is firstly described accurately by the label and secondly complies with the current legislation in force in the country where it is marketed or sold" (Lees 1998). According to this definition the geographical traceability (provenance) is part of food authenticity though consumers and mass media link it with adulteration almost exclusively and also with quality issues. Food adulteration, however, has been practiced since ancient times though it has become more technologically evolved in the recent past due to the higher efficiency of control methods. This is the case of virgin olive oil, which is a preferred target for fraudsters because of its high price and its recently gained popularity as a healthy and delectable oil. The illicit activities of the fraudsters result in considerable monetary losses worldwide and erode the confidence of consumers who buy olive oil for its health benefits (García-

---
[1] Instituto de la Grasa (CSIC), Ctra. de Utrera, km. 1, Campus Universitario Pablo de Olavide building 46, 41013 - Sevilla, Spain.
[2] Institute for Global Food Security, Queen's University Belfast, David Keir Building, BT95BN, Belfast, United Kingdom.
* Corresponding author

González et al. 2009a) and are surprised to receive an oil that does not fulfil their expectation.

The high number of possible adulterants gives an idea of all the fraud possibilities. These adulterants are seen as opportunities by fraudster and they are usually other oils such as refined olive oil, raw olive pomace oil, synthetic olive oil-glycerol mixtures, and other oils obtained from avocado, corn, cottonseed, hazelnut, rapeseed, soybean, and sunflower among others. The adulterations with many of these oils are, however, well controlled thanks to methods that have been perfected over years and new authenticity issues are coming into play. Over the last years, the new agricultural practices and consumer sensitivities have been gaining importance in the market and have contributed to the redefining of olive oil authenticity. Thus, the agricultural practices are changing, slowly but inexorably, from the classical orchards to the super intensive orchards and they are moving the production from the Mediterranean Basin to regions as far as New Zealand and Argentina. This slow evolution is leading to a loss of biodiversity (only a few varieties are adapted to super intensive agricultural practices) and the reduction in the price of the final product. The second great factor that is contributing to a new concept of authenticity is that the consumers are far better informed today, which is resulting in a demand for virgin olive oils with higher quality. In consequence, the classical problems focused on the adulteration with cheap oils are now compounded with aspects related to geographical provenance and quality, both being parts of the olive oil authenticity and integrity.

In the context of a more complex concept of authenticity, exporting and importing countries are compelled to address some improvements in the regulation to obtain tighter controls, clearer definitions for olive oil products and uniform labelling guidelines and directives. In addition to these challenges, the control labs require the development of more rapid and accurate instrumental techniques and methodologies. Implementing the latter, which is decisive for the success of the previous ones, is not an easy matter. In fact, analytical advances have led to great success in the fight against adulteration. Nevertheless, the new techniques and knowledge are also used by fraudsters to invalidate the usefulness of standard methods. Since the perceived risk of detection is one of the considerations that fraudsters keep in mind when committing fraud, novel analytical methodologies are pivotal to uncover new olive oil frauds that are designed for passing over the traditional controls. Such competition between fraudster and control methods requires a considerable investment in perfecting techniques or developing new ones as well as a rapid pace of R&D to obtain methods for the detection of malpractices (Aparicio et al. 2007).

No rapid and universal method exists that can be officially recognized for being applied to all the authenticity issues. The authenticity is a multifaceted concept that covers many aspects including adulteration, characterization, mislabelling, and misleading origin among many others. Consequently, a single and easy approach cannot be applied to assuring all these aspects at the same time and each authenticity issue needs fit-for-purpose strategies. With this plethora of possible issues, the ample number and characteristics of olive oil categories (IOC 2016), and the numerous edible oils that can potentially be used in spiking as already mentioned, it is not surprising that numerous methods have been used or suggested to protect olive oil authenticity during the past ten years. Many methods have been proposed, evaluated and adopted in the trade standards and regulations of international institutions (e.g., International Olive Council) that are actively involved in anti-fraud measures. The standards, which are periodically revised and upgraded, contain those provisions needed to ensure fair trade, to prevent fraud and to avoid any image of a hypothetical uncontrolled distribution of adulterated olive oil into the market.

This chapter analyzes the different analytical methods used to verify authenticity, which would be described and their strength and weaknesses discussed. The current chapter describes a set of analytical methods that are able to detect classical adulterations at very low percentages, which make those adulterations non-profitable for the fraudsters. Furthermore, fraudsters have developed new adulterations, as the case of the addition of soft-deodorized virgin olive oil to extra virgin olive oils, that are more difficult to be detected and new strategies need to be developed.

2. The Authenticity of Olive Oil

When the definition of a foodstuff is fairly complicate and exhaustive, which is the case of olive oil and its numerous categories (IOC 2016), the legal detection of adulteration is much more difficult due to the wide variety of chemical characteristics of the different categories and the numerous possibilities of sophisticated fraudulent practices (Aparicio et al. 2013). On the other hand, the strict definition of each one of the olive categories requires an armoury of analytical techniques that have to be capable of distinguishing adulteration by picking up minor differences on some chemical compounds present, even at trace levels. Thus, olive oil authenticity is faced with the problem of numerous kinds of adulterations and the need of sensitive analytical methods that can detect them (Fig. 1).

The methods for olive oil analysis are internationally reviewed by means of expert panels though some differences do exist between the standard methods proposed by different international regulations. Furthermore, there are some in-house proposals that are put in practice in some laboratories. The development of standard methods normally results from an industrial or commercial need to implement a particular control and through the validation of the method by international collaborative studies. This

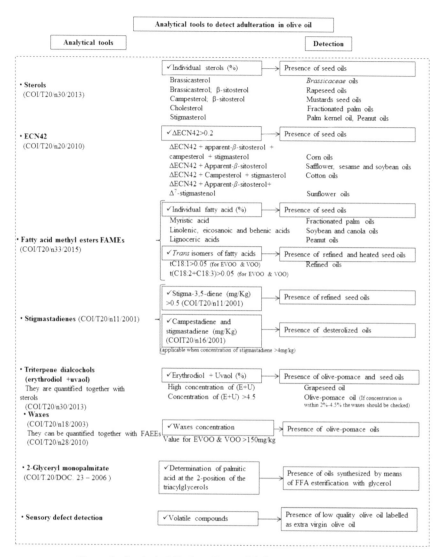

Figure 1. Analytical Tools to Detect Adulterations in Olive Oil.
Note: The COI/T20 documents can be downloaded from the website: http://www.internationaloliveoil.org/estaticos/view/224-testing-methods. EVOO, extra virgin olive oil; VOO, virgin olive oil.

latter requirement is strictly necessary for international bodies to adopt standard methods. That is the case of International Olive Council (IOC), the American Oils Chemists' Society (AOCS), the International Union of Pure and Applied Chemistry (IUPAC), and International Organization for Standardization (ISO).

Olive oil authenticity is commonly addressed through an international approach, which is a direct consequence of the worldwide trade of virgin olive oil that requires an agreement between all stakeholders involved in commercialization and control. Olive oil production and international trading have been traditionally ruled by three different normative and legal sources. EU regulations are in force for EU countries since 1966 (Regulation 136/66), while the standards of IOC—since its foundation in 1956—and *Codex Alimentarius*—since 1981 (Codex STAN 33-1981)—are agreements that signatory countries voluntarily accept to comply with the limits established for each quality and purity criterion, including the precision values.

From an analytical viewpoint, the current methods basically correspond to IOC trade standards for olive oil and olive-pomace oil, ISO methods, and some alternative methods fostered by other associations.

2.1 Analytical methods in authentication

Different analytical methods have been approved by the regulatory bodies to detect the presence of adulterant edible oils in virgin and refined olive oils over the years. Since a standard method requires to be validated prior to a definitive approval, several years are usually spent till the method is endorsed as a standard method. The validation step should be carried out under the strictest conditions, with a right application of the statistical procedures and an adequate selection of representative validation samples to avoid the approval of methods that may lack of a causal relationship with authenticity or quality (i.e., diacylglycerols to determine the presence of soft-deodorized virgin olive oils, or fatty acid alkyl esters to assess sensory assessment). Fortunately, most of these methods met these requirements during validation and the panoply of methods described in IOC trade standards are good enough to detect the presence of undesirable edible oils in olive oils even at low percentages (Table 1).

The chemical compounds, whose contents point out differences between genuine and adulterated olive oils with regard to their categories, are traditionally clustered into major and minor compounds. The former compounds are mostly responsible for the olive oil's general characteristics while the latter are markers for their peculiarities (characteristics associated with cultivar, quality, geographical provenance, etc.). A critical analysis of the information provided by each chemical compound means to analyze this information from three successive perspectives: the scientific reason for using it in olive oil authentication, the pros and cons of the standard analytical method by which it is determined and quantified, and the potential alternatives to standard methods.

Table 1. Chemical Parameters that are Useful in Olive Oil Authenticity and Approximate Detection Limits for Some Adulterant Oils.

Chemical parameters	Standard method	Informing variables	Adulterant oil	Approx. % detection
Sterols	COI/T.20/Doc. No. 30	Brassicasterol Apparent β-sitosterol	Rapeseed oil	1.0–1.5%
		Apparent β-sitosterol Campesterol	Mustard seed oil	1.0–1.5%
Sterols + ΔECN42	COI/T.20/Doc. No. 30 COI/T.20/Doc. No. 20	Apparent β-sitosterol ΔECN42 Campesterol Stigmasterol	Corn oil	1.0–1.5%
		Apparent β-sitosterol ΔECN42	Safflower oil Sesame oil Soybean oil	1.0–4.0%
		Campesterol Stigmasterol ΔECN42	Cotton oil	3.5–5.0%
		Apparent β-sitosterol ΔECN42 Δ⁷-stigmastenol	Sunflower oil	0.5–2.5%
FAMES + Sterols	COI/T.20/Doc. No. 33 COI/T.20/Doc. No. 30	Behenic acid Stigmasterol	Peanut oil	3.0–5.0%
		Miristic acid Stigmasterol	Palm oil Palm kernel oil	3.5–10.0%

Note: The COI/T20 documents can be downloaded from the website: http://www.internationaloliveoil.org/estaticos/view/224-testing-methods.

2.1.1 Detection of edible oils in olive oil

The composition of the major compounds provides basic information about the genuineness. The major compounds of any edible oil are usually the fatty acids. A small amount of them can be present as free fatty acids and they can form esters with glycerol to produce glycerides and phospholipids. They can also form esters with aliphatic alcohols of linear structure (waxes) or terpenic structure (terpene and sterol esters). The molecular structure of each individual trihydric alcohols esterified with three fatty acids (triacylglycerol, TAG) is another source of information on genuineness. In fact, differences between the experimental composition of TAGs and their theoretical composition determined from fatty acids according to their 1, 3-random 2-random distribution was a milestone in 1990s (Cortesi et al. 1990) because the difference between both values should be zero for an authentic olive oil sample. However, the analytical error—for example, only some theoretical TAGs show good mathematical significance (García-Pulido and Aparicio 1993)—directed the attention to the difference between the real and theoretical ECN42 (equivalent carbon number 42) triacylglyceride content (ΔECN42).

The composition of the minor compounds provides information about the genuineness of edible oils that major compounds cannot overall do after the genetic improvement of seeds for producing high oleic oils. The most important of those minor compounds are the sterols, which comprise a major portion of the unsaponifiable matter.

All these chemical compounds, which are quantified by chromatographic techniques, are detectable at minimum addition of less than 10% of other edible oils (Table 1). These techniques are based on target analysis and they are associated usually to particular adulterants, which can make the detection of unexpected adulterations difficult as it happened in the past regarding the addition of refined hazelnut oil to refined olive oil or the addition of soft-deodorized virgin olive oil to extra virgin olive oil, the latter case being still non-detectable at low percentages.

2.1.1.1 Major compounds in olive oil authenticity

Fatty acid composition has been extensively used in authenticity since the first chromatographic methods were proposed for authenticity control. Thus, a high percentage of myristic acid in labelled olive oils may indicate the presence of seed oils, mainly fractionated palm oil, while high percentages of linolenic, eicosanoic, eicosenoic and behenic acids, are associated with the presence of soybean and low erucic rapeseed oils; and a high content in lignoceric acid is associated with peanut oils.

From an analytical viewpoint, the standard method based on gas-chromatography for the quantification of fatty acids is widely accepted by analysts though some attention should be paid to the methylation step (Table 2, Part I). The methylation should be acid in case of oils with high concentration of free fatty acids. There is, however, an alternative to the GC technique based on NMR (nuclear magnetic resonance) though it is not a standard method (Table 3).

The best NMR methodology for the quantification of fatty acids is based on ^{1}H-NMR despite signal overlaps and mathematical algorithms being needed for determining the appropriate signal intensities of saturated, monounsaturated, and polyunsaturated fatty acids. The individual determination of fatty acids cannot be carried out like GC-methods with the exception of linolenic acid whose methyl protons are detected at δ 0.96. The isomers *cis* and *trans* have also been determined by using the allylic methylene protons adjacent to *cis* and *trans* double bonds (Sedman et al. 2010). The alternative to proton NMR is ^{13}C-NMR, which is suggested when information about *trans* fatty acids is required because there is enough distance (5 ppm) between signals of the allylic carbons of *cis* and *trans* double bonds (Gao et al. 2009). NMR instruments, however, require a large

Table 2. The Most Common Methods for Quantifying Major and Minor Components in Olive Oil (Part I).

Compounds	Technique	Sample pre-treatment	Method details
Fatty acids *trans* fatty acids	GC–FID	Purification with silica gel extraction cartridge. Cold methylation with KOH.[a]	Column: Capillary 50–100 m × 0.25–0.32 mm × 0.1–0.3 μm Stationary phase: Cross-linked cyanopropylsilicone Carrier gas: Hydrogen or helium Temperature program: 165°C (8 min) to 210°C at 2°C/min Injection mode: Split
Triacyglycerols	HPLC–RI	0.12 g oil in 0.5 mL hexane is charged into column (SPE: 1 g of Si) and solution pulled through then, eluted with 10 mL hexane-diethylether (87:13v/v). The purified oil is dissolved in acetone at 5%.[b]	Oven temperature: 25°C Mobile phase: acetone/acetonitrile (1:1 v/v) or propionitrile (flow-rate 0.6 to 1.0 mL/min) Column: RP-18 (25 cm × 4 mm i.d.) (5 μm) with 22%–23% carbon in form of octadecylsilane Detector: RI
	HPLC–RI	Purification by SPE (IUPAC method 2507).[c] The purified oil is dissolved in acetone at 5%.[d]	Oven temperature: 20°C Mobile phase: propionitrile (flow-rate 0.6 mL/min) Column: RP-18 (25 cm × 4.5 mm i.d.) (4 μm) Detector: RI
Triacylglycerols Diacylglycerols	GC–FID	Requires silylation. Gives information clustered into their carbon numbers.[e]	Capillary column: 3–7 m × 0.25–0.32 mm × 0.10–0.15 μm. C58 and C60 loss by thermal degradation can be avoided by short columns (5 m should be enough) Phase: SE52, SE54 (5% diphenyl dimethyl polyxilosane) Carrier gas: Hydrogen or helium Temperature program: 80°C -1 min- to 220°C at 20°C/min to 340°C -10 min- at 5°C/min Injection mode: On-column

Table 2 Part 1 contd. …

...Table 2 Part I contd.

Compounds	Technique	Sample pre-treatment	Method details
2-glyceryl monopalmitate (%)	GC-FID	GLC or SPE separation after hydrolytic reaction with pancreatic lipase. Silylation should be applied.[f]	Capillary column: 12 m × 0.32 mm × 0.10–0.30 μm Phase: Methylpolisiloxane or 5% phenylmethylpolysiloxane Carrier gas: Hydrogen or helium Temperature program: 60°C -1 min- to 180°C at 15°C/min to 340°C -13 min- at 5°C/min Injection mode: On-column
Waxes	GC-FID	Fractionation by LC on hydrated silica gel column (15 g of silica).[g]	Capillary column: 8-5 m × 0.25–0.32 mm × 0.1–0.3 μm Phase: 5% Phenylmethylpolysiloxane, liquid phase SE52, SE54 Carrier gas: Hydrogen or helium Temperature program: 80°C -1 min- to 240°C at 20°C/min to 325°C -6 min- at 5°C/min to 340°C -10 min- at 20°C/min Injection mode: On-column

Note: GC: Gas Chromatography; FID: Flame Ionization Detector; HPLC: High Performance Liquid Chromatography; RI: Refractive Index; SPE: Solid Phase Extraction; [a] COI/T20/Doc. No. 33; [b] COI/T20/ Doc. No. 20 Rev 3; [c] (IUPAC 1987); [d] COI/T20/Doc. No. 25 Rev 1; [e] COI/T.20/Doc No. 32; [f] COI/T.20/Doc No. 23; [g] COI/T.20/Doc No. 18 Rev 2. The COI/T20 documents can be downloaded from the website: http://www.internationaloliveoil. org/estaticos/view/224-testing-methods.

Table 2. The Most Common Methods for Quantifying Major and Minor Components in Olive Oil (Part II).

Compounds	Technique	Sample pre-treatment	Method details
Stigmastadienes	GC-FID	Fractionation of unsaponifiable matter on LC Si-column.[b]	Capillary column: 25–30 m × 0.25–0.32 mm × 0.15–0.30 µm. Phase: 5% Methylpolysiloxane. Carrier gas: Hydrogen with quality N-50 or helium. Temperature program: gradient (235°C -6 min- to 285°C at 2°C/min). Injection mode: Split (ratio 1:15) or on column
Aliphatic hydrocarbons and sterenes	GC-FID	Fractionation of unsaponifiable matter on LC Si-column impregnated with silver nitrate.[c]	Column: Capillary (25–30 m × 0.25–0.32 mm × 0.15–0.30 µm). Phase: 5% Phenylmethylpolysiloxane. Carrier gas: Hydrogen or helium. Temperature program: 235°C -6 min- to 285°C at 2°C/min. Injection mode: Split (ratio 1:15) or on column
Sterols[a] and triterpene dialcohols (erythrodiol+uvaol)	GC-FID	Fractionation of unsaponifiable matter on TLC or HPLC. Silylation should be applied.[d]	Capillary column: 20–30 m × 0.25–0.32 mm × 0.15–0.30 µm). Phase: 5% Diphenyl-95% dimethylpolysiloxane (SE-52 or SE-54). Carrier gas: Hydrogen or helium. Operation conditions: Isothermal 260 ± 5°C. Injection mode: Split (ratio from 1:50 to 1:100)
Aliphatic alcohols	GC-FID	Fractionation of unsaponifiable matter on TLC or HPLC. Silylation should be applied.[e]	Capillary column: 20–30 m × 0.25–0.32 mm × 0.10–0.30 µm. Phase: SE-52, SE-54. Carrier gas: Hydrogen or helium. Temperature program: 180°C -8 min- to 260°C -15 min- at 5°C/min. Injection mode: Split (ratio from 1:50 to 1:100)

Table 2 Part II contd. ....

...Table 2 Part II contd.

Compounds	Technique	Sample pre-treatment	Method details
Tocopherols	HPLC-FD (Si column)	0.10 g oil in 10 mL n-heptane.[f]	Oven temperature: 20°C Mobile phase: n-Hexane/isopropanol (99:1 v/v) (flow-rate 1.0 mL/min) Detector: Fluorescence λ_{ex}: 295 nm; λ_{em}: 330 nm Detector: Ultraviolet λ_{ex}: 292 nm (not recommended)
Waxes, Fatty Acid Methyl esters, Fatty Acid Ethyl Esters	GC-FID	Fractionation by LC on hydrated silica gel column 15 g of silica.[g] Fractionation by LC on hydrated silica gel column 3 g of silica.[h]	Capillary column: 8–12 m × 0.25–0.32 mm × 0.1–0.3 μm Phase: Liquid phase SE52, SE54 Carrier gas: Hydrogen or helium Temperature program: e.g., 80°C -1 min- to 140°C at 20°C/min to 335°C -20 min- at 5°C/min Injection mode: On-column
Free Fatty alcohols, Free Tocopherols Free Sterols, Free Triterpenic Alcohols, Methyl Sterols, Sterols, Triterpenic Esters	GC-FID	Silylation reaction. Fractionation by LC on silica gel column.[i]	Capillary column: 15 m × 0.32 mm × 0.1 μm Phase: Methylsilicone at 5% diphenyl Carrier gas: Hydrogen Temperature program: 80°C -1 min- to 180°C at 20°C/min to 330°C at 6.5°C/min Injection mode: On-column

Note: GC: Gas Chromatography; FID: Flame Ionization Detector;[a] See (Mariani et al. 2006) for the methodology of free and esterified sterols; TLC: Thin Layer Chromatography; HPLC: High Performance Liquid Chromatography; FD: fluorescence detector;[b] COI/T.20/Doc No 11 Rev 2;[c] COI/T.20/Doc No. 16 Rev 1;[d] COI/T.20/Doc No. 30 Rev 1;[e] COI/T.20/Doc No. 26 Rev 1;[f] ISO 9936;[g] COI/T.20/Doc No. 28 Rev 1;[h] COI/T.20/Doc No. 31;[i] Mariani et al. 1991. The COI/T20 documents can be downloaded from the website: http://www.internationaloliveoil.org/estaticos/view/224-testing-methods.

Table 3. Basic Characteristics of the Methods Proposed for the Current Analytical Challenges of the Olive Oil Authenticity Issues.

Issue	Addition of cheaper oils to olive oils
Objective:	Detection of the presence of crude or refined hazelnut oil in virgin or refined olive oil.
Analyte/Signal:	Free and esterified sterols.
Technique:	Gas chromatography.
Level of applicability:	Particular due to the design of the mathematical algorithm.
Standard method?:	No, but the method has been validated with blind trials by IOC.
Time of analysis[a]:	Sample pre-treatment: 120 min; measurement: 45 min; data analysis: 15 min.
Limit of detection[b]:	< 8%
Advantages:	Easy to carry out. Low cost analysis. Fine repeatability.
Disadvantages:	Time-consuming. Poor reproducibility because of the need of extensively well-trained analysts.
References:	García-González et al. 2007a,b, Mariani et al. 2006, Bowadt and Aparicio 2003.
Objective:	Detection of the presence of any edible oil (crude or refined) in virgin or refined olive oil.
Analyte/Signal:	Selected ^{13}C- &^{1}H-NMR bands of the spectrum.
Technique:	^{13}C-NMR and ^{1}H-NMR spectroscopies.
Level of applicability:	Universal although has been checked with only a few adulterants.
Standard method?:	No, but the adulteration with hazelnut oils have been validated with blind trials.
Time of analysis[a]:	Pre-treatment: No; measurements: 4 h for ^{1}H-NMR and 1.45 h for ^{13}C-NMR; data analysis: 20 min applying procedures of Artificial Neural Networks (ANN).
Limit of detection[b]:	> 10% using bands from ^{13}C-NMR and ^{1}H-NMR for adulterations with hazelnut oils. ~ 15% using bands from ^{13}C-NMR or from ^{1}H-NMR for adulterations with hazelnut oils.
Advantages:	Good repeatability.
Disadvantages:	Time-consuming. Poor reproducibility. False positives. Hyper-optimist models.
References:	Dais and Hatzakis 2013, Dais 2013, Mannina et al. 2009.
Objective:	Detection of the presence of any edible oil (crude or refined) in virgin or refined olive oil.
Analyte/Signal::	Infrared or Raman bands.

Table 3 contd. ...

…Table 3 contd.

Table 3. Basic Characteristics of the Methods Proposed for the Current Analytical Challenges of the Olive Oil Authenticity Issues.

Issue	Addition of cheaper oils to olive oils (contd.)
Technique:	FTIR or FT-Raman.
Level of applicability:	Universal although has been checked with only a few adulterants.
Standard method?:	No, some kinds of adulteration have been validated with blind samples.
Time of analysis[a]:	FTIR: Pre-treatment: 5 min[c]; measurement: 5 min; data analysis: 5 min applying ANN. FT-Raman: Pre-treatment: nil[c]; measurement: 10 min; data analysis: 5 min applying ANN.
Limit of detection[b]:	> 10%
Advantages:	Rapid and easily implementable method.
Disadvantages:	Full checked with hazelnut oils only. A large set of spectra is required. Unstable mathematical equations.
References:	García-González et al. 2013, Baeten et al. 2005.
Issue	**Addition of refined oils to virgin olive oils**
Objective:	Detection of the presence of any refined edible oil in virgin olive oils.
Analyte/Signal:	Hydrocarbons: Stigmastadienes.
Technique:	Gas chromatography.
Level of applicability:	Universal.
Standard method?:	Yes (Table 2).
Time of analysis[a]:	Pre-treatment: 90 min; measurement: 20 min; data analysis: No.
Limit of detection[b]:	2%
Advantages:	Global method. Excellent LOQ (0.1 ppm) & LOD (2%).
Disadvantages:	May fail detecting refined oils obtained under soft conditions (< 175°C).
References:	IOC 2001b, Lanzón et al. 1989.
Objective:	Detection of the presence of any refined edible oil in virgin olive oils.
Analyte/Signal:	*cis/trans* FTIR or FT-Raman bands.
Technique:	Spectroscopy by FTIR or FT-Raman.
Level of applicability:	Universal.
Standard method?:	No, but the method has been validated with blind samples.

Table 3 contd.…

…*Table 3 contd.*

Table 3. Basic Characteristics of the Methods Proposed for the Current Analytical Challenges of the Olive Oil Authenticity Issues.

Issue	Addition of refined oils to virgin olive oils (contd.)
Time of analysis[a]:	Pre-treatment: nil; measurement: 10 min; data analysis: 10 min.
Limit of detection[b]:	> 8%
Advantages:	Rapid method.
Disadvantages:	Limit of detection. The method does not work properly with less unsaturated oils.
References:	García-González et al. 2013.
Issue	**Geographical traceability of VOOs**
Objective:	Determination of the geographical provenance (country, region, county, PDO, PGI) of VOOs.
Analyte/Signal:	Several: fatty acids, alcohols, sterols, hydrocarbons, etc.
Technique:	Gas chromatography for chemical analysis and expert system (SEXIA®) for data analysis[e].
Level of applicability:	Whole Spain and partially the other EU producer countries.
Standard method?:	No, but SEXIA® has been validated with hundreds of samples for years.
Time of analysis[a]:	Pre-treatment: 180 min; measurement: 300 min; Data analysis: 10 min using expert system.
Correct classification[b] (%):	Average certainty factors (CF): 92% for Andalusian PDOs, 95% for Spanish regions, and 96% for the identification of major EU producing countries/varieties among others.
Advantages:	Results are associated to high certainty factors. It is based on the largest VOO database.
Disadvantages:	Time-consuming. Several different chemical analyses. It constantly needs to be updated.
References:	Aparicio and García-González 2013, Aparicio and Alonso 1994.
Objective:	Determination of the geographical provenance of VOOs.
Analyte/Signal:	δ^2H, $\delta^{13}C$ or $\delta^{18}O$.
Technique:	EA-IRMS or NMR
Level of applicability:	Universal.
Standard method?:	No.
Time of analysis[a]:	Pre-treatment: nil; measurement: few minutes; data analysis: 5 min.
Correct classification[b] (%):	Not reported.

Table 3 contd. …

...Table 3 contd.

Table 3. Basic Characteristics of the Methods Proposed for the Current Analytical Challenges of the Olive Oil Authenticity Issues.

Issue	Geographical traceability of VOOs (contd.)
Advantages:	Rapid method.
Disadvantages:	Reproducibility. Need of a previous large database. Harmonization calibration procedure.
References:	Dais and Hatzakis 2013, Dais 2013, Camin et al. 2010, Chiavaro et al. 2011, Alonso-Salces et al. 2010.
Objective:	Determination of the geographical provenance of VOOs.
Analyte/Signal:	Multi-elements.
Technique:	ICP-MS or ICP-AES.
Level of applicability:	Universal.
Standard method?:	No.
Time of analysis[a]:	Pre-treatment: 75–90 min with digestions in microwave; measurement: 3–5 min; data analysis: 15 min using ANN.
Correct classification[b] (%):	Not reported by authors.
Advantages:	Causal relationship between soil and oil. A large number of variables (elements). Repeatability.
Disadvantages:	Low concentration of elements in the oils. Need of information of soils for training the model. Intereference of fertilizers and fungicides[d].
References:	Beltrán et al. 2015, Llorent-Martínez et al. 2011, Benincasa et al. 2007, Zeiner et al. 2005.

Note:
[a] checked by the authors at their labs and in the course of collaborative analyses of European funded projects.
[b] the best figure reached in the course of collaborative analyses with blind samples.
[c] the measurement is carried with the entire oil but if the measurement is of the unsponifiable matter, 60 min has to be added to the total analytical procedure.
[d] foliar fertilizers can contain K, Fe, Mg, Mn, P and Zn in different proportions, together with other elements (i.e., B, Ca), which can be presented complexed with amino acids such in the cases of Ca, Fe, Mg, Mn and Zn. Fungicides can contain Cu among other elements.
[e] other authors have proposed the study of particular geographical production zones by diverse series of compounds, and data are analyzed by a high number of statistical procedures, either unsupervised (e.g., PCA, MDS) or supervised (e.g., LDA, PLS).

initial outlay and their methods need longer time for the analysis than GC methods (Table 3).

The study of other chemical compounds as triacylglycerols (TAG) shows that their biosynthetic pathways in vegetable oils do not allow high concentration of saturated fatty acids to be present at 2-position of the molecule; 1% of such acids esterified at the 2-position of glycerol is enough to conclude that the olive oil is not genuine. The overlaps of individual TAGs in the chromatographic determination led researchers to calculate them as joined by their same equivalent carbon numbers (ECN), which have no distinction between TAGs that are positional isomers. The determination of TAGs by RP-HPLC (Table 2, Part I) is not free of problems (Aparicio et al. 2013). For example, some difficulties in the analysis is due to the formation of complex mixtures of varying molecular structure like "critical pairs" with close retention time when applying this technique. Thus, the focus was on HPLC mobile phase. A mobile phase of acetone/acetonitrile (1:1 v/v) in isocratic elution allows an effective separation of critical pairs though saturated long-chain TAGs are not sufficiently dissolved. The use of propionitrile in isocratic elution (Fiebig 1985) results in an improved separation of the groups of peaks clustered as ECN42, which is the parameter selected for preventing olive oil adulteration.

Furthermore, the content of 2-glyceryl mono-palmitate is of usefulness to determine the presence of strange oils in olive oil with high content of palmitic acid (EC 2007) like re-esterified edible oils by synthetic means or by the addition of animal fat. The method for this determination (Table 2, Part I) can be tedious if the sample has a free acidity upper 3% because it has to be previously neutralized or simply because the separation between phases after lipase pancreatic digestion is not automatic (Tena et al. 2015). Other aspects as the use of dangerous chemicals, tailing peaks in the chromatograms and the low repeatability are inherent aspects to the method.

2.1.1.2 Minor compounds in olive oil authenticity

Sterols are the main series of the unsaponifiable matter. They are not a homogeneous group but three groups with analogous chemical structure that are named 4-demethylsterols or phytosterols, 4, 4-dimethylsterols or triterpenic alcohols, and 4-monomethylsterols or methylsterols. The research has been mainly focused on phytosterols because the ranges of concentration of some of their constituents are characteristic of the genuineness of many edible oils (Table 4); for example, rapeseed oils contain significant concentrations of brassicasterol, safflower and sunflower seed oils contain high concentration of Δ^7-stigmastenol, and olive oils have high concentrations of β-sitosterol and Δ^5-avenasterol and also low concentrations of campesterol and stigmasterol. The phytosterol concentrations are, generally speaking, sum of both possible forms (free and esterified) as their sterol acetates (IOC 2001a) differences between samples do not change if free and esterified sterols are determined independently (Mariani et al. 2006).

These chemical differences in the botanical origin of the oils made decisive to use the sterol composition in olive oil authenticity since 1985.

Table 4. Basic Composition of Sterols Determined in Several Oils, According to Codex Alimentarius (CAC 2009) and (Firestone and Reina 1996). The concentrations are expressed in percentage for individual sterols, and in mg/kg for total sterols.

Sterolic compounds	Olive	Safflower High Linoleic	Safflower High Oleic	Rapeseed Low Erucic	Sunflower High Linoleic	Sunflower High Oleic	Hazelnut	Soybean	Palm Stearin	Corn	Cottonseed	Palm kernel	Groundnut	Sesame
Cholesterol	≤0.5	nd / 0.7	nd / 0.5	nd / 1.3	nd / 0.7	nd / 0.5	0.8	0.2 / 1.4	2.5 / 5.0	0.2 / 0.7	0.7 / 2.9	0.6 / 3.8	nd / 4.2	0.1 / 0.2
Brassicasterol	≤0.1	nd / 0.4	nd / 2.2	nd / 13.0	nd / 0.2	nd / 0.3	<0.15	nd / 0.3	nd	nd / 0.2	nd / 0.9	nd / 0.8	nd / 0.2	0.1 / 0.2
Campesterol	≤4.0	9.2 / 13.3	8.9 / 19.9	24.7 / 38.6	6.5 / 13.0	5.0 / 13.0	3.8 / 5.6	15.8 / 24.2	15.0 / 26.0	17.8 / 24.3	5.8 / 15.2	8.2 / 13.0	12.1 / 20.4	9.8 / 20.7
Stigmasterol	≤Camp.	4.5 / 9.6	2.9 / 8.9	0.2 / 1.0	6.0 / 13.0	4.5 / 13.0	0.8 / 1.4	14.9 / 19.1	9.0 / 15.0	3.9 / 8.2	2.0 / 7.1	12.1 / 16.9	5.1 / 13.2	3.0 / 7.3
β-sitosterol	≥93.0	40.2 / 50.6	40.1 / 66.9	45.1 / 57.9	50 / 70	42.0 / 70.0	87.0 / 94.0	47.0 / 60.0	50.0 / 60.0	54.0 / 67.5	75.8 / 88.2	61.8 / 73.5	46.7 / 69.2	57.1 / 61.8
Δ⁵-avenasterol	--	0.8 / 4.8	0.2 / 8.9	2.5 / 6.6	nd / 6.9	1.5 / 6.9	1.1 / 5.2	1.5 / 3.7	nd / 3.0	4.0 / 8.1	1.8 / 7.4	1.0 / 9.2	8.1 / 19.2	5.8 / 7.8
Δ⁷-stigmastenol	≤0.5	13.7 / 16.4	3.4 / 16.4	nd / 1.3	6.5 / 24.0	6.5 / 24.0	0.9 / 3.7	1.4 / 5.2	nd / 3.0	0.9 / 4.1	nd / 1.4	nd / 2.0	nd / 5.0	2.0 / 8.4
Δ⁷-avenasterol	--	2.2 / 6.3	nd / 8.3	nd / 0.8	3.0 / 7.5	nd / 9.0	0.3 / 1.6	1.0 / 4.6	nd / 3.0	0.7 / 3.2	0.8 / 3.2	nd / 1.4	nd / 6.2	1.0 / 6.3
Others	--	0.5 / 6.4	4.4 / 11.9	nd / 4.2	nd / 5.3	3.5 / 9.5	--	nd / 1.8	nd / 5.0	nd / 2.1	nd / 1.6	nd / 3.2	nd / 1.5	0.7 / 9.0
Total sterols	≥1000	2100 / 4600	2000 / 4100	4500 / 11300	2400 / 5000	1700 / 5200	1000 / 2000	1800 / 4500	250 / 500	7810 / 22300	2600 / 6500	750 / 1470	850 / 2912	4400 / 19200

Note: Camp.: Campesterol; nd: not detected.

Brumley et al. (1985) detected the presence of brassicasterol in olive oil by GC-EIMS.

The first standard method based on sterols for detecting the presence of different crude and refined seed oils in crude and refined olive oil was published by IOC in 2001a (COI/T.20/Doc. No. 10/Rev.1). The method needs lengthy and tedious sample preparation (Table 2, Part II) on silica gel plate followed by the recovery of the band of phytosterols. Trimethylsilyl ethers are obtained after a silylation step, which are not exempted of difficulties and problems like, for example, the need of experienced analysts (Tena et al. 2015) and poor sensitivity and reproducibility for phytosterols present at low concentrations (i.e., Δ^7-stigmastenol).

Analysts, however, have simplified the analysis to avoid the problems of using TLC. Thus, the most reliable and widely used technique is the off-line HPLC-GC coupling that comprises separation of the fractions of sterols or other unsaponifiable components (e.g., branched and terpenic hydrocarbons) by HPLC with a silica gel column, collection of the fraction, elimination of the solvent, further derivation, and injection onto GC (Arjona 2013).

The success in the application of sterol composition to detect fraudulent practices, however, encouraged the fraudsters to remove sterols from the adulterants, using a bleaching step at forced conditions, prior to be added to olive oil. However, an intense refining process produced olefinic degradation products and this fraudulent practice was detected by an anomalous increment in the amount of *trans* unsaturated C_{18}-fatty acids (Grob et al. 1994).

Individual sterols were then expressed as percent of the total content but the illegal process of removing sterols without forming fatty acid *trans*-isomers suggested to give the information in absolute concentration as well. Thus, authentic VOOs cannot have a total amount of sterols less than 1000 mg/kg (IOC 2016). Below this value, it is reasonable to think that olive oil might have been spiked with "desterolized" seed oils. However, with the aid of computer software and compositional databases of edible oils, fraudulent recipes including several edible seed oils for sophisticated adulterations may remain undetectable (e.g., 65% refined olive oil, 15% refined hazelnut oil, 15% desterolized sunflower, and 5% palm olein without free sterols) though it should firstly yield a profit for fraudsters.

More recently, the detection of refined olive oil spiked with refined hazelnut oil at low concentrations was the target of researchers. One of the most successful proposals was based on the quantification of sterols although it is not a standard method yet. A double mathematical model based on three free and esterified sterols (campesterol, Δ^7-stigmastenol and Δ^7-avenasterol) was able to detect the presence of hazelnut oil in olive oil at percentages in the range 6–8% (Mariani et al. 2006, García-González et al. 2007a).

In addition to qualitative differences in the sterol profiles according to the vegetable origin of the edible oil, some researchers have pointed out that the pedoclimate (soil, climate, latitude, altitude, etc.) of the orchards and the cultivars affect the olive oil sterol composition (Aparicio and García-González 2013). With the increase of olive tree growing areas as far from the latitude and hemisphere of the Mediterranean basin as Argentina, Australia and South Africa, the studies on the influence of the pedoclimatic parameters on the sterol composition has gone further than initially was thought, up to the point that it has been observed some authentic olive oils can be qualified as non-genuine (Mailer et al. 2010, Ceci and Carelli 2007). This problem was known in particular cultivars (e.g., Cornicabra, Koroneiki) and in small producing areas (e.g., some olive oils from Syria, Palestine and Eastern Greek islands showed levels of Δ^7-stigmastenol slightly exceeding the legal limit, 0.5%).

A compromise solution to include those anomalous varieties within the limits established in the regulation has been made in the so-called decision trees (IOC 2016), which are applied to those olive oils with values of some sterol that do not fit to the current regulations in force. The objective of the decision trees is to overcome the number of false positives (genuine olive oil qualified as adulterated) with the implementation of equations that combine values of several chemical parameters (Table 5). However, it is required to provide statistical values (i.e., percentages of false positives and false negatives) associated to each branch of the tree in order to determine the possibility or probability with which a theoretically anomalous olive oil is classified as genuine. The absence of these statistical parameters, which should be mandatory in a regulation or trade standard of these characteristics, can lead to the increment in the number of false negatives (a non-genuine classified as genuine).

Table 5. Decision Trees to Protect Genuine Virgin and Extra Virgin Olive Oils from False Positives As Currently Published in the IOC Trade Standard (IOC 2016).

Rule 1: IF 4.0% < Campesterol ≤ 4.5%, THEN Stigmaterol ≤ 1.4% AND Δ^7-Stigmastenol ≤ 0.3%
Rule 2: IF 0.5% < Δ^7-Stigmastenol ≤ 0.8%, THEN Campesterol ≤ 3.3% AND Stigmaterol ≤ 1.4% AND apparent β-Sitosterol ≥ 25 AND ΔECN42 ≤ |0.1|

2.1.2 Detection of refined and desterolized edible oils in virgin olive oils

The main physical difference in producing refined and crude edible oils is the use of temperature. Thus, the research on protecting virgin olive oil from fraudulent mixtures was focused on knowing the chemical compounds that are affected by the action of that source of energy that is forbidden in the production of virgin olive oil (IOC 2016). Today, we know that the sterol dehydration products and *trans* isomers are good markers of the presence of refined edible oils, either deodorized or desterolized, but it was not until the mid-1990s that the importance of stigmastadienes (Lanzón et al. 1989) with a certainty factor higher than 0.99 was discovered. Thousands of oil samples were analysed to verify that refined oils have some chemical compounds that virgin olive oils have not, such as n-alkanes, n-hexacosadiene, stigmasta-3, 5-diene, isomerization products of squalene, isoprenoidal polyolefins produced from hydroxy derivatives of squalene and steroidal hydrocarbons from 24-methylene-cycloarthanol. The discovery was later followed by another study that pointed out that the desterolized process converted Δ^7-sterols into $\Delta^{8(14)}$- and Δ^{14}-sterols (Mariani et al. 1995) and hence allowing the detection of low percentages of a desterolized high oleic sunflower oil in virgin olive oil. The standard method (Table 2, Part II) is not exempt of problems like some overlaps in the chromatogram peaks that can result in false results, which can be avoided with a pre-separation by HPLC as used in the proposal for the determination of sterols.

The other sensitive process used to detect chemical changes resulting from heat treatment of edible oils is the *cis-trans* isomerisation. Thus, the presence of *trans* isomers of oleic, linoleic and linolenic acids in percentages above the approved levels for virgin olive oil (IOC 2016) alerts the analyst about the presence of refined oils that could have been obtained from esterified olive oils or (partially-) hydrogenated seed oils or seed oils desterolized at high temperatures among others.

2.1.3 Detection of olive-pomace oil in olive oil.

The protection of olive oil against adulteration is not only circumscribed to detect the presence of other edible oils but also of olive-pomace oils, an oil obtained from the residual paste left over from the production of VOO (IOC 2016). As this kind of oil is obtained by treating the paste with solvents, its main characteristic is higher concentration of those chemical compounds mostly present in the olive pulp and skin though the concentrations can vary depending on the cultivar. Triterpene dialcohols (percentages of erythrodiol and uvaol) and waxes (ppm of C40–C46) have been traditionally used to determine this kind of adulteration.

Triterpene dialcohols are determined by scraping their TLC band with the band of 4-demethylsterols or phytosterols (IOC 2013b) and subsequent GC analysis with capillary columns (Table 2, Part II). There are several methods for determining waxes though the simultaneous quantification of alkyl esters and waxes (Table 2, Part II) is the most promising as it saves time and allows knowing if the content of waxes C36 and C38 is lower than the sum of C40, C42, C44 and C46 which would mean the presence of olive-pomace oil in olive oil. The combined action of the ratio of waxes and the concentration of triterpene dialcohols seems to work very well in terms of false positives.

2.1.4 Alternatives being studied for detection of low quality olive oil labelled as extra virgin olive oil

Another issue in the authenticity of virgin olive oil is related to the mislabelling of categories (e.g., a virgin olive oil labelled as "extra virgin" while it is actually of "virgin" category). The sensory assessment, the so-called panel test (IOC 2015), is one of the methodologies applied for the category classification of the oils. This method is based on the score given by sensory assessors according to their perception of the oils as a consequence of the chemical compounds present in VOO that are responsible for their VOO flavour and aroma (Aparicio et al. 2012). There is a demand for new tools for assessing the sensory quality by an objective measurement which will also support the standard method. Different analytical methods based on the determination of volatile compounds have been proposed to support the panel tests. The separation and quantification of volatiles is usually carried out by GC, although chemosensors is gaining in popularity (García-González and Aparicio 2010). In general, the analytical techniques require improvements to perfect the chromatographic resolution in order to indentify a higher number of volatiles that may have a significant sensory impact (García-González et al. 2011). A complete information obtained by these techniques would give coherence and unity to the sensory information, thereby avoiding the current confusion that sometimes affect the sensory assessment of VOOs, in particular when the oils are located in the boundaries between categories.

3. The Traceability of Olive Oil

The different characteristics of olive cultivars and production practices around the world coupled with the clear increase in demand for higher quality olive oils led to the emergence of olive oil elaborated with specific composition and quality attributes such as oils from specific well-known regions such as Cretan, Andalusian, and Tuscan olive oils, which are usually linked with specific olive cultivars, such as Koroneiki, if they are monovarietal. Monovarietal olive oils have certain specific characteristics related to the olive variety from which they are elaborated (Montealegre et al. 2010). This led the European Union and other constitutions around the world to legislate the appearance of products with denominations and protected indications of origin on their label. EU Regulation 2081/92 introduced the Protected Designation of Origin (PDO), Protected

Geographical Indication (PGI) and the "Traditional Speciality Guaranteed" (TSG) schemes to promote and protect food products.

A virgin olive oil with PDO registration requires precise definition of several parameters, with the most important to be the cultivar (botanical origin) and the geographical origin followed by agronomic practice, production/extraction, and organoleptic attributes. All of these parameters have to be investigated to certify its origin and quality (Giménez et al. 2010). There is a slight yet important difference between the botanical and the geographical traceability: the botanical traceability is aiming to find the olive (i.e., which cultivar(s)) that was used to produce olive oil whereas the geographical traceability is investigating the geographical location of the olive tree that the oil came from. Often these two terms are considered as the same but there are clear cases that they are not. Independently, both the botanical and the geographical traceability are studied with the selection of markers, i.e., compounds within olive oil with a discriminating power to belong to one group or the other.

Botanical traceability is particularly relevant when monovarietal olive oils and oils with PDO are concerned because these high-quality olive oils may be adulterated by other 'lower quality' olive oils, with anonymous or low added-value cultivars (Breton 2004). It has to be noted however that one olive variety can be cultivated in multiple geographical locations, which makes the distinction of olive varieties a complex problem (Montealegre et al. 2010).

Traditionally, differentiation among olive cultivars has been supported by numerous morphological traits but their subjectivity, inherit difficulty in the evaluation and their reliance of environmental factors have led to the investigation for most robust markers (Sanz-Cortes et al. 2003). One of the most common markers that are employed for this type of traceability is the compositional markers. The composition of the oil includes specific levels of sterols, chlorophyll, carotenoid and pheophytin pigments, hydrocarbons, and fatty acids among others. Many of them may provide basic information on olive cultivars. In particular, minor components can provide more useful information and have been more widely used to differentiate the botanical origin of olive oils (Bianchi et al. 2001, Karabagias et al. 2013, Portarena et al. 2014). Physical parameters, especially the colour of a virgin olive oil as associated with the different pigment levels, have also been employed.

An appropriate traceability system can be designed by means of a large database built with major and minor compounds that allows determining the geographical origin of the most representative European VOOs with fine certainty factors. That was the purpose of the SEXIA project, which allowed characterizing samples with an expert system (Aparicio and Alonso 1994). Thus, the geographical identification of samples can take the advantage of chemical analysis and chemometric procedures (García-González et al. 2009b). The high number of VOO PDOs has raised the interest for a major control of the oils protected under this geographical identification (García-González et al. 2012). Considering these challenges, future work on geographical traceability involves building an olive oil map including information from chromatographic, spectroscopic, and isotopic techniques properly organized by geographical areas and PDOs.

Molecular markers, also known as DNA-based markers, have been the topic of significant research interest over the last 10–15 years especially as these methods are widely available and affordable. The genomic analysis of olive oil involves two main obstacles: extraction of DNA from an olive oil and selection of appropriate molecular markers that can provide a reliable result. Most of the research in the area has focused on tackling these two issues.

Good quality DNA from an oil matrix such as olive oil, even if no chemical and physical refining has taken place, is hard to extract due to the filtration step that eliminates most of the useful plant material that are mechanically transferred from the olive drupe to the oil at the crushing stage. However, several researchers have overcome the problem of low quality/ high degradation grade DNA by introducing wide range of protocols on DNA extraction from olive oil, either cold press or refined. Pioneering role corresponded to Pafundo et al. (2007), Pasqualone et al. (2007), Spaniolas et al. (2008), and more recently to Raieta et al. (2015), who successfully amplified the DNA extracted from olive oil using a variety of techniques involving PCR amplification. To date, however, no consensus among researchers has been reached on the most suitable extraction method to apply for olive oil DNA extraction (Muzzalupo et al. 2015).

All the common classes of molecular markers have been investigated for their suitability for olive cultivar identification and geographical traceability. These include both non species-specific markers such as randomly amplified polymorphic DNA (RAPD), inter-simple sequence repeat (ISSR) and amplified fragment length polymorphism (AFLP) marker (i.e., they do not require prior knowledge of the genome sequence for primer design) and species-specific markers such as microsatellites, DNA barcoding, sequence-characterized amplified region (SCAR), qualitative real-time PCR (qRT-PCR) and single-nucleotide polymorphism. The later techniques selectively amplify the target nucleotide sequences that are present in the species of interest and require the support of a large database and prior knowledge of the same sequence (Rabiei and Enferadi 2012, Scollo et al. 2016). The essential criteria for the selection of molecular markers in terms of reliability were ordered hierarchically as followed: (i) the discrimination power; (ii) the correlation between the fruit DNA (the olive drupe and leaf in this case) and oil DNA profiles; (iii) reproducibility and repeatability of results; (iv) simplicity of analysis (Agrimonti et al. 2011). All these findings helped to build enough data for genomics databases such as the olive genomics information present on the National Centre for Biotechnology Information (NCBI) database.

It is beyond the scope of this section to extensively cite all the work carried out over the years. In summary, RAPDs and AFLPs are found to give complex profiles that can be applicable to monovarietal olive oils. As with many techniques, it is much harder to identify mixtures of more than three cultivars, which is apparently the case in some PDO olive oils (Pafundo et al. 2007). Simple Sequence Repeat (SSR) and Single Nucleotide Polymorphic (SNP) markers are preferred to multilocus markers (AFLP, RAPD) because they are simpler and faster to perform, giving clear results. Some researchers raised concerns about the reliability of the methods altogether. Bracci et al. (2011) claimed that the DNA extracted from olive oil contains fruit pulp tissues of the tree as well as alleles of the seed embryo which may contain exogenous alleles from the pollinator. Thus, DNA purity should be another factor to be considered when interpreting DNA profiles for resolving botanical and geographical traceability. Following the trend of multimethod analysis, Rotondi et al. (2011) demonstrated that the genetic (SSR analysis) component and the selected fatty acids (eicosenoic, linoleic, oleic, stearic, palmitic and linolenic acids) seems to represent a possible tool for inter- and intra-varietal characterisation and for monovarietal traceability.

4. Conclusions

The new challenges in olive oil authenticity can be focused on, firstly, perfecting the current protocol for applying the methods in authentication (EU 2014); secondly, validating the usefulness of the limits established in the trade standard with mathematical parameters (e.g., probability, certainty factor, false positives, etc.); and thirdly, making the effort to diminish the number of analyses without reducing the information of chemical compounds.

Relevant problems today are the authenticity of olive oils added to canned fish or inside bottles labelled as spiced VOOs, the presence of deodorized oil in VOO, or even the geographical origin of olive oil and the resulting overlap between quality and authenticity. Regarding the traceability of olive oil, the DNA analysis represents an attractive and viable alternative to the more classical chemical methods that focus on macromolecules and metabolite as targets, due to DNA's independence of environmental and processing conditions. More work however needs to be carried out before this technique is mature to a standard method for regulatory purposes to be adapted as a reliable method for provenance and authenticity of olive oil.

One of the requirements of an efficient authenticity method is to be based on markers that have a causal relationship with the authenticity issue. For that reason, it is important to avoid markers selected only because they show a mere strong mathematical relationship with the authenticity issue without any chemical interpretation. The inappropriate application of statistical procedures sometimes leads to overoptimistic results that later fails in the validation step or in the daily application of the method. Thus, this aspect should be considered when evaluating new proposals of methods to prevent a casual relationship between chemical compounds and authenticity. The paradigm is the implication of chemical compounds that, being related to fermentation processes, and not having sensory impact, are used to explain the presence of sensory defects because of a simple correlation. Authenticity issues that are complex to solve, such as geographical identification, can be also affected by proposals of markers with non-causal relationship that are not directly associated to the genuineness of the oils.

Keywords: Olive oil, authenticity, traceability, adulteration, standard methods, classification, extra virgin, detection

References

Agrimonti, C., Vietina, M., Pafundo, S. and Marmiroli, N. 2011. The use of food genomics to ensure the traceability of olive oil. Trends in Food Science & Technology 22: 237–244.

Alonso-Salces, R.M., Moreno-Rojas, J.M., Holland, M.V., Reniero, F., Guillou, C. and Heberger, K. 2010. Virgin olive oil authentication by multivariate analyses of ^{1}H-NMR fingerprints and δ^{13}C and δ^2H data. Journal of Agriculture and Food Chemistry 58: 5586–5596.

Aparicio, R. and Alonso, V. 1994. Characterization of virgin olive oils by SEXIA expert system. Progress in Lipid Research 33: 29–38.

Aparicio, R., Aparicio-Ruiz, R. and García-González, D.L. 2007. Rapid methods for testing authenticity: the case of olive oil. pp. 163–188. *In*: van Amerongen, A., Barug, D.M. and Lauwars, M. (eds.). Rapid Methods for Food and Feed Quality Determination. Wageningen Academic, Wageningen.

Aparicio, R., Morales, M.T., Aparicio-Ruiz, R., Tena, N. and García-González, D.L. 2013. Authenticity of olive oil: Mapping and comparing official methods and promising Alternatives. Food Research International 54: 2025–2038.

Aparicio, R., Morales, M.T. and García-González, D.L. 2012. Towards new analyses of aroma and volatiles to understand sensory perception of olive oil. European Journal of Lipid Science and Technology 114: 1114–1125.

Aparicio, R. and García-González, D.L. 2013. Olive oil characterization and traceability. pp. 431–478. *In*: Aparicio, R. and Harwood, J. (eds.). Handbook of Olive Oil: Analysis and Properties. Springer Science, New York.

Arjona, J. 2013. Oral communication: Characterization of monovarietal virgin olive oils: Implications of the geographical origin (In Spanish). Master Thesis. University of Sevilla.

Baeten, V., Fernández-Piernas, J.A., Dardenne, P., Meurens, M., García-González, D.L. and Aparicio-Ruiz, R. 2005. Detection of the presence of hazelnut oil in olive oil by FT-Raman and FT-MIR spectroscopy. Journal of Agriculture and Food Chemistry 53: 6201–6206.

Beltrán, M., Estudillo, M., Aparicio, R. and García-González, D.L. 2015. Geographical traceability of virgin olive oil from South Western Spain by their multi-elemental composition. Food Chemistry 165: 350–357.

Benincasa, C., Lewis, J., Perri, E., Sindona, G. and Tagarelli, A. 2007. Determination of trace element in Italian virgin olive oils and their characterization according to geographical origin by statistical analysis. Analytica Chimica Acta 585: 366–370.

Bianchi, G., Giansante, L., Shaw, A. and Kell, D.B. 2001. Chemometric criteria for the characterisation of Italian protected denomination of origin (DOP) olive oils from their metabolic profiles. European Journal of Lipid Science and Technology 103: 141–150.

Bowadt, S. and Aparicio, R. 2003. The detection of the adulteration of olive oil with hazelnut oil: A challenge for the chemists. Informatica 14: 342–344.

Breton, C., Claux, D., Metton, I., Skorski, G. and Bervillé, A. 2004. Comparative study of methods for DNA preparation from olive oil samples to identify cultivar SSR alleles in commercial oil samples: possible forensic applications. Journal of Agricultural and Food Chemistry 52(3): 531–537.

Bracci, T., Busconi, M., Fogher, C. and Sebastiani, L. 2011. Molecular studies in olive (*Olea europaea* L.): overview on DNA markers applications and recent advances in genome analysis. Plant Cell Report 30: 449–462.

Brumley, W.C., Sheppard, A.J., Rudolf, T.S., Shen, C.S., Yasaei, P. and Sphon, J.A. 1985. Mass spectrometry and identification of sterols in vegetable oils as butyryl esters and relative quantitation by gas chromatography with flame ionization detection. Journal of Association of Analytical Chemistry 68: (4) 701–709.

Camin, F., Larcher, R., Perini, M., Bontempo, L., Bertoldi, D., Gagliano, G., Nicolini, G. and Versini, G. 2010. Characterization of authentic Italian extra-virgin olive oils by stable isotope ratios of C, O and H and mineral composition. Food Chemistry 118: 901–909.

Ceci, L.N. and Carelli, A.A. 2007. Characterization of monovarietal Argentinian olive oils from new productive zones. Journal of American Oil Chemist Society 84: 1125–1136.

Chiavaro, E.C., Cerratini, L., Di Mateo, A., Barnaba, C., Bendini, A. and Iacumin, P. 2011. Application of a multidisciplinary approach for the evaluation of traceability of extra virgin olive oil. European Journal of Lipid Science and Technology 113: 1509–1519.

Codex Alimentarius Commssion (CAC). 2009. Distribution of the report of the 21st session of the Codex Committee on fats and oils (ALINORM 09/32/17). FAO, Roma.

Cortesi, N. Rovellini, P. and Fedeli, E. 1990. I trigliceridi degli oli naturali. Nota I. Rivista Italiana Delle Sostanze Grasse 67: 69–73.

Dais, P. and Hatzakis, E. 2013. Quality assessment and authentication of virgin olive oil by NMR spectroscopy: a critical review. Analytica Chimica Acta 765: 1–27.

Dais, P. 2013. Nuclear magnetic resonance: methodologies and applications. pp. 395–430. *In*: Aparicio, R. and Harwood, J. (eds.). Handbook of Olive Oil: Analysis and Properties. Springer Science, New York.

European Communities (EC). 2007. Official Journal of the European Union. Regulation No. 702/2007, 22 June 2007.

European Commission (EU). 2014. EU Programmes Horizon 2020. H2020-SFS-2014-2. Authentication of olive oil. http://ec.europa.eu/research/participants/portal/desktop/en/opportunities/h2020/topics/sfs-14a-2014.html.

Fiebig, H.J. 1985. HPLC-Trennung von Triglyceriden. Fette Seifen Anstrichm 87: 53–57.

Firestone, D. and Reina, R.J. 1996. Authenticity of vegetable oils. pp. 198–258. *In*: Ashurst, P.R. and Dennis, M.J. (eds.). Food Authentication. Blackie Academic & Professional, London.

Gao, L., Sedman, J., García-González, D.L., Ehsan, S., Sprules, T. et al. 2009. ^{13}C NMR as a primary method for determining saturates, *cis* and *trans*-monounsaturates and polyunsaturates in fats and oils for nutritional labelling purposes. European Journal of Lipid Science and Technology 11: 612–622.

García-González, D.L. and Aparicio, R. 2010. Coupling MOS sensors and gas chromatography to interpret the sensor responses to complex food aroma: application to virgin olive oil. Food Chemistry 120: 572–579.

García-González, D.L., Aparicio-Ruiz, R. and Aparicio, R. 2009a. Olive oil. pp. 33–72. *In*: Moreau, R.A. and Kamal-Eldin, A. (eds.). Gourmet and Health-Promoting Speciality Oils. AOCS Press, Urbana.

García-González, D.L., Baeten, V., Fernández-Piernas, J.A. and Tena, N. 2013. Infrared, Raman and fluorescence spectroscopies: methodologies and applications. pp. 335–394. *In*: Aparicio, R. and Harwood, J. (eds.). Handbook of Olive Oil: Analysis and Properties (2nd ed.). Springer Science, New York.

García-González, D.L., Luna, G., Morales, M.T. and Aparicio, R. 2009b. Stepwise geographical traceability of virgin olive oils by chemical profiles using artificial neural network models. European Journal of Lipid Science and Technology 111: 1003–1013.

García-González, D.L., Viera-Macías, M., Aparicio-Ruiz, R., Morales, M.T. and Aparicio, R. 2007a. Validation of a method based on triglycerides for the detection of low percentages of hazelnut oil in olive oil by column liquid chromatography. Journal of AOAC International 90: 1346–1353.

García-González, D.L., Viera, M., Tena, N. and Aparicio, R. 2007b. Evaluation of the methods based on triglycerides and sterols for the detection of hazelnut oil in olive oil. Grasas Aceites 58: 344–350.

García-González, D.L., Vivancos, J. and Aparicio, R. 2011. Mapping brain activity induced by olfaction of virgin olive oil aroma. Journal of Agriculture and Food Chemistry 59: 10200–10210.

García-González, D.L., Tena, N. and Aparicio, R. 2012. Describing the chemical singularity of the Spanish protected designations of origin for virgin olive oils in relation to oils from neighbouring areas. Grasas Aceites 63(1): 26–34.

García-Pulido, J. and Aparicio, R. 1993. Triacylglycerol determination based on fatty acid composition using chemometrics. Analytica Chimica Acta 271: 293–298.

Giménez, M.J., Pistón, F., Martín, A. and Atienza, S.G. 2010. Application of real-time PCR on the development of molecular markers and to evaluate critical aspects for olive oil authentication. Food Chemistry 118: 482–487.

Grob, K., Giuffré, M.A., Biedermann, M. and Bronz, M. 1994. The detection of adulteration with desterolized oils. Fat Science and Technology 9: 341–345.

International Olive Council (IOC). 2001a. COI/T.20/Doc. No. 10/Rev. 1. Determination of the composition and content of sterols by capillary-column gas chromatography. Madrid, Spain.

International Olive Council (IOC). 2001b. COI/T.20/Doc. No. 11/Rev. 2. Determination of stigmastadienes in vegetable oils. Madrid, Spain.

International Olive Council (IOC). 2013b. COI/T.20/Doc. No. 30/Rev. 1. Determination of the composition and content of sterols and triterpene dialcohols by capillary column gas chromatography. Madrid, Spain.

International Olive Council (IOC). 2015. Document COI/T.20/Doc. No. 15/Rev. 8. Sensory analysis of olive oil: method for the organoleptic assessment of virgin olive oil. Madrid, Spain.

International Olive Oil (IOC). 2016. COI/T.15/NC No. 3/Rev.11. Trade standard applying to olive oils and olive-pomace oils. Madrid, Spain.

International Union of Pure and Applied Chemistry (IUPAC). 1987. Method 2.507, Determination of polar compounds in frying fats. *In*: Paquot, C. and Haufenne, A. (eds.). Standard Methods for the Analysis of Oils, Fats and Derivatives, 7th edition. Blackwell Scientific Publications, Oxford.

Karabagias, I., Michos, C., Badeka, A., Kontakos, S., Stratis, I. and Kontominas, M.G. 2013. Classification of Western Greek virgin olive oils according to geographical origin based on chromatographic, spectroscopic, conventional and chemometric analyses. Food Research International 54(2): 1950–1958.

Lanzón, A., Albi, T. and Cert, A. 1989. Detección de la presencia de aceite de oliva refinado en el aceite de oliva virgen. Grasas y Aceites 40: 385–388.

Lees, M. 1998. Introduction. pp. 11–17. *In*: Lees, M. (ed.). Food Authenticity: Issues and Methodologies. Eurofins Scientific, Nantes.

Llorent-Martínez, E.J., Ortega-Barrales, P., Fernández-de Córdova, M.L., Domínguez-Vidal, A. and Ruiz-Medina, A. 2011. Investigation by ICP-MS of trace elements in vegetable edible oils produced in Spain. Food Chemistry 1127: 257–1262.

Mailer, R.J., Ayton, J. and Graham, K. 2010. The influence of growing region, cultivar and harvest timing on the diversity of Australian olive oil. Journal of American Oil Chemists Society 87: 877–884.

Mannina, L., D'Imperio, M., Capitani, D., Rezzi, S., Guillou, C., Mavromoustakos, T., Molero Vilchez, M.D., Fernandez, A.H., Thomas, F. and Aparicio, R. 2009. 1H NMR-based protocol for the detection of adulterations of refined olive oil with refined hazelnut oil. Journal of Agriculture and Food Chemistry 5: 11550–11556.

Mariani, C., Fedeli, E. and Grob, K. 1991. Valutazione dei componenti minori liberi ed esterificati nelle sostanze grasse. Rivista Italiana Delle Sostanze Grasse 68: 233–242.

Mariani, C., Venturini, S. and Grob, K. 1995. Individuazione dell'olio di girasole alto oleico desterolato nell'olio d'oliva. Rivista Italiana Delle Sostanze Grasse 72: 473–482.

Mariani, C., Bellan, G., Lestini, E. and Aparicio, R. 2006. The detection of the presence of hazelnut oil in olive oil by free and esterified sterols. European Food Research and Technology 223: 655–661.

Montealegre, C., Marina Alegre, M.L. and García-Ruiz, C. 2010. Traceability markers to the botanical origin in olive oils. Journal of Agricultural and Food Chemistry 58: 28–38.

Muzzalupo, I., Pisani, F., Greco, F. and Chiappetta, A. 2015. Direct DNA amplification from virgin olive oil for traceability and authenticity. European Food Research and Technology 241(1): 151–155.

National Centre for Biotechnology Information (NCBI) database. https://www.ncbi.nlm.nih.gov/.

Pafundo, S., Agrimonti, C., Maestri, E. and Marmiroli, N. 2007. Applicability of SCAR markers to food genomics: olive oil traceability. Journal of Agricultural and Food Chemistry 55: 6052–6059.

Pasqualone, A., Montemurro, C., Summo, C., Sabetta, W., Caponio, F. and Blanco, A. 2007. Effectiveness of microsatellite DNA markers in checking the identity of PDO extra virgin olive oil. Journal of Agricultural and Food Chemistry 55: 3857–3862.

Portarena, S., Gavrichkova, O., Lauteri, M. and Brugnoli, E. 2014. Authentication and traceability of Italian extra-virgin olive oils by means of stable isotopes techniques. Food Chemistry 164: 12–16.

Rabiei, Z. and Enferadi, S.T. 2012. Traceability of origin and authenticity of olive oil. INTECH Open Access Publisher. doi:10.5772/28458.

Raieta, K., Muccillo, L. and Colantuoni, V. 2015. A novel reliable method of DNA extraction from olive oil suitable for molecular traceability. Food Chemistry 172: 596–602.

Rotondi, A., Beghe, D., Fabbri, A. and Ganino, T. 2011. Olive oil traceability by means of chemical and sensory analyses: A comparison with SSR biomolecular profiles. Food Chemistry 129(4): 1825–1831.

Sanz-Cortes, F., Parfitt, D.E., Romero, C., Struss, D., Llacer, G. and Badenes, M.L. 2003. Intraspecific olive diversity assessed with AFLP. Plant Breeding 122: 173–177.

Scollo, F., Egea, L.A., Gentile, A., La Malfa, S., Dorado, G. and Hernandez, P. 2016. Absolute quantification of olive oil DNA by droplet digital-PCR (ddPCR): Comparison of isolation and amplification methodologies. Food Chemistry 213: 388–394.

Sedman, J., Gao, L., García-González, D.L., Ehsan, S. and van de Voort, F.R. 2010. Determining nutritional labeling data for fats and oils by 1 H NMR. European Journal of Lipid Science and Technology 112: 439–451.

Spaniolas, S., Bazakos, C., Ntourou, T., Bihmidine, S., Georgousakis, A. and Kalaitzis, P. 2008. Use of lambda DNA as a marker to assess DNA stability in olive oil during storage. European Food Research and Technology 227: 175–179.

Tena, N., Wang, S.C., Aparicio-Ruiz, R., García-González, D.L. and Aparicio, R. 2015. In-depth assessment of analytical methods for olive oil purity, safety, and quality characterization. Journal of Agriculture and Food Chemistry 63: 4509–4526.

Zeiner, M., Steffan, I. and Cindric, I.J. 2005. Determination of trace elements in olive oil by ICP-AES and ETA-AAS: a pilot study on the geographical characterization. Microchemistry Journal 81: 171–176.

14

Analytic Aspects of Mycotoxin Traceability

Nolwenn Hymery

1. Introduction

Mycotoxins: Origin and surveillance

Mycotoxins are secondary metabolites produced by molds that contaminate various agricultural commodities at any stage in the food production chain: before harvesting, between harvesting and drying, drying and during storage. The Food and Agriculture Organization (FAO) estimated that 25% of the world's food crops are affected by mycotoxins to a certain degree (Stepien et al. 2007). Mycotoxins are natural contaminants produced by a range of fungal species. Their common occurrence in food and feed poses a serious threat to the human and animal health. This threat is caused either by the direct contamination of plant material and derived products or by a "carry-over" of mycotoxins and their metabolites into animal tissues, eggs and milk after feeding of contaminated matrix. The most important agro-economic and public health classes of mycotoxins are aflatoxins (AFT), ochratoxins (OTA), zeralenone (ZEA), trichothecenes, and fumonisins (FUMs) that are produced by species of *Fusarium, Penicillium* and *Aspergillus*.

Due to their diverse chemical structures and their differing physical properties, mycotoxins show a wide range of biological effects. Mycotoxins cause adverse health effects such as genotoxic, mutagenic, carcinogenic, teratogenic, and oestrogenic damages alone or in combinations (Bennett

Brest University, EA 3882, Laboratoire Universitaire de Biodiversité et Ecologie Microbienne, ESIAB, Technopôle Brest-Iroise, 29280 Plouzané, France; E-mail: nolwenn.hymery@univ-brest.fr and Klich 2003, Cano et al. 2016, Smith et al. 2016). Detection, identification and surveillance of mycotoxin in food and feed have become major issues for producers, regulatory authorities and scientific community to reduce economic loss and to protect consumers from mycotoxin-derived health risks. There are key management tools and traceability procedures which should be used to limit mycotoxin contamination. It is essential that Good Agricultural Practices are in place. Prevention strategies pre- and post-harvest can be effective for mycotoxins. For pre-harvest identification of mycotoxigenic fungal species and strains, it is essential to develop effective strategies for control. Concerning post-harvest, several parameters should be monitored such as humidity, temperature.

Traceability required effective diagnostic tools which can be used to monitor and quantify mycotoxins. The main pillars combating mycotoxins on the consumer's plate are adequate agricultural practices (including storage, plant breeding, pesticides, proper varieties resistant to the local environment, non-toxigenic fungal spraying, etc.), decontamination and analytical mycotoxin monitoring of products. Decontamination is only allowed for feed and under stringent rules as the quality of the product will also suffer (e.g., vitamin losses). The most practiced consumer protection action is the regular testing of products for mycotoxins, which have no structural similarity between different groups of toxins. As a result they exhibit a very broad range of physic-chemical properties. It requires therefore specific methods that must be used for specific groups of toxins (or fractions of different groups). This monitoring is often targeted to the most likely mycotoxins present. However recent efforts have been made to screen for a vast array of mycotoxins with a "single" method (Berthiller et al. 2016).

However it seems impossible to test for all mycotoxins on a routine bases as the analytical effort (incl. calibration) will limit such a method to (research) applications where the whole mycotoxin spectra is of interest. As a result analytical methods for routine target to a large degree only regulated mycotoxins for a specific product, while the number of different mycotoxins analyzed for in a single run can be more than a dozen.

2. Methodology for Mycotoxin Determination

Increasing demands for fast, reliable, inexpensive and food matrices specific method is still a challenge and has led to many scientific papers describing a vast number of analytical methods developed with an attempt to validate demands (Table 1). Because distribution of mycotoxins in food matrices is heterogeneous, upstream stages of an accurate analysis requires a sampling and a rigorous homogenization of samples (Whitaker et al. 1974, 1976, 1979, 2009, Whitaker 2006). Sampling must be representative of the lot. Number

Table 1. Example of Methods of Determination of the Most Common Mycotoxin in Different Food Matrix.

Mycotoxins	Methods	Matrix	References
Aflatoxin B1	ELISA*	Rice	Reddy and Reddy 2009
	LC-MS/MS	Seafoods	Huang et al. 2011
	UHPLC-MS/MS	Milk and milk powder	Wang and Li 2015
Aflatoxin M1	ELISA	Milk	Decastelli et al. 2007
	UHPLC-MS/MS	Milk and milk powder	Wang and Li 2015
Aflatoxins	HPLC	Peanut	Njumbe Ediage et al. 2015
	LC-MS/MS	Sorghum	Wan Ainiza and Jinap 2015, Shi et al. 2014
	HPLC-FLD	Spices, peanut	Hashemi and Taherimaslak 2014
Citrinin	LC-MS/MS	Rice	Ji et al. 2015
Ergot alkaloids	LC/MS	Cereals	Bryla et al. 2015
Fumonisins	ELISA	Spanish beer	Torres et al. 1998
	LC-MS/MS	Maize	Oliveira et al. 2015
OTA	HPLC	Dried figs, coffee, wine	Bircan 2009
	ELISA	Wine	Novo et al. 2013
	LC-MS/MS	Sorghum, wine	Njumbe Ediage et al. 2015, Cao et al. 2013
	UHPLC-MS/MS	Milk and milk powder	Wang and Li 2015
	HPLC-FLD	Spices	Wan Ainiza and Jinap 2015
	TLC	Corn	Keller et al. 2013
Patulin	HPLC	Apple juice	Li et al. 2007
	UHPLC-UV	Fruit leather	Maragos et al. 2015
Trichothecenes	GC	Cereals	Pereira et al. 2015
	LC-MS/MS	Wheat	Bertuzzi et al. 2014
Zearalenone	UPLC, HPLC	Cereals	Ok et al. 2014
	LC-MS/MS	Seafoods	Woźny et al. 2013
	UHPLC-MS/MS	Milk and milk powder	Wang and Li 2015

*ELISA: Enzyme Linked Immunosorbent Assays; GC: Gas Chromatography; HPLC: High-Performance Liquid Chromatography; HPLC-FLD: High-Performance Liquid Chromatography-Fluorescence detector; TLC: Thin Layer Chromatography; LC-MS: Liquid chromatography-mass Spectrometry; UHPLC-MS/MS: Ultra performance liquid chromatography-tandem mass spectrometer; UHPLC-UV: ultra-high performance liquid chromatographic (UHPLC) method coupled with an ultraviolet detector (UV).

of incremental samples of sub-lots to be taken shall depends on the size of the lot and kind of food such as cereals and cereal products, dried fruits, nuts, and spices that have been tabulated in EC401/2006. For example concerning aflatoxin, 39 countries have regulations for aflatoxin B1 in feed with the most common limit concentration of 5 µg/kg, which is equivalent to 0.005 g/ton or 0.05 g for a container (10 tons).

Concerning extraction, purification of the extract is an essential step, the desired compounds can be concentrated in a solvent, and interferences can be removed. Briefly it can be performed by solid phase extraction (SPE) using a solid and a liquid phase and by multifunctional clean-up columns, e.g., MycoSep®, and immunoaffinity columns (IACs) which are frequently used to clean-up the extracts and rarely by liquid–liquid extraction (LLE) using 2 immiscible liquid-phases (Rahmani et al. 2009). In most cases, and this being due to the strong sensitivity in analytical technique like mass spectrometric, no clean-up is carried out beforehand and samples are just diluted and shoot.

Classical analytical methods for mycotoxins include thin layer chromatography (TLC), enzyme linked immunosorbent assays (ELISA), high-performance liquid chromatography (HPLC) coupled to diode array (DAD), fluorescence (FLD), single mass spectrometric (MS) or tandem mass spectrometric (MS/MS) detection and gas chromatography (GC) or MS detection coupled with electron capture (ECD), flame ionisation (FID). Among this wide range of analytical techniques, Tandem Mass Spectrometry Hyphenated with HPLC (HPLC-MS/MS) or multiple-stage MS (also called tandem MS or MS/MS or MSn) using atmospheric pressure chemical ionization (APCI) or electrospray ionization (ESI) interfaces has become the performing method for mycotoxin quantification in food matrices, due to the sensitivity and reliability of this setup (Sforza et al. 2006, Zöllner and Mayer-Helm 2006, Berthiller et al. 2007, Songsermsakul and Razzazi-Fazeli 2008).

Furthermore it is possible to separate and detect all relevant mycotoxins mentioned above in a single run without derivatization. For the sum of these reasons, HPLC-MS/MS is constantly gaining importance in multi mycotoxin analysis (Sagawa et al. 2006, Ren et al. 2007, Sulyok et al. 2007, Spanjer et al. 2008, Di Mavungu et al. 2009, Monbaliu et al. 2009). GC based methods were used for quantitative determination of mycotoxins as well. However, some mycotoxin specific problems, such as non-linearity of calibration curves, poor repeatability, matrix induced over-estimation, and memory effects from previous sample injections have led to the increasing use of HPLC-MS/MS (Pettersson and Langseth 2002a,b).

2.1 Classical methods: chromatographic techniques

In combination with the diversity of chemical structures, it is still impossible to use a unique technique for the determination and quantification of all toxins. Chromatography analysis is based on distribution or partition of a sample solute between two phases: stationary phase and mobile phase. Most common chromatography techniques used to analyze food matrices are **gas chromatography** (GC), HPLC, and supercritical fluid chromatography (SFC). These methods, when connected to another instrument such as mass spectrometer, represent a good separation method.

2.1.1 Gas chromatography

Gas chromatography coupled to different detection techniques is widely used in food analysis. Typically, GC is suited for semi or not polar, semi or volatile and thermostable elements. The majority of mycotoxins are small size compounds, they are non-volatile and polar, and so derivatization must be made before GC analysis with pre-column (e.g., fumonisins, aflatoxins, trichothecenes). Nevertheless some toxins may be analyzed directly (e.g., ZEA, OTA, ergot alkaloids, citrinin) (Tanaka et al. 2007, Köppen et al. 2010).

2.1.2 High-performance liquid chromatography coupled to classical detectors

HPLC is the most popular method for the analysis of mycotoxins in food and feed. During these last few years, HPLC coupled with MS and, especially HPLC-MS/MS have become the most popular techniques for mycotoxin analysis.

For getting an efficient separation for a good quantification of mycotoxins in HPLC analysis, separation of the individual mycotoxins on the stationary phase (in most cases RP-C18) is essential. Liquid chromatography with fluorescence detection (**LC–FLD**) is one of the most widely used techniques for the analysis of mycotoxins for mycotoxins with native fluorescence (e.g., Af, OTA, ZEN) or those that can easy be derivatized (e.g., fumonisins). Some of the mycotoxins such as fumonisins or trichothecenes are not fluorescent, so, prior derivatization of these compounds is needed to make fluorescent derivatives. Aflatoxins can also be detected in food commodities using LC-FLD, even though the native fluorescence emission of aflatoxins is significantly quenched by aqueous mixtures used for reversed-phase chromatography (Maragos 2008). This technique is mainly used for the detection of ZEA and OTA in different food matrices such as red wine (Aresta et al. 2006), dried fruits and rice (Zinedine et al. 2007), coffee (La Pera et al. 2008).

2.1.3 *High performance liquid chromatography with mass spectrometric detection (HPLC-MS)*

Liquid chromatography hyphenated to mass spectrometry (LC-MS) or **tandem mass spectrometry (LC-MS/MS)** has in the last ten years advanced to the status of the reference and definitive method in the field of mycotoxin analysis. On different food commodities, LC-MS (or HPLC-MS/MS) methods using ionization interfaces (ESI or APCI) can simultaneously run with positive and negative mode, which means the best possible detection conditions for all metabolites and offers new ways for fast mycotoxin analysis. It means that simultaneous determination of different chemical mycotoxins can be performed without the need for derivatization. Nevertheless, type and nature of a matrix can strongly affect the suitability of a method (Caprotti et al. 2012). Methods are developed on various food matrices like cheese (Fontaine et al. 2015), for example. Recently, the method based on **QuEChERS** (quick, easy, cheap, effective, rugged, and safe) extraction coupled for example with HPLC-MS/MS or Ultra HPLC-ESI-MS/MS increased attention in the research field of mycotoxins due to its simplicity and effectiveness for isolating mycotoxins from complex matrices. This method has been described in many products, like cereals, spices, nuts, seeds, human breast milk, sesame butter, noodles, eggs, baby food (Sun et al. 2016).

In the last years a large number of studies described multi-mycotoxin methods, showing the special interest for high-throughput multi-mycotoxins routine analysis and therefore, new methods will be developed in the coming years (Di Mavungu et al. 2009, Sulyok et al. 2010, Habler and Rychlik 2016).

2.2 *Immunochemical and rapid techniques*

Among screening methods, immunochemical analyses with high sensitivity and selectivity provided by specific antibodies are the most popular. Contrary to chromatographic methods, immunoassays usually (but not necessary) need a simple extraction, concentration and clean-up procedure.

Screening methods to detect mycotoxins are mostly used to determine the approximate level of toxins in a sample. They are generally less sensitive and precise than laboratory methods like chromatographic techniques. Nevertheless, rapid methods allow the analysis of many samples, even in a non-laboratory environment. Many screening tests are available for main mycotoxins (AFTs, ZEA, DON, OTA, T-2, FMs, citrinin) (Zheng et al. 2006, Maragos 2009). They can be divided in immunochemical and non-immunochemical tests, whereas the first type clearly dominates. The most frequently used methods for mycotoxin immunoassay are binding of specific antibodies to a solid support (direct competitive enzyme-linked immunosorbent assay ELISA) or coated antigens (indirect competitive ELISA). This last technique involves the reaction of antigen and antibody in micro-plate wells. ELISA tests for ZEA (Thongrussamee et al. 2008, Burmistrova et al. 2009) and main mycotoxins in various matrices have been described such as DON in barley, OTA in swine or in green coffee (Matrella et al. 2006) for example.

Another important immunochemical method for rapid mycotoxin detection is based on **fluorescence polarization** that differs from ELISA in that it is a homogeneous assay conducted in solution phase. It is a homogeneous method; in this format, the analyte is labeled with a fluorophore, usually fluorescein. Several publications described this technique for DON, OTA and were reviewed (Maragos 2009).

A relatively recent technology for rapid mycotoxin detection is the imaging **surface plasmon resonance biosensor (iSPR)** (Joshi et al. 2016), an optical phenomenon used to measure changes on the surface of thin metal films. This technique is based on the detection of a change of the refractive index of the medium when an analyte binds to an immobilized element (antibody) on the surface of the sensor chip (reusable). The number of molecules bonded by antibodies to a thin metal layer correlates with the changing of the resonance angle. Rapid multi-mycotoxins detection is described with this method (AFTs, ZEA, OTA, FB1, HT-2, DON) and portable equipment may be used for this type of analysis (Zheng et al. 2006, Turner et al. 2015). This application has several advantages such as small sample volumes (μL) and sensitivity comparable to ELISA.

2.3 *Non-instrumental rapid screening techniques*

These techniques could be used outside the laboratory environment, at the place of sampling, and hence are becoming more and more important. Non-instrumental antibody based applications are lateral flow tests, dipstick tests or flow-through tests. In contrast to ELISA, these applications use antigens or antibodies that are immobilized on carrier membranes instead of microtiter plates. This test is a visual evaluation, giving qualitative results presence/absence by appearance or absence of colored lines on the analysis strip. Compared to instrumental applications like ELISA results appear quickly. Applications have been published recently for several different mycotoxins (De Saeger 2007).

Semi-quantitative rapid tests were designed for screening. These tests are required by the food and agricultural industry for a continuous monitoring of their products. However, to obtain a reliable data, results should be confirmed by chromatographic methods (Burmistrova et al. 2009).

3. Validated Methods

The European Union (EU) has set maximum levels for some mycotoxins presenting a health risk (Directive 2002/32/EC, Regulation (EC) No. 1881/2006). The levels of some mycotoxins in food and animal feed especially aflatoxins, ochratoxin A, patulin, trichothecenes, zearalenone are therefore controlled. Criteria to evaluate the performance of a method are the limit of detection and quantification (LOD/LOQ), linearity, precision (repeatability and reproducibility), selectivity (interference of other compounds and/or matrix effect), robustness, and working range representative (Josephs et al. 2004).

Several analytical methods concerning mycotoxin analysis are available and validated by recognized authorities such as CEN (European Committee for Standardization), AOAC (Association of Official Analytical Chemists), and ISO (International Organization for Standardization). Performance tested methods require validation (in a multi laboratory validation study) followed by standardization according to a protocol, such as done by CEN. A number of CEN methods are also AOAC and ISO-approved because there is a close working relationship between CEN and ISO and standards are developed jointly under what is known as the Vienna Agreement.

Standardized methods for AFTs (EN12955 1999, EN14123 2001), OTA (EN14132 2003), FMs (EN13585 2001, EN14352 2004), and patulin (EN14177 2003) in various foods are available, next to a number of other mycotoxins. Inter-laboratory comparisons or proficiency tests organized by the European Union Reference Laboratory (EURL) for the National Reference Laboratories (NRLs) enable them to assess their performance and allow improving the overall correctness of analytical results for mycotoxins.

Despite great progress in terms of new analytical techniques, several problems remain:

- Conjugated mycotoxins in sample may be lost during extraction and/or clean-up, due to difference in polarity compared to the parent toxins.
- In chromatographic and immunological analytical methods, structure of altered mycotoxins is, in some cases, not recognized because of changed physicochemical properties.

4. Masked Mycotoxins

Gareis et al. (1990) referred to a zearalenone-glucoside as a *'masked mycotoxin'* (now called "modified mycotoxin") to emphasize the fact that this conjugate mycotoxin was not detected by routine analysis of food or feed, but probably contributed to the total mycotoxin content and subsequent effects. Mycotoxin derivatives that are undetectable by conventional analytical techniques because their structure has been changed in the plant are designated masked mycotoxins. The group of masked mycotoxins comprises both extractable conjugated and bound (non-extractable) varieties. Non-availability of analytical standards or calibrants (only one compound—deoxynivalenol–glucoside—is commercially available) is indeed an obstacle to the development of dedicated analytical methods to estimate accurately occurrence of mycotoxin-conjugates in the food and feed, and then evaluate their potential effects in humans and animals.

Nevertheless, major methodological approaches described to detect masked mycotoxins are indirect determination (hydrolysis of the conjugated mycotoxins by enzymatic, acidic or basic treatment), direct analysis by chromatography and direct analysis by ELISA (Bertiller et al. 2013).

5. Conclusion

Two main groups of methods exist for mycotoxin analysis: laborious methods for determination of mycotoxins with high sensitivity and precision, and screening methods for rapid detection in a non-laboratory environment. One of new challenge is to perform available data for modified mycotoxin with accurate analytical methods because modified mycotoxins might also be more toxic than their parent compounds. Tracing mycotoxin by analysis is still challenging with regard to accurate quantization and newly identified compounds. The cost of analysis stays high and the performant equipment are expensive. New research has to be done on rapid and accurate methods.

Keywords: AFTs: aflatoxins, DON: deoxynivalenol, FB1: fumonisin B1, FUMs: fumonisins, OTA: ochratoxin, ZEA: zeralenone

References

AOAC. 2016. Official Methods of Analysis of AOAC International. Appendix f. AOAC International. Available: http://www.eoma.aoac.org.

Aresta, A., Vatinno, R., Palmisano, F. and Zambonin, C.G. 2006. Determination of ochratoxin A in wine at sub ng/mL levels by solid-phase microextraction coupled to liquid chromatography with fluorescence detection. Journal of Chromatography A 1115: 196–201.

Bennett, J.W. and Klich, M. 2003. Mycotoxins. Clinical Microbiology Reviews 16: 497–516.

Berthiller, F., Sulyok, M., Krska, R. and Schuhmacher, R. 2007. Chromatographic methods for the simultaneous determination of mycotoxins and their conjugates in cereals. International Journal of Food Microbiology 119: 33–37.

Berthiller, F., Crews, C., Dall'Asta, C., Saeger, S.D., Haesaert, G., Karlovsky, P., Oswald, I.P., Seefelder, W., Speijers, G. and Stroka, J. 2013. Masked mycotoxins: a review. Molecular Nutrition of Food Research 57: 165–86.

Berthiller, F., Brera, C., Iha, M.H., Krska, R., Lattanzio, V.M.T., MacDonald, S., Malone, R.J., Maragos, C., Solfrizzo, M., Stranska-Zachariasova, M., Stroka, J. and Tittlemier, S.A. 2016. Developments in mycotoxin analysis: an update for 2015–2016. World Mycotoxin Journal in press.

Bertuzzi, T., Camardo, L.M., Battilani, P. and Pietri, A. 2014. Co-occurrence of type A and B trichothecenes and zearalenone in wheat grown in northern Italy over the years 2009–2011. Food Additives and Contaminants Part B 7: 273–281.

Bircan, C. 2009. Incidence of ochratoxin A in dried fruits and co-occurrence with aflatoxins in dried figs. Food and Chemical Toxicology 47: 1996–2001.

Bryla, M., Szymczyk, K., Jedrzejczak, R. and Roszko, M. 2015. Application of liquid chromatography/ion trap mass spectrometry technique to determine ergot alkaloids in grain products. Food Technology and Biotechnology 53: 18–28.

Burmistrova, N.A., Goryacheva, I.Y., Basova, E.Y., Franki, A.S., Elewaut, D., Van Beneden, K., Deforce, D., Van Peteghem, C. and De Saeger, S. 2009. Analytical and Bioanalytical Chemistry 39: 1301–1307.

Cano, P. M., Puel, O. and Oswald, I.P. 2016. Mycotoxins: Fungal Secondary Metabolites with Toxic Properties. Fungi: Applications and Management Strategies 318.

Cao, J., Kong, W., Zhou, S., Yin, L., Wan, L. and Yang, M. 2013. Molecularly imprinted polymer-based solid phase clean-up for analysis of ochratoxin A in beer, red wine, and grape juice. Journal of Separation Sciences 36: 1291–1297.

Capriotti, A.L., Caruso, G., Cavaliere, C., Foglia, P., Samperi, R. and Lagan, A. 2012. Multiclass mycotoxin analysis in food, environmental and biological matrices with chromatography/mass spectrometry. Mass Spectrometry Reviews 31: 466–503.

De Saeger, S. 2007. Immunochemical methods for rapid mycotoxin detection: evolution from single to multiple analyte screening. A review. Food Additives and Contaminants 24: 1169–1183.

Decastelli, L., Lai, J., Gramaglia, M., Monaco, A., Nachtmann, C., Oldano, F., Ruffier, M., Sezian, A. and Bandirola, C. 2007. Aflatoxins occurrence in milk and feed in Northern Italy during 2004–2005. Food Control 18: 1263–1266.

Di Mavungu, J.D., Monbaliu, S., Scippo, M.L., Maghuin-Register, G., Schneider, Y.J., Larondelle, Y., Callebaut, A., Robbens, J., Van Peteghem, C. and De Saeger, S. 2009. LC-MS/MS multi-analyte method for mycotoxin determination in food supplements. Food Additives and Contaminants A 26: 885–895.

[EC] European Community. 2006. EC401/2006, laying down the methods of sampling and analysis for the official control of the levels of mycotoxins in foodstuffs. Official Journal of EU 70: 12–34.

EN 12955. 1999. Foodstuffs. Determination of aflatoxin B1, and the sum of aflatoxins B1, B2, G1 and G2 in cereals, shell-fruits and derived products; High performance liquid chromatographic method with post column derivatization and immunoaffinity column clean-up.

EN 13585. 2001. Foodstuffs. Determination of fumonisins B1 and B2 in maize; HPLC method with solid phase extraction clean-up.

EN 14123. 2003. Foodstuffs. Determination of aflatoxin B1 and the sum of aflatoxin B1, B2, G1 and G2 in peanuts, pistachios, figs, and paprika powder; High performance liquid chromatographic method with postcolumn derivatization and immunoaffinity column clean-up.

EN 14132. 2003. Foodstuffs. Determination of ochratoxin A in barley and roasted coffee; HPLC method with immunoaffinity column clean-up.

EN 14177. 2003. Foodstuffs. Determination of patulin in clear and cloudy apple juice and puree. HPLC method with liquid/liquid partition clean-up.

EN14352. 2004. Foodstuffs - Determination of fumonisin B1 and B2 in maize based foods - HPLC method with immunoaffinity column clean up.

Fontaine, K., Passeró, E., Vallone, L., Hymery, N., Coton, M., Jany, J.L., Mounier, J. and Coton, E. 2015. Occurrence of roquefortine C, mycophenolic acid and aflatoxin M1 mycotoxins in blue-veined cheeses. Food Control 47: 634–640.

Gareis, M., Bauer, J., Thiem, J., Plank, G., Grabley, S. and Gedek, B. 1990. Cleavage of zearalenone-glycoside, a 'masked' mycotoxin, during digestion in swine. Journal of Veterinary Medicine Series B 37: 236–240.

Habler, K. and Rychlik, M. 2016. Multi-mycotoxin stable isotope dilution LC-MS/MS method for Fusarium toxins in cereals. Analytical and Bioanalytical Chemistry 408: 307–17.

Hashemi, M. and Taherimaslak, Z. 2014. Separation and determination of aflatoxins B1, B2, G1 and G2 in pistachio samples based on magnetic solid phase extraction followed by high performance liquid chromatography with fluorescence detection. Analytical Methods 6: 7663–7673.

Huang, Y., Zhu, X.M., Yang, Y.X., Jin, J.Y., Chen, Y.F. and Xie, S.Q. 2011. Response and recovery of gibel carp from subchronic oral administration of aflatoxin B1. Aquaculture 319: 89–97.

Ji, X., Xu, J., Wang, X., Qi, P., Wei, W., Chen, X., Li, R. and Zhou, Y. 2015. Citrinin determination in red fermented rice products by optimized extraction method coupled to liquid chromatography tandem mass spectrometry (LC-MS/MS). Journal of Food Science 80: 1438–1480.

Josephs, R.D., Derbyshire, M., Stroka, J., Emons, H. and Anklam, E. 2004. Trichothecenes: reference materials and method validation. Toxicology Letters 153: 123–132.

Joshi, S., Segarra-Fas, A., Peters, J., Zuilhof, H., van Beek, T.A. and Nielen, M.W. 2016. Multiplex surface plasmon resonance biosensing and its transferability towards imaging nanoplasmonics for detection of mycotoxins in barley. The Analyst 141: 1307–1318.

Keller, L.A.M., Pereyra, M.L.G., Keller, K.M., Alonso, V.A., Oliveira, A.A., Almeida, T.X., Barbosa, T.S., Nunes, L.M.T., Cavaglieri, L.R. and Rosa, C.A.R. 2013. Fungal and mycotoxins contamination in corn silage: monitoring risk before and after fermentation. The Journal of Stored Products Research 52: 42–47.

Köppen, R., Koch, M., Siegel, D., Merkel, S., Maul, R. and Nehls, I. 2010. Determination of mycotoxins in foods: current state of analytical methods and limitations. Applied Microbiology and Biotechnology, New York 86: 1595–1612.

La Pera, L., Avellone, G., Lo Turco, V., Di Bella, G., Agozzino, P. and Dugo, G. 2008. Influence of roasting and different brewing processes on the ochratoxin A content in coffee determined by high-performance liquid chromatography-fluorescence detection. Food Additives and Contaminants Part A 25: 1257–1263.

Li, J.K., Wu, R., Hu, Q.H. and Wang, J.H. 2007. Solid-phase extraction and HPLC determination of patulin in apple juice concentrate. Food Control 18: 530–534.

Maragos, C.M. 2008. Extraction of aflatoxins B1 and G1 from maize by using aqueous sodium dodecyl sulfate. Journal of AOAC International 91: 762–767.

Maragos, C. 2009. Fluorescence polarization immunoassay of mycotoxins: a review. Toxins 1: 196–207.

Maragos, C.M., Busman, M., Ma, L. and Bobell, J. 2015. Quantification of patulin in fruit leathers by ultra-high-performance liquid chromatography-photodiode array (UPLC-PDA). Food Additives and Contaminants Part A 32: 1164–1174.

Matrella, R., Monaci, L., Milillo, M.A., Palmisano, F. and Tantillo, M.G. 2006. Ochratoxin A determination in paired kidneys and muscle samples from swines slaughtered in southern Italy. Food Control 17: 114–117.

Monbaliu, S., Van Poucke, C., Van Peteghem, C., Van Poucke, K., Heungens, K. and De Saeger, S. 2009. Development of a multimycotoxin liquid chromatography/tandem mass spectrometry method for sweet pepper analysis. Rapid Commun. Mass Spectrom. 23: 3–11.

Njumbe Ediage, E., Van Poucke, C. and De Saeger, S. 2015. A multi-analyte LC-MS/MS method for the analysis of 23 mycotoxins in different sorghum varieties: the forgotten sample matrix. Food Chemistry 177: 397–404.

Novo, P., Moulas, G., Prazeres, D.M.F., Chu, V. and Conde, J.P. 2013. Detection of ochratoxin a in wine and beer by chemiluminescence-based ELISA in microfluidics with integrated photodiodes. Sensors and Actuators B 176: 232–240.

Ok, H., Choi, S.W., Kim, M. and Chun, H.S. 2014. HPLC and UPLC methods for the determination of zearalenone in noodles, cereal snacks and infant formula. Food Chemistry 163: 15: 252–257.

Oliveira, M.S., Diel, A.C.L., Rauber, R.H., Fontoura, F.P., Mallmann, A., Dilkin, P. and Mallmann, C.A. 2015. Free and hidden fumonisins in Brazilian raw maize samples. Food Control 53: 217–221.

Pereira, V.L., Fernandes, J.O. and Cunha, S.C. 2015. Comparative assessment of three cleanup procedures after QuEChERS extraction for determination of trichothecenes (type A and type B) in processed cereal-based baby foods by GC–MS. Food Chemistry 182: 143–149.

Pettersson, H. and Langseth, W. 2002a. Intercomparison of trichothecene analysis and feasibility to produce certified calibrants and reference material-method studies. BCR Information, EUR 20285/1 EN. Office for Official Publications of the European Communities, Luxembourg.

Pettersson, H. and Langseth, W. 2002b. Intercomparison of trichothecene analysis and feasibility to produce certified calibrants and reference material-homogeneity and stability studies, final intercomparison. BCR Information, EUR 20285/2 EN. Office for Official Publications of the European Communities, Luxembourg.

Rahmani, A., Jinap, S. and Soleimany, F. 2009. Qualitative and quantitative analysis of mycotoxins. Comprehensive Reviews in Food Science and Food Safety 8: 202–251.

Reddy, K.R., Reddy, C.S. and Muralidharan, K. 2009. Detection of *Aspergillus* spp. and aflatoxin B1 in rice in India. Food Microbiology 26: 27–31.

Ren, Y., Zhang, Y., Shao, S., Cai, Z., Feng, L., Pan, H. and Wang, Z. 2007. Simultaneous determination of multi-component mycotoxin contaminants in foods and feeds by ultra-performance liquid chromatography tandem mass spectrometry. Journal of Chromatograhy A 1143: 48–64.

Sagawa, N., Takino, T. and Kurogochi, S. 2006. A simple method with liquid chromatography/tandem mass spectrometry for the determination of the six trichothecene mycotoxins in rice medium. Bioscience, Biotechnology and Biochemistry 70: 230–236.

Sforza, S., Dall'Asta, C. and Marchelli, R. 2006. Recent advances in mycotoxin determination in food and feed by hyphenated chromatographic techniques/mass spectrometry. Mass Spectrometry Reviews 25: 54–76.

Shi, Y.G., Zhong, W., Wen, H., Ma, J. and Sun, P. 2014. Applied technology in using HPLC for determining aflatoxin contents in peanuts. Advanced Materials Research 1022: 76–80.

Smith, M.C., Madec, S., Coton, E. and Hymery, N. 2016. Natural co-occurrence of mycotoxins in foods and feeds and their *in vitro* combined toxicological effects. Toxins 8: 94.

Songsermsakul, P. and Razzazi-Fazeli, E.A. 2008. Review of recent trends in applications of liquid chromatography-mass spectrometry for determination of mycotoxins. Journal of Liquid Chromatography and Related Technologies 31: 1641–1686.

Spanjer, M.C., Rensen, P.M. and Scholten, J.M . 2008. LC-MS/MS multimethod for mycotoxins after single extraction, with validation data for peanut, pistachio, wheat, maize, cornflakes, raisins and figs. Food Additives and Contaminants 25: 472–489.

Stepien, M., Sokol-Leszczynska, B. and Luczak, M. 2007. Mycotoxins, food products and human health. Postepy Mikrobiol 46: 167–177.

Sulyok, M., Krska, R. and Schuhmacher, R. 2007. A liquid chromatography/tandem mass spectrometric multi-mycotoxin method for the quantification of 87 analytes and its application to semiquantitative screening of moldy food samples. Analytical and Bioanalytical Chemistry 389: 1505–1523.

Sulyok, M., Krska, R. and Schuhmacher, R. 2010. Application of an LC-MS/MS based multi-mycotoxin method for the semi-quantitative determination of mycotoxins occurring in different types of food infected by moulds. Food Chemistry 119: 408–416.

Sun, J., Li, W., Zhang Y., Sun, L., Dong, X., Hu, X. and Wang, B. 2016. QuEChERS purification combined with ultrahigh-performance liquid chromatography tandem mass spectrometry for simultaneous quantification of 25 mycotoxins in cereals. Toxins 8: 375–393.

Tanaka, K., Sago, Y., Zheng, Y.Z., Nakagawa, H. and Kushiro, M. 2007. Mycotoxins in rice. International Journal of Food Microbiology 119: 59–66.

Thongrussamee, T., Kuzmina, N.S., Shim, W.B., Jiratpong, T., Eremin, S.A., Intrasook, J. and Chung, D.H. 2008. Monoclonal-based enzyme-linked immunosorbent assay for the detection of zearalenone in cereals. Food Additives and Contaminants 25: 997–1006.

Torres, M.R., Sanchis, V. and Ramos, A.J. 1998. Occurrence of fumonisins in Spanish beers analyzed by an enzyme-linked immunosorbent assay method. International Journal of Food Microbiology 3: 139–43.

Turner, N.W., Bramhmbhatt, H., Szabo-Vezse, M., Poma, A., Coker, R. and Piletsky, S.A. 2015. Analytical methods for determination of mycotoxins: An update (2009–2014). Analytica Chimica Acta 901: 12–33.

Wan Ainiza, W.M. and Jinap, S. 2015. Simultaneous determination of aflatoxins and ochratoxin A in single and mixed spices. Food Control 50: 913–918.

Wang, X. and Li, P. 2015. Rapid screening of mycotoxins in liquid milk and milk powder by automated size-exclusion SPE-UPLC-MS/MS and quantification of matrix effects over the whole chromatographic run. Food Chemistry 173: 897–904.

Whitaker, T.B. 2006. Sampling food for mycotoxins. Food Additives Contaminants 23: 50–61.

Whitaker, T.B., Dickens, J.W. and Monroe, R.J. 1974. Variability of aflatoxins test results. Journal of American Oils Chemists' Society 51: 214–218.

Whitaker, T.B., Whitten, M.E. and Monroe R.J. 1976. Variability associated with testing cottonseed for aflatoxin. Journal of American Oils Chemists' Society 53: 502–507.

Whitaker, T.B., Dickens, J.W. and Monroe, R.J. 1979. Variability associated with testing corn for aflatoxin. Journal of American Oils Chemists' Society 56: 789–794.

Whitaker, T.B., Trucksess, M.W., Weaver, C.M. and Slate, A. 2009. Sampling and analytical variability associated with the determination of aflatoxins and ochratoxin A in bulk lots of powdered ginger marketed in 1-lb bags. Analytical and Bioanalytical Chemistry 395: 1291–1299.

Woźny, M., Obremski, K., Jakimiuk, E., Gusiatin, M. and Brzuzan, P. 2013. Zearalenone contamination in rainbow trout farms in north-eastern Poland. Aquaculture 416: 209–211.

Zheng, M.Z., Richard, J.L. and Binder, J. 2006. A review of rapid methods for the analysis of mycotoxins. Mycopathologia 161: 261–273.

Zinedine, A., Soriano, J.M., Juan, C., Mojemmi, B., Molto, J.C., Bouklouze, A., Cherrah, Y., Idrissi, L., El Aouad, R. and Manes, J. 2007. Incidence of ochratoxin A in rice and dried fruits from Rabat and Sale area, Morocco. Food Additives Contaminants 24: 285–291.

Zöllner, P. and Mayer-Helm, B. 2006. Trace mycotoxin analysis in complex biological and food matrices by liquid chromatography atmospheric pressure ionisation mass spectrometry. Chromatography A 1136: 123–169.

15

Animal Diet Authentication from Food Analysis

Erwan Engel* and *Jérémy Ratel*

1. Introduction

"A response to rising society: wide demand for guarantees on quality"

—Authors

There is rising consumer demand for guarantees on the quality and safety of their food, especially for farmed animal products. This growing need is largely driven by frequent widely publicized food safety scares, and cases of inappropriate risk management practices (van Rijswijk et al. 2008). Consumers are aware that the quality of an animal product is largely determined by production conditions, and are pressing for guaranteed traceability on food production. EU regulation 178/2002 making traceable documents compulsory for all food/feed businesses lays the foundations to the EU-wide regulatory architecture framing food safety. From 2006, a new draft of 'hygiene legislation' composed of five further regulations (EC 853/2004, EC 882/2004, EC 852/2004, EC 854/2004, EC 183/2005) consolidated, supplemented and modernized the foundation EU legislation on foodstuffs. However, documents can be faked, making it essential to develop additional robust methods that use analyses run on the finished

INRA, UR370 QuaPA, MASS Group, 63122 Saint-Genès-Champanelle, France.
  E-mail: jeremy.ratel@inra.fr
* Corresponding author: erwan.engel@inra.fr

product to authenticate the key information channelled through the paper-based traceability system. Foods of animal origin, which have been at the centre of a number of recent food scares, are now first in line for research, and priority should be given to address certain issues related to critical control points in the production chain that have heavy repercussions on farmed-product quality. High on the list is the animal production systems used to farm livestock, particularly diet and the feedstuffs consumed by the animals which impact the quality of animal-origin foods by altering their composition and/or structure. Certain types of animal diet add significant market value to the animal farmed products as they are considered—whether proven or perceived—nutritionally superior. There is rising demand for authentication of these products obtained with particular animal diet, even if other rearing parameters (breeds, rearing practices, geographical location...) can make it difficult the authentication diagnosis by influencing also the structure and composition of animal-derived products. The development of robust analytical means of characterizing the key variables of change in foodstuffs will ultimately serve to reliably authenticate the key steps in how animals have been fed.

Here we review the main potentially implementable means of authenticating information about diet of animals from analysis of animal-origin foods.

2. Markers in food for Authenticating Animal Diet—The Candidates

When dealing with livestock production systems, by far the most critical information to authenticate is the type of ration fed, as it decisively shapes the nutritional, sensory and food health properties of the animal-source product and its brand image in the consumer's mind. From a sensory standpoint, the diet received by an animal on the farm has effects on the colour, flavour and texture of its meat (Wood et al. 2004) and dairy products (Chilliard et al. 2000, Chouinard et al. 2001). From a nutritional standpoint, it is the diet ration that modulates the fatty acid, antioxidant and mineral contents of the end-product (Chilliard et al. 2000, Scollan et al. 2006, Biesalski 2005, Valsta et al. 2005, Williamson et al. 2005). From a food-safety standpoint, environmental pollutants or certain pathogenic microorganisms can contaminate component ingredients of the animal ration, accumulate in the animal's body tissue, and transfer through to meat and dairy end-products, where they can potentially represent a human health risk (Bernard et al. 2002, Kabak et al. 2006). Finally, farming systems allowing ruminants to feed on pasture have a positive brand image, as pasture-feeding is associated with good animal welfare, ties to home soil, and the natural, ethically-sound dimension attached to the food produced. The development of analytical methods for authenticating the diet of farmed animals therefore represents a major challenge for food product authentication research.

There are two types of components of animal-source foods that can currently be considered potential candidate markers of the animal's on-farm diet. The first possibility is molecular or elemental components introduced directly by the feeds consumed by the animal on the farm, and that show little if any degradation during ingestion or digestion. Candidates would include metabolites introduced by the plants the animals have fed on, like carotenoids, polyphenols, terpenes or fatty acids. Ratios of stable isotopes of the main atoms forming molecular fractions of animal-source products (C, N, H, O, S, Sr) also make relevant candidates as they are dependent on the isotopic composition of certain features of the farm environment, like the plants and water they ingest.

The second possibility also involves metabolites produced by the animal and whose concentrations vary according to type of diet. The main candidates are hydrocarbons, ketones, aldehydes, lactones, sulphur compounds, and fatty acids, among others. Note that this compound group includes a very broad range of 'volatile' compounds whose molecular weight is less than 300 daltons and whose composition is extremely sensitive to animal farming system conditions. These tiny molecules, as end-products of metabolism, make the best-option tracers of metabolic activity in animal tissues or fluids. Recent research has confirmed their position as strong candidate markers for authenticating the type of diet given to farmed animals (Vasta and Priolo 2006).

3. Candidate Markers from Animal Diet

Here we start by reviewing the main candidate markers from animal diet, i.e., carotenoids, polyphenols, terpenes, fatty acids, and stable isotopes, with special focus on stable isotopes as the most intensively researched candidates in the literature to date. We then move on to focus on the promising potential of low-molecular-weight or 'volatile' compounds as candidate markers produced by animal metabolism in response to different livestock production systems.

3.1 Carotenoids

Carotenoids form the major class of natural pigments. Carotenoid concentration in animal body tissues and animal products is directly linked to the animal's carotenoid intake (Calderón et al. 2007, Dian et al. 2007). Type of feed therefore affects the amount of carotenoids found in the animal food produced, which is why this pigment class is proposed to discriminate between feeding regimes used for livestock. The effect of the animal diet on the presence of carotenoids and vitamin A in fluids, tissues and products has been widely evaluated in ruminants, e.g., in sheep, in cattle, and to a lesser extend in goats (Alvarez et al. 2015). Pasture feeding provides the highest dietary carotenoid supply, as fresh green grass is particularly rich in carotenoids (430–700 mg/kg DM), whereas a dry forage-based diet provides far less input, as the carotenoids are temperature-sensitive and photodegrade. Independent of wilting, the ensiling process decreases carotenoid concentration to various degrees, with a concentration loss generally less than 20% of the initial concentration, but which can reach 80% in improper silages (Nozière et al. 2006). Corn silage is equally carotenoid-poor (70–80 mg/kg DM), and zeaxanthin, the major carotenoid in corn kernels, is not even stored in ruminant body tissues. Carotenoid contents are also very low in concentrate feeds.

Several teams have investigated carotenoid content analysis in ruminant body tissues and fluids as a way to authenticate diet, chiefly grass-based diet. Numerous studies have managed, at least at experimental level, to successfully discriminate pasture-based and concentrate-based diet regimes based on milk (Martin et al. 2005), cheese (Verdier-Metz et al. 1998), blood (Prache et al. 2003a) and adipose tissue (Prache et al. 2003b, Röhrle et al. 2011). A common denominator to all of this research is that β-carotene and lutein were analyzed by high-performance liquid chromatography or by spectroscopy. Spectrometry offers the advantage of providing fast and cheap yet reproducible analyses (Martin et al. 2005), and direct-at-slaughterhouse measurements are even in reach with portable spectro-colorimeters (Prache and Thériez 1999). In contrast with ruminants, there are few reports on the presence of carotenoids and retinoids in monogastric livestock and their use as diet biomarker. Álvarez et al. (2014) proposed retinol levels in pig liver as biomarkers for feeding traceability purposes. They discriminated Iberian pigs from two types of feeding systems, Montanera (typical outdoor rearing system with a nutritional strategy based on pasture and acorns) and Cebo (intensive system with a diet based on concentrate). For poultry, carotenoids are supplied in feed for proper pigmentation since chickens do not produce carotenoids naturally (Bortolotti et al. 2003). For example, maize is the major source in chicken feed of carotenoids, pigmenting egg yolks and meat. However, carotenoids have been little used as biomarkers of feeding systems, even if Van ruth et al. (2011) have used them for authentication of eggs according to organic and conventional systems.

Some rapid instrumental techniques have been also used to predict the quality of agricultural products and foods, and particularly to predict carotenoid composition in animal tissues (Osborne et al. 1993). Near-infrared (NIR) spectroscopy has emerged as a rapid method for traceability purposes, mainly to discriminate grass-feeding systems. NIR spectroscopy allows to assess chemical structures through the analysis of the molecular bonds in the near-infrared spectrum and to build a characteristic "fingerprint" of the sample. Animal tissue analysis by NIR has been used to separate grass-feeding systems by quantifying the signature of carotenoid pigments mainly in ruminants. This approach was used for example to separate pasture and concentrate feedings regimes for lambs (Dian et al. 2008) or pasture and corn silage feeding regimes for beef (Cozzolino et al. 2002).

However, studies attempting to discriminate animals reared in less carotenoid-contrasted settings have shown the limits of carotenoids as markers, as they proved highly sensitive to numerous other farm factors, including climate, season, and production factors as diet switching, as well as the common practice of using carotenoids supplements in the concentrates fed to non-pasture-system animals. For example, farmers may elect to add carotenoid-rich alfalfa or spirulina to the concentrate (Kandlakunta et al. 2008) as the vitamin A formed by β-carotene degradation brings added nutritional value for consumers (Nozière et al. 2006). In short, carotenoids carry information that can be workably used for livestock ration authentication purposes, but diagnostic tests based on quantifying carotenoid content alone would appear venturesome at best.

3.2 Polyphenols

Certain fodder plants used in animal diet are very rich in polyphenols, particularly flavonoids and other intracellular phenolic compounds. Polyphenol content and composition vary with plant species and stage of maturity—dicots contain much more polyphenols than grasses. When ingested by ruminants, these hydrophilic compounds get metabolized in the rumen, liver and muscle tissues. Depending on which metabolic pathway they follow, the polyphenols are stored more or less metabolized in different body compartments, while a fraction gets excreted in milk (Besle et al. 2005, O'Connell and Fox 2001). Various teams have demonstrated the ability of certain metabolites to discriminate particular feeding regimes. For example, Besle et al. (2005) found that equol, which is an isoflavone metabolite, formononetin and free aglycones were able to discriminate a red clover-rich ration using high-performance liquid chromatography on cow's milk. However, just like carotenoids, polyphenol content and composition in a plant vary with botanical family, stage of maturity, and environmental factors (Fraisse et al. 2007), which again limits their potential use alone as markers of the diet given to animals and, crucially, herbivores.

3.3 Terpene

Terpenes count among the many low-molecular-weight substances found in meat and dairy products. They are almost exclusively produced by plants.

The 10-carbon-atom monoterpenes and 15-carbon-atom sesquiterpenes, along with their oxygenated derivatives, make up the volatile fraction of terpenes. The big determinant of forage volatile terpene content is botanical composition: volatile terpenes are much less abundant and less diversified in grass-based forage than in diversified praireland forage which counts a diverse mosaic of terpene-rich aromatic dicots (Cornu et al. 2001). Plant terpene content also varies with stage of maturity and is generally higher in late-stage plants than early-stage plants (Cornu et al. 2001, Tornambé et al. 2006). Back in 1978, Dumont and Adda showed that Comté and Beaufort cheeses contained radically substantially more terpene-rich when produced with upland milk than plain land milk. Later work found evidence that the volatile terpene fraction of forages is deposited rapidly and relatively unaltered in the animal tissues (Vasta and Priolo 2006). However, as a grassland's terpene profile is highly dependent on its botanical composition, the specificity of these candidate markers does not fit with the genericity needed for animal diet authentication models. Note, however, that certain individual terpenes like *p*-cymene, β-caryophyllene and trans-cadina-1(6), 4-diene have repeatedly been identified as markers of pasture feeding (Vasta and Priolo 2006, Engel et al. 2007).

3.4 Fatty acids

The type of certain fatty acids (FA) secreted into milk or stored in fat is heavily diet-dependent. This means that the FA composition of meat or milk can be used to trace animal diet. 60% of milk FA (counting 18-plus carbon atoms) come from diet, with the remaining 40% coming from *de novo* synthesis in the mammary glands, predominantly using the acetate and β-hydroxybutyrate produced by ruminal fermentation of sugars (Chilliard et al. 2007). Odd-chain and/or branched-chain FA represent a minor fraction of total milk FA and are synthesized by ruminal bacteria. The polyunsaturated FA (PUFA, primarily linoleic and linolenic acid) introduced through diet undergo microbial activity-driven ruminal biohydrogenation, which culminates in stearic acid and produces intermediates such as the *cis* and *trans* C18:1, C18:2 and C18:3 isomers. As ruminants are unable to synthesize PUFA, the PUFA concentrations in their milk are entirely dependent on the amounts of PUFA that are absorbed in the intestine and thus managed to escape ruminal digestion. Ration factors (forage type and conservation method, forage-to-concentrate ratio, starch content of the ration) and lipid supplementation (supplement type, form of supply, dose and duration) are the main dietary factors capable of altering the ruminal metabolism of PUFA and consequently the profile of the FA leaving the rumen (Chilliard et al. 2007). These same factors also partly alter the bacterial synthesis of minor odd-chain and/or branched-chain FA in the rumen (Vlaeminck et al. 2006). Corn silage, cereals, concentrates, and oilseeds such as soybean and sunflower meal are rich sources of linoleic acid, whereas green grass and linseed are the main sources of linolenic acid. Green grass grazed at early-stage growth yields a milk that is richer in C18:1 *cis*-9 and rumenic acid (the main conjugated linoleic acid isomer in milk) and poorer in linoleic acid that are in concentrate-rich or corn silage-based rations (Ferlay et al. 2006). Mellado et al. used the C18:2 and C18:3 content of perinal fat of goat in order to separate successfully animals with a fattening diet based exclusively on goat milk, from animals fed with milk replacer or with milk-based starter (Mellado et al. 2009). Engel et al. (2007) showed that two FA ratios—*trans*-11, *cis*-15 C18:2/*trans*-11 C18:1 and *cis*-9 C16:1/*iso* C16:0—were able to correctly class all 49 bulk milk samples according to the proportion of corn silage in the basic fodder rations (Fig. 1). The evidence thus suggests that PUFA, intermediates in ruminal biohydrogenation of PUFA, and minor bacterially-synthesized FA could well contribute valuable information for tracing animal diet. As for carotenoids, spectroscopy techniques were also used to predict fatty composition of animal adipose tissues and to generate a

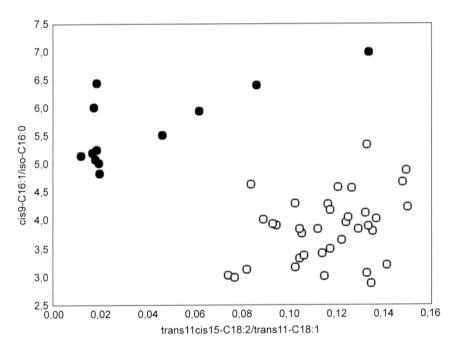

Figure 1. Discrimination of bulk milk samples from plainland-reared cows (proportion of corn in the ration > 30%, black circles) and upland-reared cows (proportion of corn in the ration < 25%, white circles) based on two fatty acid ratios.

Source: Engel et al. (2007).

fingerprint useful for separation of feeding systems. For example, Arce et al. (2009) used the combination of near- and mid-infrared spectral information from pig fat samples to separate satisfactorily Iberian pigs reared with different fattening diets. These authors showed that even when pigs are fed with a special formulation, which has increased oleic acid content in a way that mimics the fatty acid profile of free-range pigs, discrimination is possible based on the spectral characteristics of the samples. Fat tissue fingerprint obtained by Fourier transform MIR spectroscopy was also used to trace back the feeding of suckling lambs, fed with a milk replacer or with mother milk (Osorio et al. 2009). However, there are factors that can skew the diagnostic picture, one is grass wilt, which reduces total FA and linolenic acid concentrations, and another is silage, which produces intermediate responses (Chilliard et al. 2007). Consequently, any diagnostic test of farmed-animal diet based on FA cannot be robust without also combining other types of markers.

3.5 Stable isotopes

Each environment contains different forms of the elements that make up all biological molecules, and these isotopes are used to discriminate between environments by a specific number of neutrons. The distribution profile between stable isotope forms traditionally expressed as the relative abundance of two isotopes—termed "isotopic ratios"—can be influenced by local environment-related factors, such as geographic origin and/or livestock system and feeding regime. The elements involved—chiefly hydrogen, oxygen, carbon, nitrogen, sulphur or strontium—can be used to characterize animal farming conditions (Kelly et al. 2005, Piasentier et al. 2003). For example, $^2H/^1H$ ratio and $^{18}O/^{16}O$ ratio inform on the latitude and altitude at which the animal was farmed. $^{87}Sr/^{86}Sr$ ratio translates the age and composition of the subsoil bedrock. $^{13}C/^{12}C$ ratio is modulated by the type of photosynthesis employed by the plants the animal has fed on, while $^{15}N/^{14}N$ ratio informs on the use of nitrogen fertilizer and $^{34}S/^{32}S$ ratio informs on the use of ammonium sulphate fertilizer.

Stable isotope ratios are generally measured using a specialization of mass spectrometry called isotope-ratio mass spectrometry (IRMS). IRMS is able to differentiate compounds that share the same chemical structure but different relative content of stable isotopes. A study by Boner and Förstel (2004) demonstrated that the IRMS technique can be used to determine the stable isotope ratios of carbon ($^{13}C/^{12}C$), nitrogen ($^{15}N/^{14}N$), oxygen ($^{18}O/^{16}O$), hydrogen ($^2H/^1H$) and sulphur ($^{34}S/^{32}S$). This is achieved by heating the product to a high temperature to transform all its component compounds into gases, generally by combustion or pyrolysis, that are then purified by gas-phase chromatography before being injected into a magnetic sector mass spectrometer. NMR with compound-specific isotopic enrichment cascades can also be used to assess $^2H/^1H$ ratios (Renou et al. 2004). Rather than working on the full product, this method can be applied to target components extracted from the product beforehand, including water, lipid fractions or even molecules pre-extracted from the meat matrix and separated by gas-phase chromatography (Ehtesham et al. 2013). The heavier isotope ratios of strontium ($^{86}Sr/^{87}Sr$) are detected using inductively coupled plasma-mass spectrometry. This 'plasma torch' technique consists in ionizing the sample by injecting it into a high-temperature argon plasma and using mass spectrometry to selectively quantify the ions of the target element. The element isotope ratio of a given sample (R_{sample}) is conventionally compared relative to the standard value measured on a reference product (R_{ref}), and expressed as an index, $\delta_{‰} = (R_{sample}/R_{ref}-1) \times 1000$, reported in *per mil* units.

The challenge of authenticating geographic origin and type of diet have been advanced and addressed in abundant literature (Kelly et al. 2005, Engel et al. 2007). Indeed, the challenge is a completely different issue for products from ruminants like goats, sheep and cattle, which can be strongly tied to the home soil due to local diet composition, and for products from monogastric animals like pigs and poultry, whose diet is essentially off-ground and relatively standardized across countries and regions. This is probably why the majority of research into isotope signature-based authentication of livestock production conditions has essentially focused on products from ruminants. Approaches based on measuring a single stable isotope ratio has generally proven unable to categorically discriminate between products sourced from geographically disperse countries. For example, Hegerding et al. (2002) showed that the $\delta^{18}O_{‰}$ ratio measured in beef meat was unable to differentiate samples sourced from Germany or Argentina from samples sourced from the UK. Franke et al. (2008) further showed that simply measuring $\delta^{18}O_{‰}$ on chicken breast muscle could not discriminate French-raised chickens from Brazilian-raised chickens. In response to this first set of studies, several teams sought to capitalize on the informational complementarity between several isotopic ratios through multi-tracer diagnostics solutions. Boner and Förstel (2004) managed to differentiate beef samples sourced from Germany, Argentina and Chile by measuring $\delta^2H_{‰}$, $\delta^{18}O_{‰}$, $\delta^{13}C_{‰}$, $\delta^{14}N_{‰}$, $\delta^{34}S_{‰}$. More recently, Nakashita et al. (2008) showed that combining $\delta^{13}C_{‰}$ measurements with $\delta^{18}O_{‰}$ measurements made it possible to discriminate between Japanese, Australian and American-sourced beef samples. Along similar lines, Osorio et al. (2011) showed that a model built from $\delta^2H_{‰}$, $\delta^{13}C_{‰}$, $\delta^{14}N_{‰}$, $\delta^{34}S_{‰}$ measurements demonstrated promising discriminant classification performances (> 80%) on beef muscles samples sourced from 9 different countries inside and outside Europe. However, the multi-tracer isotope-based approach still does not give a generic response to the question of authenticating geographic origin. As underlined by Kelly et al. (2005), seasonal and annual variations in $\delta^{18}O_{‰}$ and $\delta^2H_{‰}$ ratios together with other fluctuations in cattle diet sources or fodder production driven by economic, zootechnical or climatic factors can all significantly alter the performance of these diagnostics solutions. These arguments become stronger with increasingly similar climates and farming practices in the geographical zones compared. Heaton et al. (2008) attempted to improve the robustness of these discriminatory methods by coupling isotopic and elemental tracers. However, the best discrimination between South-American, European and Australian beef, which was obtained with a linear model built from 6 variables ($\delta^{13}C_{‰}$, Sr, Fe, $\delta^2H_{‰}$, Rb and Se), was still unable to satisfactorily classify all the samples. It is probably by continuing on this path of studying the informational complementarity between certain types of geographic origin-targeted tracers (isotopes, trace elements, volatile compounds, microbial ecology-related data, etc.) that it will eventually become possible to give a categorical answer on the geographical source of meat products. For the sharper challenge of animal diet authentication, $\delta^{13}C$ and $\delta^{14}N$ ratios look to be the most promising candidates. The stable isotope ratio of ^{13}C in an animal-source product is highly dependent on the composition of the on-farm feed ration given to the animals, particularly proportion of C3 and C4-pathway plants (Bahar et al. 2005, 2009). Several authors have used presence of corn in the ration to discriminate organically-farmed meat from conventionally-farmed meat (Fig. 2). The premise is that corn, which is a C4-metabolism plant, is an energy-packed plant practically only ever used in conventional livestock farming as a way to dope animal growth performances (Boner and Förstel 2004, Schmidt et al. 2005). Studies led in Germany and Ireland showed that conventionally-farmed beef gave a less negative $\delta^{13}C$ isotope ratio than organically-farmed beef (Boner and Förstel 2004, Schmidt et al. 2005). $\delta^{15}N$ isotope ratio is equally informative for discriminating conventionally-farmed from organically-farmed meat. Schmidt et al. (2005) showed that $\delta^{15}N$ values were slightly higher in conventionally-farmed Irish beef than organically-farmed Irish beef. As both sets of animals were reared in essentially pastoral systems, the hypothesis put forward was that soil-plant uptake of ^{15}N would be greater in conventional than organic farming due to the higher nitrogen inputs to land on conventional farms. However, further investigation is warranted on larger samples of animals and broader geographies of livestock farming. Furthermore, it has been shown that isotopic compositions of beef meat are subject to seasonal variation, likely due to seasonal patterns of change in feeding regimes combined with variations in animal tissue turnover rate

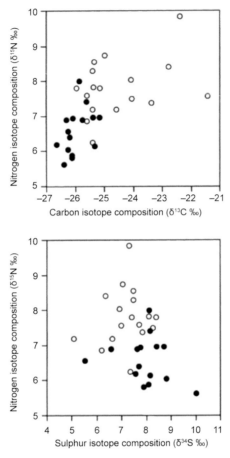

Figure 2. Authentication of meat derived from conventional production systems (white circles) and organic production systems (black circles) by measuring isotope ratios. Distribution of stable isotope composition—nitrogen, carbon and sulphur.
Source: Schmidt et al. (2005).

(Bahar et al. 2008). Similar studies have also investigated stable isotope analysis in lamb and pork to authenticate diet (Piasentier et al. 2003, Gonzalez-Martin et al. 2001) and ^{13}C stable isotope ratio has been measured in cow's milk to characterize and reconstruct diet history (Kornexl et al. 1997, Molkentin and Giesemann 2007). Furthermore, less-negative $\delta^{13}C$ values were found in milk for more intensive dairy operations where diet is more heavily based on cereal crops like corn than on pasture grazing.

3.6 Conclusions on screening for ration components

All the candidate tracers that come from diet ration sources unfortunately have little prospect for workable use in robust livestock diet authentication models as they are extremely specific to plant diversity, easily flawable by supplementation, or too sensitive to numerous farm factors other than diet. It is in an effort to address these limits that research is turning to a different strategy aimed at improving authentication diagnostics on key steps in the production of animal-source foods. With the emergence of high-throughput analytical techniques generating a huge number of variables and also able to generate dense and complex fingerprints, the idea of capturing all the complexity of living material—and with the least restrictive methods possible—has spurred a new paradigm of "omics" approaches. Metabolomics, which studies the set of metabolites left behind in response to a given set of environmental factors, is already making huge strides and could well overcome the hurdles plaguing authentication efforts. The study of low-molecular-weight or 'volatile' metabolites is undoubtedly one of the most promising ways forward.

4. Candidate Markers from Animal Metabolism: A Look at Volatolomics

Recent studies have shown that metabolomics is a particularly promising approach by learning about the subtle changes in cell composition resulting from the action of some factors (Engel et al. 2015). The principle of these approaches is thus to understand the biological system complexity as a whole, with the least restrictive methodologies, searching biologic compounds specifically produced in response to a given situation (Nicholson and Lindon 2008). Many studies have been carried out on the value of the low-molecular-weight or 'volatile organic compounds' to reveal particularly pathologies in humans (Hakim et al. 2012). The literature has also clearly demonstrated the ability of volatile compound profile in livestock organs, tissues or fluids to authenticate certain key stages that shape the quality of animal produce, including animal feeding regime (Vasta and Priolo 2006). Berge et al. (2011) posited that the production of low-molecular-weight secondary metabolites in animal products could be modulated by certain events equating to metabolic 'stresses', like diet shifts, diseases, or an on-farm exposure to environmental pollution. The outcome would be a differential and stress-dependent expression of certain compounds, which could thus serve for authentication purposes. The authors proposed that an animal tissue's compositional profile in volatiles is effectively a metabolic signature able to trace and reconstruct key stages shaping the quality of the resulting animal produce (Fig. 3).

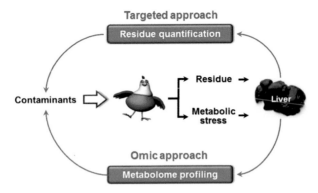

Figure 3. Analytical strategies for back–tracing animal exposure to contaminants.
Source: adapted from Berge et al. (2011).

4.1 Detection methods

To characterize volatile compounds that are commonly found at ultra-low concentrations (often less than ng/g) and possess a broad spectrum of polarities, solubilities and volatilities, the first step is to efficiently extract them from the sample matrix. Solventless headspace extraction techniques like solid-phase microextraction (SPME) are now widely used as they enable automated volatile compound extraction from the media without significantly denaturing the sample analyte. SPME revolves around the adsorption of volatile compounds onto a polymer-coated fused silica fiber extracting phase. The fiber adsorbs the analytes present in the gas phase floating just above the liquid or solid sample. In most analytical systems, the compounds adsorbed onto the SPME fiber are then being desorbed thermally. However, there is still room to improve the repeatability of this extraction step via new technological and/or chemometric advances (Zhang et al. 2014). Once the volatile fraction has been extracted from the matrix and concentrated, the technique best-geared to separating the rich and complex volatile compound fraction from the products is gas chromatography (GC). The compounds are separated according to their volatility by a temperature gradient programmed into the GC step, and according to their affinity for the stationary phase coating the inside walls of the chromatography column. Once separated by gas chromatography, the volatile components are generally identified and quantified by mass spectrometry (MS). The chemical compound is impacted by a high-power (70 eV) electron beam until it fragments into ions, and the abundances of each ion present can then be measured after a separation step. The component is identified by analyzing the relative abundance of the different ions, called a 'mass spectrum', which gives a characteristic picture for each component. The absolute abundance of the signal of one or more of these ions, where each

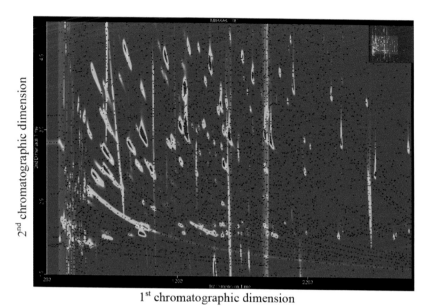

Figure 4. 2D chromatogram obtained from the volatolome of lamb adipose tissue by two–dimensional gas chromatography coupled with time–of–flight mass spectrometry (GC×GC–TOFMS). The second chromatographic dimension is able to highlight volatile compounds that could not be separated through the first chromatographic dimension.
Source: taken from Ratel and Engel (2009).

signal is identified through its specific mass-to-charge ratio (m/z), serves to quantify the component. Recent advances in GC–MS are set to make it the gold standard in the search for markers of quality. The lead-candidate emerging technology is two-dimensional gas chromatography coupled with mass spectrometry (GC×GC–MS), a technique developed to overcome persistent coelution problems hampering classic one-dimensional GC as the volatile fractions of foods are so rich and complex that many components stay mixed, or 'co-eluted', after the chromatography step. The premise behind GC×GC is to re-separate the eluate from the first column on a second column with a stationary phase that has entirely different physical-chemical properties to the first column. This 2D GC approach brings substantially greater resolving power over more traditional GC–MS techniques, making it possible to identify a multitude of volatile components that would be impossible to resolve otherwise (Tranchida et al. 2004, Ratel and Engel 2009). Figure 4 shows a 2D chromatogram acquired after GC×GC–MS analysis of the volatile fraction extracted from lamb fat as part of research to identify compounds discriminating pasture-based or concentrate-based feed rations (adapted from Ratel and Engel 2009).

Ion mobility spectrometry (IMS) is another technique which could be interesting to analyse these volatile promising markers. Ion mobility spectrometry (IMS) is an analytical technique for the determination of volatile and semivolatile compounds based on the gas-phase separation of the resulting ions under a weak electric field at ambient pressure (Eiceman 2002). This technique has been showed suitable for application in the field of food quality and safety (Armenta et al. 2011). Alonso et al. (2008) showed that characteristic ion mobility spectra for volatile compounds present in fat were useful to authenticate the feeding regime of Iberian pigs. These authors obtained a promising classification of fat samples either from pigs reared under the free-range regime and fed mainly on pasture and acorns, or from pigs reared in confinement facilities and fed with commercial pig feed, or from pigs fed with standard pig feed or fed with a special feed possessing an increased oleic acid content.

4.2 Applications

The challenge of authenticating ruminant diet is probably the point that the quality sector has worked on the most, and the bulk of research in this field has focused on discriminating pasture-based and concentrate-based diets (Vasta and Priolo 2006). Recent breakthroughs in the analysis of volatolome in body fluids and tissues have made it possible to identify a large number of tracers that act as signatures distinguishing pastured-based or concentrate-based diet in ruminants fed exclusively on either feed. Since these volatile compounds are generally lipophilic, it is adipose tissue that has proven the main target for screening (Engel and Ratel 2007). Indeed, a recent study has demonstrated that running parallel analyses in several different adipose tissues makes the diet authentication tests more robust (Sivadier et al. 2008). Vasta et al. (2007) also demonstrated the feasibility of using volatolome analysis to authenticate exclusive diets in lamb meat. Whatever tissue is studied, the tracers identified all belong to different chemical families (fatty acids, alkanes, alcohols, aldehydes, ketones, lactones, terpenes, or others) and all carry complementary information on animal diet regimen. Building on this initial research confirming the ability of volatile compounds to differentiate contrasting dietary regimens, several recent studies have investigated the possibilities for authenticating diets closer to real-world farm conditions. Many real-world farms run 'diet-switching', where zootechnical factors often dictate alternating periods of exclusively pasture or concentrate-based diet, and it is common farm practice to finish pasture-fed animals on concentrates in order to achieve carcass fattening targets, but some farms may fraudulently attempt to simulate an exclusively pasture-based system by finishing stall-reared lambs on pasture. Recent research has been able to characterize the appearance and disappearance rates of volatile tracers of pasture-feeding in the adipose tissue of either pasture-fed lambs finished for different-length periods on a concentrate diet (Sivadier et al. 2010) or lambs finished on pasture following different-length periods of a concentrate-based diet (Sivadier et al. 2009). This information was combined into chemometrics run on appropriate data to successfully discriminate between the four different finishing periods studied and under both types of diet-switching system. Figure 5 presents the outcome of this exploratory data analysis which visually evidences successful discrimination between the four degrees of grass finishing. Variations in herd management systems are not the only possible sources of bias in diet authentication tests constructed from volatile compounds. There are changes in muscle tissue composition that occur after slaughter and after the sample are taken, which adds a further source of bias. This is because in the time between slaughter and consumer retail of the cut of meat, the meat muscle goes through a maturation step. By tracking the patterns of grass-diet tracers in muscle tissue during ageing, Lehallier et al. (2012) demonstrated that signature compounds of pasture feeding can be identified whatever the stage in the muscle-to-meat process. Applying novel chemometrics processing based on reconstructing the ratio of volatile

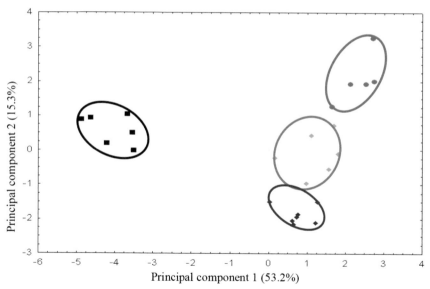

Figure 5. Authentication of meat farming systems: discrimination between lambs whose diets have been switched during farming by probing the volatile fraction of their adipose tissue using gas chromatography-mass spectrometry. First principle component analysis performed on dietary regimen markers identified in caudal subcutaneous and perirenal adipose tissue: clear discrimination between the 4 groups of lambs raised either exclusively on concentrate (■) or raised on a concentrate diet and 'finished' on pasture for 17 days (♦), 51 days (♦) or 85 days (•). Source: adapted from Sivadier et al. (2009).

compounds enabled correct classification of 100% of the meat samples according to animal diet, regardless of how long the meat had aged (Fig. 6).

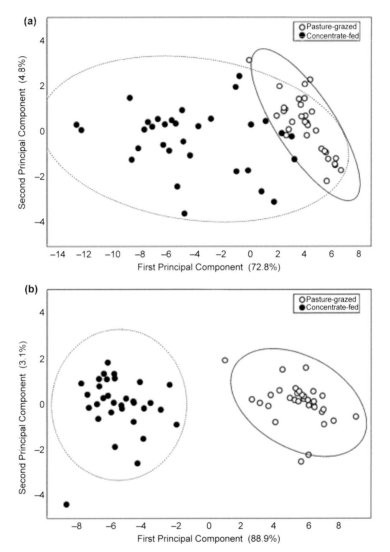

Figure 6. Authentication of ruminant diet regime (pasture-grazed lambs in white circles vs concentrate-fed lambs in black circles) from analysis of meat samples aged to different degrees (ageing for 1, 2, 3, 4 or 7 days) by gas chromatography-mass spectrometry. Ability of novel chemometrics approaches to improve sample discrimination and biomarker identification. (a) Discrimination obtained before applying systematic ratio normalization to reprocess the GC–MS signal data. (b) Discrimination obtained after applying systematic ratio normalization to reprocess the GC–MS signal data.
Source: Lehallier et al. (2012).

Even if these volatile markers from animal metabolism hold great promise for authenticating animal diet, there are two point factors that still need to be improved or completed in order to lend these tracers infallible robustness. The first point concerns the critical step of volatile compound extraction from the animal products. Headspace extraction techniques, particularly SPME which is now the mainstay for this process-critical step, still carry big limitations in terms of volatile trapping capacity, analytical repeatability and coefficient of variation. Alternative techniques have been developed but are not yet as widely accessible as SPME. New sorbents have also been developed to address these limitations, and the development effort may ultimately produce materials that enable efficient and repeatable recovery of in-food volatiles (Gallidabino et al. 2014, Marquez et al. 2014). The second point concerns the robustness of the candidate-marker volatiles identified up to date in what is essentially experimental research and which will need to get through field tests. Validation of the robustness of these markers could well consist in reproducing exactly the same sampling protocol that served to first identify these candidate markers and re-running the analyses but, this time, only ruling in those markers that demonstrate the strongest cross-agreement and the weakest context-bias. In those cases where the samples come from controlled experimental studies, this first step will need to be completed by re-sampling in real-world conditions. The second step could then be to take a candidate marker ruled in by the first validation step and performance-test it for relevancy. Understanding the biochemical and metabolic pathways that produce these markers will be relevant here for explaining where quantitative differences observed between feeding regimes come from. Radiolabelling the precursors of candidate markers and tracking these isotope fluxes is almost certainly the most appropriate analytical solution for capturing this information.

5. Conclusion and Perspectives

The increase in international scientific literature on feed authentication probably reflects a growing interest generated by analytical breakthroughs in the ability to not only provide objective guarantees of transparency in food production practices but also to legitimize and safeguard distinctive signs of food quality in an increasingly segmented marketplace. There is expected to be significant spin-off, as much for customer-perceived trustworthiness of animal-product quality and safety claims as for animal producers and institutions that will be given fully objective methods for vindicating the quality of their products. Authentication challenges do not stop and start at animal diet. Authentication includes also, among others, the origin (species, geographical or genetic), production method (conventional, organic, traditional procedures, free range), or processing technologies (irradiation, freezing, microwave heating). One of the big analytical challenges facing the scientific community today—and strongly tying into diet authentication—is arguably the authentication of "sustainable" agriculture, typically organically-farmed produce. Another challenge could be to determine if a food product is originated from an animal fed with genetically modified feeds, but the control of transgenic materials in animal-derived products seems to be needless since transgenic proteins and DNA have been shown rapidly degraded by normal digestive processes. Authenticating produce from non-cloned animals could fast become the next big challenge. Finally, food authentication on health-and-hygiene credentials is also anticipated as major challenge on the near horizon. Consumers, once again spurred by frequent and high-profile "chemical" scares on food safety, are growing increasingly attentive to this dimension of food quality, which could fast become a key criterion dictating decision to buy. Providing consumers assurances that livestock has never been exposed, especially via diet, to toxicants while being farmed is inescapably a very real challenge given today's growing awareness of the multiple sources of environmental pollution. Research in this direction is going to be pushed to come up with analytical methods enabling large-scale standard-procedure system-wide food safety tests on all animal-source foods put on sale. As things stand, there is little prospect that this kind of large-scale testing will actually see the light of day given the price of routine contaminant assay methods and the multitude of potential animal produce contaminants. Any and all alternatives developed by research teams to meet these key challenges will have to be robust yet fraud proof methods that can be used for regulatory purposes. Research is set to follow two main paths based on in-food analysis: the development of poly-residue analytical approaches targeted at a handful of *a priori* relevant components, or the development of multi-marker "omics" approaches based on identifying metabolic markers capable of flagging up potential health-hazard situations. An approach based on multivariate analysis combined with powerful chemometrics is set to be the pivotal strategy to achieve this robustness objective for diet authentication tests. For industry sectors and institutions, the challenge is to update the regulations so they can quickly incorporate these methodological advances and thereby enable more systematic control-tests to provide consumers with solid guarantees on the quality of meat products, and possibly also their unique selling points.

Keywords: Authentication, traceability, farmed-animal diet, animal-derived food, markers, carotenoids, polyphenols, terpenes, fatty acids, stable isotopes, volatolomics, metabolomics, volatile organic compounds, GC×GC–TOF/MS, chemometrics

References

Alonso, R., Rodríguez-Estévez, V., Domínguez-Vidal, A., Ayora-Cañada, M.J., Arce, L. and Valcárcel, M. 2008. Ion mobility spectrometry of volatile compounds from Iberian pig fat for fast feeding regime authentication. Talanta 76: 591–596.

Álvarez, R., Vicario, I.M., Meléndez-Martínez, A.J. and Alcalde, M.J. 2014. Effect of different carotenoid-containing diets on the vitamin A levels and colour parameters in Iberian pigs' tissues: utility as biomarkers of traceability. Meat Science 98: 187–192.

Álvarez, R., Meléndez-Martínez, A.J., Vicario, I.M. and Alcalde, M.J. 2015. Carotenoid and vitamin A contents in biological fluids and tissues of animals as an effect of the diet: A review. Food Reviews International 31: 319–340.

Arce, L., Domínguez-Vidal, A., Rodríguez-Estévez, V., López-Vidal, S., Ayora-Cañada, M.J., and Valcárcel, M. 2009. Feasibility study on the use of infrared spectroscopy for the direct authentication of Iberian pig fattening diet. Analytica Chimica Acta 636: 183–189.

Armenta, S., Alcala, M. and Blanco, M. 2011. A review of recent, unconventional applications of ion mobility spectrometry (IMS). Analytica Chimica Acta 703: 114–123.

Bahar, B., Moloney, A.P., Monahan, F.J., Harrison, S.M., Zazzo, A., Scrimgeour, C.M., Begley, I.S. and Schmidt, O. 2009. Turnover of carbon, nitrogen, and sulfur in bovine longissimus dorsi and psoas major muscles: implications for isotopic authentication of meat. Journal of Animal Science 87: 905–913.

Bahar, B., Monahan, F.J., Moloney, A.P., O'Kiely, P., Scrimgeour, C.M. and Schmidt, O. 2005. Alteration of the carbon and nitrogen stable isotope composition of beef by substitution of grass silage with maize silage. Rapid Communications in Mass Spectrometry 19: 1937–1942.

Bahar, B., Schmidt, O., Moloney, A.P., Scrimgeour, C.M., Begley, I.S. and Monahan, F.J. 2008. Seasonal variation in the C, N and S stable isotope composition of retail organic and conventional Irish beef. Food Chemistry 106: 1299–1305.

Berge, P., Ratel, J., Fournier, A., Jondreville, C., Feidt, C., Roudaut, B., Le Bizec, B. and Engel, E. 2011. Use of volatile compound metabolic signatures in poultry liver to back-trace dietary exposure to rapidly metabolized xenobiotics. Environmental Science and Technology 45: 6584–6591.

Besle, J.M., Lamaison, J.L., Dujol, B., Pradel, P., Fraisse, D., Viala, D. and Martin, B. 2005. Flavonoids and other phenolics in milk as a putative tool for traceability of dairy production systems. *In*: Hocquette, J.F. and Gigli, S. (eds.). Indicators of Milk and Beef Quality. Wageningen Academic Publishers, Wageningen, EAAP Publication 112: 345–350.

Bernard, A., Broeckaert, F., De Poorter, G., De Cock, A., Hermans, C., Saegerman, C. and Houins, G. 2002. The Belgian PCB/dioxin incident: Analysis of the food chain contamination and health risk evaluation. Environmental Research 88: 1–18.

Biesalski, H.K. 2005. Meat as a component of a healthy diet—are there any risks or benefits if meat is avoided in the diet? Meat Science 70: 509–524.

Boner, M. and Forstel, H. 2004. Stable isotope variation as a tool to trace the authenticity of beef. Analytical and Bioanalytical Chemistry 378: 301–310.

Bortolotti, G.R., Negro, J.J., Surai, P.F. and Prieto, P. 2003. Carotenoids in eggs and plasma of red-legged partridges: effects of diet and reproductive output. Physiological and Biochemical Zoology 76: 367–374.

Calderón, F., Chauveau–Duriot, B., Pradel, P., Martin, B., Graulet, B., Doreau, M. and Nozière, P. 2007. Variations in Carotenoids, Vitamins A and E, and color in cow's plasma and milk following a shift from hay diet to diets containing increasing levels of carotenoids and vitamin. European Journal of Dairy Science 90: 5651–5664.

Chilliard, Y., Ferlay, A., Mansbridge, R.M. and Doreau, M. 2000. Ruminant milk fat plasticity: nutritional control of saturated, polyunsaturated, trans and conjugated fatty acids. Annales de Zootechnie. EDP Sciences 49: 181–205.

Chilliard, Y., Glasser, F., Ferlay, A., Bernard, L., Rouel, J. and Doreau, M. 2007. Diet, rumen biohydrogenation, cow and goat milk fat nutritional quality: a review. European Journal of Lipid Science and Technology 109: 828–855.

Chouinard, P.Y., Corneau, L., Butler, W.R., Bauman, D.E., Chilliard, Y. and Drackley, J.K. 2001. Effect of dietary lipid source on conjugated linoleic acid concentrations in milk fat. Journal of Dairy Science 84: 680–690.

Cornu, A., Carnat, A.P., Martin, B., Coulon, J.B., Lamaison, J.L. and Berdagué, J.L. 2001. Solid phase micro-extraction of volatile components from natural grassland plants. Journal of Agricultural and Food Chemistry 49: 203–209.

Cozzolino, D., Martins, V. and Murray, I. 2002. Visible and near infrared spectroscopy of beef longissimus dorsi muscle as a means of dicriminating between pasture and corn silage feeding regimes. Journal of Near Infrared Spectroscopy 10: 187–193.

Dian, P.H.M., Chauveau–Duriot, B., Prado, I.N. and Prache, S. 2007. A dose-response study relating the concentration of carotenoid pigments in blood and fat reflectance spectrum characteristics to the carotenoid intake level in sheep. Journal of Animal Science 85: 3054–3061.

Dian, P.H.M., Andueza, D., Jestin, M., Prado, I.N. and Prache, S. 2008. Comparison of visible and near infrared reflectance spectroscopy to discriminate between pasture-fed and concentrate-fed lamb carcasses. Meat Science 80: 1157–1164.

Dumont, J.P. and Adda, J. 1978. Occurrence of sesquiterpenes in mountain cheese volatiles. Journal of Agricultural and Food Chemistry 26: 364–367.

Eiceman, G.A. 2002. Ion-mobility spectrometry as a fast monitor of chemical composition. TrAC Trends in Analytical Chemistry 21: 259–275.

Ehtesham, E., Hayman, A.R., McComb, K.A., Van Hale, R. and Frew, R.D. 2013. Correlation of geographical location with stable isotope values of hydrogen and carbon of fatty acids from New Zealand milk and bulk milk powder. Journal of Agricultural and Food Chemistry 61: 8914–8923.

Engel, E., Ferlay, A., Cornu, A., Chilliard, Y., Agabriel, C., Bielicki, G. and Martin, B. 2007. Relevance of isotopic and molecular biomarkers for the authentication of milk according to production zone and type of feeding of the cow. Journal of Agricultural and Food Chemistry 55: 9099–9108.

Engel, E. and Ratel, J. 2007. Correction of the data generated by mass spectrometry analyses of biological tissues: Application to food authentication. Journal of Chromatography A 1154: 331–341.

Engel, E., Ratel, J., Bouhlel, J., Planche, C. and Meurillon, M. 2015. Novel approaches to improving the chemical safety of the meat chain towards toxicants. Meat Science 109: 75–85.

Ferlay, A., Martin, B., Pradel, P., Coulon, J.B. and Chilliard, Y. 2006. Influence of grass-based diets on milk fatty acid composition and milk lipolytic system in Tarentaise and Montbéliarde cow breeds. Journal of Dairy Science 89: 4026–4041.

Fraisse, D., Carnat, A., Viala, D., Pradel, P., Besle, J.M., Coulon, J.B., Felgines, C. and Lamaison, J.L. 2007. Polyphenolic composition of a permanent pasture: variations related to the period of harvesting. Journal of the Science of Food and Agriculture 87: 2427–2435.

Franke, B.M., Koslitz, S., Micaux, F., Piantini, U., Maury, V., Pfammatter, E., Wunderli, S., Gremaud, G., Bosset, J.O., Hadorn, R. and Kreuzer, M. 2008. Tracing the geographic origin of poultry meat and dried beef with oxygen and strontium isotope ratios. European Food Research and Technology 226: 761–769.

Gallidabino, M., Romolo, F.S., Bylenga, K. and Weyermann, C. 2014. Development of a novel headspace sorptive extraction method to study the aging of volatile compounds in spent handgun cartridges. Analytical Chemistry 86: 4471–4478.

González–Martín, I., González, Pérez C., Hernández Méndez, J. and Sánchez González, C. 2001. Differentiation of dietary regimene of Iberian swine by means of isotopic analysis of carbon and sulphur in hepatic tissue. Meat Science 58: 25–30.

Hakim, M., Broza, Y.Y., Barash, O., Peled, N., Phillips, M., Amann, A. and Haick, H. 2012. Volatile organic compounds of lung cancer and possible biochemical pathways. Chemical Reviews 112: 5949–5966.

Heaton, K., Kelly, S.D., Hoogewerff, J. and Woolfe, M. 2008. Verifying the geographical origin of beef: The application of multi-element isotope and trace element analysis. Food Chemistry 107: 506–515.

Hegerding, L., Seidler, D., Danneel, H.J., Gessler, A. and Nowak, B. 2002. Oxygen isotope–ratio–analysis for the determination of the origin of beef. Fleischwirtschaft 82: 95–100.

Kabak, B., Dobson, A.D.W. and Var, I. 2006. Strategies to prevent mycotoxin contamination of food and animal feed: a review research. Critical Reviews in Food Science and Nutrition 46: 593–619.

Kandlakunta, B., Rajendran, A. and Thingnganing, L. 2008. Carotene content of some common (cereals, pulses, vegetables, spices and condiments) and unconventional sources of plant origin. Food Chemistry 106: 85–89.

Kelly, S., Heaton, K. and Hoogewerff, J. 2005. Tracing the geographical origin of food: the application of multi-element and multi-isotope analysis. Trends in Food Science and Technology 16: 555–567.

Kornexl, B.E., Werner, T., Rossmann, A. and Schmidt, H.L. 1997. Measurements of stable isotope abundances in milk and milk ingredients—a possible tool for origin assignment and quality control. Zeitschrift für Lebensmitteluntersuchung und Forschung A 205: 19–24.

Lehallier, B., Ratel, J., Hanafi, M. and Engel, E. 2012. Systematic ratio normalization of gas chromatography signals for biological sample discrimination and biomarker discovery. Analytica Chimica Acta 733: 16–22.

Marquez, A., Serratosa, M.P., Merida, J., Zea, L. and Moyano, L. 2014. Optimization and validation of an automated DHS–TD–GC–MS method for the determination of aromatic esters in sweet wines. Talanta 123: 32–38.

Martin, B., Cornu, A., Kondjoyan, N., Ferlay, A., Verdier–Metz, I., Pradel, P., Rock, E., Chilliard, Y., Coulon, J.B. and Berdagué, J.L. 2005. Milk indicators for recognizing the types of forages eaten by dairy cows. *In*: Hocquette, J.F. and Gigli, S. (eds.). Indicators of Milk and Beef Quality. Wageningen Academic Publishers, Wageningen, EAAP publication 112: 127–136.

Mellado-González, T., Narváez-Rivas, M., Alcalde, M.J., Cano, T. and León-Camacho, M. 2009. Authentication of fattening diet of goat kid according to their fatty acid profile from perirenal fat. Talanta 77: 1603–1608.

Molkentin, J. and Giesemann, A. 2007. Differentiation of organically and conventionally produced milk by stable isotope and fatty acid analysis. Analytical and Bioanalytical Chemistry 388: 297–305.

Nakashita, R., Suzuki, Y., Akamatsu, F., Iizumi, Y., Korenaga, T. and Chikaraishi, Y. 2008. Stable carbon, nitrogen, and oxygen isotope analysis as a potential tool for verifying geographical origin of beef. Analytica Chimica Acta 617: 148–152.

Nicholson, J.K. and Lindon, J.C. 2008. Systems biology: metabolomics. Nature 455: 1054–1056.

Nozière, P., Graulet, B., Lucas, A., Martin, B., Grolier, P. and Doreau, M. 2006. Carotenoids in ruminants: from forage to dairy products. Animal Feed Science and Technology 131: 418–450.

O'connell, J.E. and Fox, P.F. 2001. Significance and applications of phenolic compounds in the production and quality of milk and dairy products: a review. International Dairy Journal 11: 103–120.

Osborne, B.G., Fearn, T. and Hindle, P.H. 1993. Practical NIR spectroscopy with applications in food and beverage analysis. Longman Scientific and Technical.

Osorio, M.T., Zumalacárregui, J.M., Alaiz-Rodríguez, R., Guzman-Martínez, R., Engelsen, S.B. and Mateo, J. 2009. Differentiation of perirenal and omental fat quality of suckling lambs according to the rearing system from Fourier transforms mid-infrared spectra using partial least squares and artificial neural networks analysis. Meat Science 83: 140–147.

Osorio, M.T., Moloney, A.P., Schmidt, O. and Monahan, F.J. 2011. Multielement isotope analysis of bovine muscle for determination of international geographical origin of meat. Journal of Agricultural and Food Chemistry 59: 3285–3294.

Piasentier, E., Valusso, R., Camin, F. and Versini, G. 2003. Stable isotope ratio analysis for authentication of lamb meat. Meat Science 64: 239–247.

Prache, S., Priolo, A. and Grolier, P. 2003a. Persistence of carotenoid pigments in the blood of concentrate-finished grazing sheep: its significance for the traceability of grass-feeding. Journal of Animal Science 81: 360–367.

Prache, S., Priolo, A. and Grolier, P. 2003b. Effect of concentrate finishing on the carotenoid content of perirenal fat in grazing sheep: its significance for discriminating grass-fed, concentrate-fed and concentrate-finished grazing lambs. Animal Science 77: 225–233.

Prache, S. and Theriez, M. 1999. Traceability of lamb production systems: carotenoids in plasma and adipose tissue. Animal Science 69: 29–36.

Ratel, J. and Engel, E. 2009. Determination of benzenic and halogenated volatile organic compounds in animal-derived food products by one-dimensional and comprehensive two-dimensional gas chromatography-mass spectrometry. Journal of Chromatography A 1216: 7889–7898.

Renou, J.P., Bielicki, G., Deponge, C., Gachon, P., Micol, D. and Ritz, P. 2004. Characterization of animal products according to geographic origin and feeding diet using nuclear magnetic resonance and isotope ratio mass spectrometry. Part II: Beef meat. Food Chemistry 86: 251–256.

Röhrle, F.T., Moloney, A.P., Osorio, M.T., Luciano, G., Priolo, A., Caplan, P. and Monahan, F.J. 2011. Carotenoid, colour and reflectance measurements in bovine adipose tissue to discriminate between beef from different feeding systems. Meat Science 88: 347–353.

van Ruth, S., Alewijn, M., Rogers, K., Newton-Smith, E., Tena, N., Bollen, M. and Koot, A. 2011. Authentication of organic and conventional eggs by carotenoid profiling. Food Chemistry 126: 1299–1305.

Schmidt, O., Quilter, J.M., Bahar, B., Moloney, A.P., Scrimgeour, C.M., Begley, I.S. and Monahan, F.J. 2005. Inferring the origin and dietary history of beef from C, N and S stable isotope ratio analysis. Food Chemistry 91: 545–549.

Scollan, N., Hocquette, J.F., Nuernberg, K., Dannenberger, D., Richardson, I. and Moloney, A. 2006. Innovations in beef production systems that enhance the nutritional and health value of beef lipids and their relationship with meat quality. Meat Science 74: 17–33.

Sivadier, G., Ratel, J., Bouvier, F. and Engel, E. 2008. Authentication of meat products: determination of animal feeding by parallel GC-MS analysis of three adipose tissues. Journal of Agricultural and Food Chemistry 56: 9803–9812.

Sivadier, G., Ratel, J. and Engel, E. 2009. Latency and persistence of diet volatile biomarkers in lamb fats. Journal of Agricultural and Food Chemistry 57: 645–652.

Sivadier, G., Ratel, J. and Engel, E. 2010. Persistence of pasture feeding volatile biomarkers in lamb fats. Food Chemistry 118: 418–425.

Tornambé, G., Cornu, A., Pradel, P., Kondjoyan, N., Carnat, A.P., Petit, M. and Martin, B. 2006. Changes in terpene content in milk from pasture fed cows. Journal of Dairy Science 89: 2309–2319.

Tranchida, P.Q., Dugo, P., Dugo, G. and Mondello, L. 2004. Comprehensive two-dimensional chromatography in food analysis. Journal of Chromatography A 1054: 3–16.

Valsta, L.M., Tapanainen, H. and Mannisto, S. 2005. Meat fats in nutrition. Meat Science 70: 525–530.

Van Rijswijk, W., Frewer, L.J., Menozzi, D. and Faioli, G. 2008. Consumer perceptions of traceability: A cross-national comparison of the associated benefits. Food Quality and Preference 19: 452–464.

Vasta, V. and Priolo, A. 2006. Ruminant fat volatiles as affected by diet. A review. Meat Science 73: 218–228.

Vasta, V., Ratel, J. and Engel, E. 2007. Mass spectrometry analysis of volatile compounds in raw meat for the authentication of the feeding background of farm animals. Journal of Agricultural and Food Chemistry 55: 4630–4639.

Verdier–Metz, I., Coulon, J.B., Pradel, P., Viallon, C. and Berdague, J.L. 1998. Effect of forage conservation (hay or silage) and cow breed on the coagulation properties of milks and on the characteristics of ripened cheeses. Journal of Dairy Research 65: 9–21.

Vlaeminck, B., Fievez, V., Cabrita, A.R.J., Fonseca, A.J.M. and Dewhurst, R.J. 2006. Factors affecting odd and branched-chain fatty acids in milk: a review. Animal Feed Science and Technology 131: 389–417.

Williamson, C.S., Foster, R.K., Stanner, S.A. and Buttriss, J.L. 2005. Red meat in the diet. Nutrition Bulletin 30: 323–335.

Wood, J.D., Richardson, R.I., Nute, G.R., Fisher, A.V., Campo, M.M., Kasapidou, E., Sheard, P.R. and Enser, M. 2004. Effects of fatty acids on meat quality: a review. Meat Science 66: 21–32.

Zhang, H., Ng, B.W.L. and Lee, H.K. 2014. Development and evaluation of plunger-in-needle liquid-phase microextraction. Journal of Chromatography A 1326: 20–28.

16

Authentication and Traceability of Agricultural and Food Products Using Vibrational Spectroscopy

Philippe Vermeulen,[1] Juan A. Fernandez Pierna,[1] Ouissam Abbas,[1] Hervé Rogez,[2] Fabrice Davrieux[3] and Vincent Baeten[1,]*

1. Introduction

Food and feed safety is an increasing concern for consumers following major crises related directly or indirectly to human health. EU has created a key tool, the Rapid Alert System for Food and Feed (RASFF 2015), for reacting quickly to food and feed safety emergencies and incidents. Currently, European foods are recognized globally for their high standards of production, labelling and safety. This is not yet the case, however, for the detection of food fraud or the enforcement of the relevant legislation. There

---

[1] Food and Feed Quality Unit, Quality Department of Agricultural Products, Walloon Agricultural Research Centre (CRA-W), Henseval building, 24, Chaussée de Namur B-5030 Gembloux Belgium.
  Email: p.vermeulen@cra.wallonie.be; j.fernandez@cra.wallonie.be; o.abbas@cra.wallonie.be
[2] Centre for Agro-food Valorisation of Amazonian Bioactive Compounds (CVACBA), University Federal do Para (UFPA), Av. Perimetral, s/n Guamá 66075-900 – Belem, PA – Brasil; Email: herverogez@gmail.com
[3] UMR Qualisud, CIRAD Persyst, 7, Chemin de l'IRAT 97455 Saint Pierre Réunion, France. Email: davrieux@cirad.fr
* Corresponding author: v.baeten@cra.wallonie.be; FoodFeedQuality@cra.wallonie.be

is a clear need for an initiative that will link the major stakeholders, establish data-sharing tools and working practices, and provide rapid fit-for-purpose screening and verification methods. The European Food Integrity project (2014–2018) aims to address that need. It is not a single method based on a single technique that will address all the needs of farmers, producers, retailers, regulatory bodies and consumers, but rather a combination of methods and strict legislation. Farmers need analytical methods giving them the ability at the farm level to check that production matches the authenticity and quality criteria included in the product specifications. Food producers need analytical methods enabling them to define the authenticity criteria and check the compliance of the raw material produced. Retailers need tools enabling them to check that products reflect the criteria agreed with food producers. On the other hand, regulatory bodies need analytical methods for certifying the products in terms of the legislation on quality, safety and authenticity. Confirmatory methods are needed that provide indisputable information that could be used, if necessary, in court. The development of analytical tools is less important for consumers, who simply want enough information to feel reassured about the authenticity of a product and the likelihood of it meeting their expectations.

Currently, many methods based on various analytical techniques are used to authenticate agro-food products. Among them, vibrational spectroscopy techniques include NIR, MIR, Raman and Terahertz techniques. They are based on measuring the amount of electromagnetic radiation absorbed by a sample according to the Beer-Lambert law. They are techniques to consider when authenticity controls need to be established at the field level, at the point where products are delivered to factories or during the production process. Methods based on these techniques are indirect, rapid and do not require skilled staff. They are not confirmatory and are therefore seldom used in official control processes. Fingerprinting methods, however, are of interest to regulatory bodies because they allow rapid preventative action to be taken. It should be noted that, despite the many studies demonstrating their potential, the application of fingerprinting methods in routine analysis and food authenticity surveillance remains limited (Riedl et al. 2015). The second section of this chapter provides a general overview of the technology, main principles, instrumentation, sample presentation and new trends, as well as giving a brief overview of the chemometric tools used to extract chemically relevant information from the spectra.

In the third section of the chapter, several examples are discussed to illustrate the potential of vibrational spectroscopy in tackling authenticity challenges (e.g., the discrimination of cereal varieties), in identifying botanical origin, geographical origin and distillers' dried grains and solubles (DDGS) production process, in the traceability and authentication of fruits and in the early detection of fraud in food/feed ingredients. Several examples come from European projects focusing on authenticity: Qsaffe (2011–2014), which looked at the early detection of fraud in feed and at DDGS authentication; PhotonFruit (2014–2015), which dealt with emergent spectroscopic techniques for the quality control and traceability of fruits and fruit-based products; FoodIntegrity (2014–2018), which aims to provide industries and regulatory authorities in the food and feed sectors with information on the analytical methods available, their use, performance and cost, as well as the availability of reference data, with links to literature and other databases.

2. Vibrational Spectroscopy Methods: NIR, MIR and Raman

This section gives the main principles of near-infrared, mid-infrared and Raman spectroscopy as well as new trends regarding instrumentation and sample presentation. It gives also a brief overview of the chemometric tools used through the examples presented in Section 3.

2.1 Principle

The term 'electromagnetic spectrum' refers to the collection of radiant energy sources, from gamma rays to radio waves. These waves are characterized by wavelength λ (the length of one wave, cm), frequency v (the number of vibrations per unit time, Hz) and wave number v (the number of waves per unit length, cm^{-1}). Spectroscopy can be defined as the study of the interaction between electromagnetic radiation and matter. The electromagnetic spectrum is divided into several regions, each of which induces specific molecular or atomic transition and is therefore suited to a specific type of spectroscopy. This chapter focuses on wavelengths in the 12,500–50 cm^{-1} range within which mid-infrared (MIR), near-infrared (NIR) and Raman spectroscopies are used for traceability and authentication.

Infrared radiation, lying between visible and microwave regions of the electromagnetic spectrum, is absorbed by organic molecules and converted into energy as molecular vibration. Vibrational transitions occur in the ground state of the molecule (Li-Chan 1996). Vibrational energy is quantized. Molecules can occupy discrete energy levels defined by whole numbers 0, 1, 2, and so on. Molecules, in nature, occupy the lowest energy level, 0. A transition from levels 0 to 1, in the MIR spectral region, is referred to as a fundamental transition. Transitions from levels 0 to 2 or 3, in the NIR spectral region, are defined as first and second overtones.

Although it is beyond the scope of this chapter to present the theory of vibrational spectroscopy, it is necessary to outline some of its basic principles in order to understand how each spectroscopic technique works.

The overall objective of vibrational spectroscopic techniques is to analyze a product in order to obtain qualitative and/or quantitative information from the interaction between the electromagnetic spectrum and its constituents (Abbas et al. 2012, Baeten et al. 2016).

2.1.1 Mid-infrared spectroscopy (MIRS)

MIRS refers to the absorption measurement of different MIR frequencies by a sample positioned in the path of an MIR beam (Baranska 1987). When the frequency of a specific vibration is equal to the frequency of the infrared radiation directed at a molecule, the molecule absorbs the radiation. The method is simple, non-destructive, rapid and environmentally friendly.

The MIR spectrum lies in the 400–4,000 cm^{-1} range of the electromagnetic spectrum. It is considered as a fingerprint of the sample in that no two molecular structures produce the same infrared spectrum, making infrared spectroscopy very interesting for traceability analysis (Coates 2000). MIRS allows the quality and authenticity of a sample, as well as the quantity of its components, to be determined. Using the Beer-Lambert law, it is possible to correlate the intensity of one band with the concentration of the active group of the product. MIRS is sensitive to functional groups and to highly polar bonds. Hydroxyl, amine and carbonyl groups are very active in the MIR region and produce high spectral signals.

2.1.2 Near-infrared spectroscopy (NIRS)

NIRS is based on the same absorption phenomenon described above for MIRS, but relates to the 12,500–4,000 cm^{-1} wavelength range (equivalent to 800–2,500 nm; nm is the unit usually used in NIRS) (Baranska 1987). This energy range is high enough to promote molecules from their fundamental vibrational energy levels to second or third excited vibrational states, but it is low enough not to reach the level of electron excitation in molecules. The method is simple, non-destructive and very fast (< 30 sec analysis time), but there is a greater penetration of the radiation than in MIRS. NIRS is suitable for in-line, on-line or at-line use, with minimum sample preparation requirements. It allows highly accurate and precise multi-component analysis (C-H, NH, S-H or O-H bonds).

NIRS can be used as a qualitative method. Due to its large bands being less resolved than in the case of MIRS (Luykx and van Ruth 2008), it is used mainly for quantifying sample properties, such as chemical composition (e.g., protein, glucose, humidity), bulk properties (e.g., density, viscosity, ripeness) and physical properties (e.g., temperature, particle size).

2.1.3 Raman spectroscopy

Unlike MIRS and NIRS, Raman spectroscopy is not concerned with an absorption phenomenon. It is based on irradiation with an intense monochromatic light source (usually a laser), which raises the energy of the system by inducing polarization in the chemical species. The polarized condition referred to a 'virtual state'. The vibrational energy levels in the molecules rise from the ground state to a short-lived, high-energy collision state, which returns to a lower-energy state by emitting a photon that has a lower frequency than the laser light (Stokes Raman scattering). A Raman spectrum between 4,000 and 50 cm^{-1} is a plot of the intensity of Raman scattered radiation as a function of its frequency difference from the incident radiation (Baranska 1987). This difference is called the Raman shift.

Raman spectroscopy has the advantage of requiring little or no sample preparation and allows samples to be measured through a glass container (samples can be analyzed directly inside a glass bottle or plastic bag). In addition, it is not affected by water band interference (ease of aqueous solutions analysis) or atmospheric perturbation, such as CO_2 and humidity (no need to purge the instrument).

As in the case of MIRS, Raman spectroscopy provides structural information and can be considered as a fingerprint. It also provides information from backbone structures and symmetric bonds (e.g., carbon double and triple bonds, and aromatic groups), as well as qualitative and quantitative information. No two molecules have exactly the same Raman spectrum, and the intensity of the scattered radiation is proportional to the amount of material present.

A brief overview of the three techniques is given in Table 1. For each technique, the table shows the spectral region, radiation source, excitation conditions, origin of bands, vibrational modes, band shape, particle size, drawbacks and a selection of applications.

2.2 Instrumentation

Because of the increasing use of infrared and Raman spectroscopies as screening and quality control methods, spectrometers are evolving rapidly at the laboratory, industry, field and farm levels.

2.2.1 Mid-infrared spectrometers

Dispersive spectrometers were the first infrared instruments. The energy emitted from an infrared source is separated into individual frequencies by the use of a prism or grating system. The detector measures the quantity of energy at each frequency that has passed through or been reflected from a sample, resulting in a spectrum that is a plot of intensity vs. frequency.

Table 1. Some Characteristics of NIR, MIR and Raman Spectroscopies.

	NIR	MIR	Raman
Spectral region	4,000–12,500 cm^{-1} (800–2,500 nm)	400–4,000 cm^{-1}	50–4,000 cm^{-1}
Radiation source	Polychromatic near-infrared light from globar tungsten light source	Polychromatic mid-infrared light from globar tungsten light source	Monochromatic visible or near-infrared light from a laser
Excitation conditions	Change in dipole moment	Change in dipole moment	Change in polarizability
Band origin	Radiation absorption	Radiation absorption	Radiation scattering
Vibrational modes	Overtones and combinations of vibrational modes	Stretching and bending fundamental vibrations	Stretching and bending fundamental vibrations
Band shape	Broad peaks arising from overlapping absorption bands	Well resolved, assignable to specific chemical groups	Well resolved, assignable to specific chemical groups
Particle size	Dependent	Dependent	Independent
Interference	Water	Water	Fluorescence
Main applications	Quantification	Structural elucidation and compound identification	Structural elucidation and compound identification

Due to the limitations of these dispersive instruments, however (e.g., slow scanning and lack of reproducibility), they were replaced by Fourier transform (FT) spectrometers.

FT spectrometers enable all infrared frequencies to be measured rapidly and simultaneously. They are equipped with a simple optical device called an 'interferometer'. It contains a beam splitter, which divides an incoming infrared beam into two parts. One beam is reflected off a fixed mirror and the other off a moving mirror, and then both beams are recombined at the beam splitter. Due to changes in the position of the moving mirror in relation to the fixed one, an interference pattern is generated that results in a signal called an 'interferogram'. This signal, a function of time, cannot be interpreted directly. It is converted mathematically by FT, resulting in a frequency spectrum. The detectors used are deuterated-triglycine sulfate (DTGS), based on measuring temperature changes, and the nitrogen-cooled Mercury cadmium telluride (MCT) photon detector.

The advantages of MIR spectrometers include their speed (all frequencies are scanned simultaneously), sensitivity (high optical throughput and sensitive detectors), mechanical simplicity (only one mirror of the interferometer moves) and internal calibration (an internal wavelength calibration standard using a HeNe reference laser), making FT infrared analysis very accurate and reproducible. On-line MIR spectrometers are used for quantification at low analyte levels because they are very sensitive, the main issue being the strong absorption of water (Bellon-Maurel et al. 1994).

2.2.2 Near-infrared spectrometers

The characteristics of near-infrared instruments make them ideal for industrial applications, especially because of their robustness, simplicity and humidity resistance. Their main elements are a radiation source (thermal or non-thermal) (Osborne et al. 1993, McClure 2001), wavelength selectors, sampling accessories and detectors (Single-channel or Multichannel) (McClure 2001).

Pasquini (2003) classified spectrometers according to the technology used for wavelength selection.

- Filter-based instruments using filters as wavelength selectors.
- Light-emitting diodes (LED)-based instruments using LED as a source of narrow bands of NIR radiation or to produce a polychromatic, highly stable source in which radiation is dispersed by using a monochromator based on gratings or filter optics.
- Acousto-optical tunable filters (AOTF)-based instruments using AOTF, which allows constructing instruments that have no moving parts and can reach very high scan speeds over a broad NIR spectral range. The wavelength precision is about ± 0.05 nm and the resolution depends on the wavelength, with typical values of 5–15 nm for a wavelength range of 1,000–2,500 nm.
- Dispersive optics-based instruments using diffraction gratings. Initially, these instruments suffered from slow scan speed, lack of wavelength precision and presence of moving parts, making them difficult to use. In the past decade, the construction of dispersive optics based on the concave grating and sensor array usually used in spectrophotometers with non-moving parts has meant that spectra can now be collected in a few milliseconds.
- FTIR using interferometer technology. These spectrometers combine most of the best characteristics in terms of wavelength precision and accuracy (wavelength accuracy is higher than 0.05 nm), high signal-to-noise ratio and scan speed.

On-line NIRS is well developed and widely implemented. Huang et al. (2008) conducted a review of NIR on-/in-line analysis of foods such as meat, fruit, grain, dairy products and beverages, covering the previous 10 years of research in this field. The tendency is now to use miniaturized spectrometers adapted to specific conditions of measurement in fields, greenhouses and on-line agro-food industrial production. They are flexible enough for a wide range of optical fibers and measurement accessories to be connected. These spectroscopic devices are currently the subject of extensive research and development (Crocombe 2013) aimed at the improvement of detector technologies, microelectro-mechanical systems (MEMS) and high-precision optical components. Some of the new instruments are based on MEMS technology, on the Fabry–Perot-Based MidWave InfraRed (MWIR) microspectrometer (Ebermann et al. 2009) and others such as the NIR grating spectrometer for mobile phone applications (Pügner et al. 2016).

2.2.3 Raman spectrometers

Raman spectroscopy measures the shift in frequency from the photons emitted by the excitation laser. Because it can be performed using any range from UV to NIR, there are two types of Raman instruments, dispersive Raman spectrometers and FT-Raman spectrometers, each one with advantages for specific types of analysis.

With dispersive instruments, the scattered light is collected through a filter and focused onto a monochromator, which allows the separation of the different energies of the Raman scattering. The radiation is directed onto a silicon charge-coupled device (CCD). Visible laser excitation is usually done with these instruments (lasers emitting at 473 nm, 532 nm, 633 nm and 780 nm). Irradiation at these wavelengths enables obtaining of high Raman signals because the intensity of the Raman scatter is proportional to the fourth power of the Raman excitation frequency. A problem that can occur here is the fluorescence phenomenon, which saturates the CCD detector and makes it difficult to conduct Raman measurements.

The near-infrared laser radiation range corresponding to less energy can provide a solution to the fluorescence problem, using FT-Raman spectrometers. These instruments have a neodymium-doped yttrium aluminium garnet (Nd3+:YAG) laser irradiating at 1,064 nm and sensitive, single-element, near-infrared detectors, such as indium gallium arsenide (InGaAs) or liquid nitrogen-cooled germanium (Ge) detectors. FT-Raman spectrometers use an interferometer that functions in the same way as FT-IR spectrometers and has the same advantages.

Depending on the nature of the sample and the objective of the analysis, dispersive or FT-Raman spectrometers can be used. Although the Raman instrument market is growing rapidly, the use of these devices in the agro-food industry remains limited.

2.3 Sample presentation

MIR, NIR and Raman instruments enable the analysis of a great variety of feed and food samples in their liquid or solid form. In order to obtain the best quality spectrum and have confidence in the results, it is important to use the best handling technique for the analyzed sample. Table 2 lists the most

Table 2. Comparison of Sample Presentation and Most Common Measurement Modes for MIRS and NIRS.

Mode	Principle	Type of samples	Accessories	Advantages	Drawbacks
Transmission	IR beam passes through the sample, and the transmitted energy is measured	Powders, liquids and gases	No accessory, but need to prepare a sample in a pellet or film	Low cost of cells Qualitative and quantitative measurements	Difficulty of ensuring good repeatability of deposited films or thickness of KBr pellets
Attenuated total reflection (ATR)	IR radiation enters the crystal in which it is reflected. It penetrates the sample a finite amount via the 'evanescent' wave. At the output end of the crystal, the beam is directed out of the crystal and back into the normal beam path of the spectrometer	Strongly absorbing and thick sample solids that can be ground into powders; Flat materials liquids	Crystal with a high refractive index	No sample preparation (no need to heat, press or grind samples) Fast and easy clean-up	Difficulty of analyzing heterogeneous samples
Diffuse reflection	IR radiation interacts with the particles and then reflects off their surface. Light is diffused or scattered as it moves throughout the sample	Samples that can be ground into a fine powder; Granular samples	Reflection accessory	Little sample preparation (grinding the sample, no need to press) Fast and easy clean-up	Difficulty of defining the path length
Transflection	The IR radiation collected results from light reflected by the measuring cell and light transmitted from the sample	Partially transparent samples	Cells with reflective surface Probes		Difficulty of obtaining the same repeatable optical path between samples

widely used infrared measurement modes and their principles, types of samples, accessories, advantages and drawbacks. Raman analysis requires only a glass container for liquids or a rotation sample holder for solids.

Combining NIRS, MIRS and Raman spectroscopy with imaging technology enables obtaining spectral and spatial information simultaneously. Analyses are achieved in a very short time by recording sequential images of the analyzed sample with each image plane being collected at a single wavelength band. Taking the example of NIR imaging, the compilation of the reflected energy images, taken sequentially at each wavelength, produces a hyperspectral cube. For each pixel, the compilation of absorbances at each wavelength produces a spectrum (Abbas et al. 2012). It should be noted that for feed and food applications, NIR hyperspectral imaging is far more widely used than MIR or Raman imaging, which are more suited to polymer and pharmaceutical applications.

MIRS and Raman spectroscopy, and to some extent NIRS, have evolved from interesting research techniques and now provide valuable analytical tools that can be used on farms, in industries and at production sites. Using them for authentication purposes is of great interest.

2.4 Chemometrics

As these methods are indirect, they require the use of chemometrics to extract chemically relevant information from spectra with statistical, mathematical and computer tools (Massart et al. 1988). In the authenticity and traceability examples presented in the third section, several multivariate techniques were used to explore data patterns (principal component analysis, PCA) or build discrimination models for correctly identifying the products of different functional classes of any given food ingredient (linear discriminant analysis, LDA; soft independent method of class analogy, SIMCA; partial least squares discriminant analysis, PLS-DA). More sophisticated indicators/tools (e.g., GH for calculating Mahalanobis distance; weighted principal components analysis, WPCA) can be used (see Section 3.5). The results of the discrimination models can be expressed in different ways. Sensitivity refers to the percentage of samples from the class studied that have been correctly classified by the model. Specificity refers to the percentage of samples not from the class studied that have been correctly classified by the model. Classification error is the sum of the false positive results (percentage of samples predicted to belong to the class studied, when they do not) and false negative results (percentage of samples predicted not to belong to the class studied, when they do).

3. Traceability and Authenticity: Food and Feed Examples

This section gives several examples to illustrate the potential of vibrational spectroscopy regarding authenticity challenges such as the discrimination of barley varieties, the origin identification of distillers' dried grains and solubles (DDGS), the authentication of fruits and finally, the early detection of fraud in food/feed ingredients.

3.1 Discrimination among varieties: the case of barley

Breeders, farmers and consumers are showing increasing interest in *Triticum* species other than common wheat (*T. aestivum* L.), such as emmer (*T. dicoccum* L.) and spelt (*T. spelta* L.), which are ancestral hulled wheat species characterized by higher resistance to stress and by specific nutritional qualities (Escarnot et al. 2012). Effective species discrimination among grains is increasingly important not only for the food industry in terms of the characteristics and qualities required, but also for the protection of Protected Geographical Indication (PGI)-certified species (e.g., farro della Garfagnana emmer). Discussions with representatives of the food industry involved in the cereal chain has shown that NIRS is undertaken in many cereal producer sites in order to rapidly check quality and safety parameters, with the potential to use fingerprinting approaches for detecting possible fraud (Suman 2016). In order to solve some authentication issues, the discrimination of species/varieties is needed at the kernel level. For instance, only 3% of common wheat (*T. aestivum* L.) and durum wheat (*T. durum* L.) crops are authorized to produce pasta in Italy. As the durum wheat price is generally higher than the price of common wheat (up to 35% higher), there is a considerable risk of economic fraud by mixing common and durum wheat. Food production companies also often need to discriminate species/varieties at the particle level in the flour. For example, oats (*Avena sativa* L.) are widely used in breakfast cereals because of its high nutritional value. It is economically profitable to add wheat flour to oat flour, which can lead to health problems because wheat contains gluten, whereas oats are gluten-free (Wang et al. 2014).

A previous review (Vermeulen et al. 2010) showed that the use of NIRS in wheat analysis in the 1980s focused on discriminating among wheat varieties based on flour quality. In the decade following the year 2000, more sophisticated NIR hyperspectral imaging systems (NIR-HIS) were developed, combining spatial and spectral information. The initial studies showed that applying chemometric tools in NIR-HIS offered new prospects to the agro-food industry for classifying kernels according to quality criteria (e.g., bread-baking quality, hardness) (Williams 2009a,b). The major advantages of NIR-HIS are that recognition does not depend on the expertise of the analyst and that it is possible to automate all procedures and analyze a large number of samples.

An original discrimination study of 176 barley samples representing 24 varieties tested in trials for registration in the Belgium catalogue conducted over 3 years (2004–2006) in eight Belgium locations (Monfort et al. 2006) was performed by the Walloon Agricultural Research Centre (CRA-W) research team (Vermeulen et al. 2007). The study sought to develop a fast and reliable method for varietal discrimination, essential for the efficient traceability and quality control system required by the seed sector as well as by the food and feed sectors. In this study, the use of NIR imaging technology was investigated in order to classify barley varieties at the single kernel level. The results were compared with those obtained using classical NIR methods based on bulk analysis or classical field data (e.g., earliness, yield) and technological data (e.g., kernel size, thousand kernel weight [TKW], specific weight, protein, humidity) collected by breeders.

The 24 varieties were grouped into three classes: 6-row winter barley class (6RW); 2-row winter barley class (2RW); and 2-row spring barley class (2RS). The samples were selected to represent variations in climate, geographical location and agronomy in Belgium. Two sets of samples were selected for the study:

- Sample set 1 was used for discriminating the three barley classes (6RW, 2RW, 2RS). A set of 96 samples, including 24 varieties tested at four Belgian locations (Enghien, Gembloux, Havelange and Poperinge) and obtained from the 2005 harvest, was created. This dataset contained 11 varieties of 6RW, seven varieties of 2RW and six varieties of 2RS.
- Sample set 2 included only 6RW samples and was used for discriminating the 6RW varieties. A set of 112 samples, including eight varieties tested at seven Belgian locations (Enghien, Gembloux, Havelange, Poperinge, Leffinge, Dommartin and Bassevelde) and obtained from the 2004 and 2005 harvests, was created. The varieties studied were Carola, Nikel, Seychelles, Sumatra, Palmyra, Jolival, Mandy and Pelican.

In this study, two types of NIR instruments were used: the classical NIR spectrometer and the NIR imaging spectrometer.

The classical NIR spectrometer was an NIR TECATOR Infratec 1241 (FOSS, Hillerod, Denmark) working in the 850–1,050 nm range and used mainly at the cereal collection sites. With this instrument, one spectrum by sample (bulk of kernels) was obtained. Initially, the spectra were used to predict some technological parameters (protein content and humidity) by using the calibration equations developed from barley databases built at CRA-W over more than 30 years. The mean spectra per sample were then used to build the discrimination equations among barley classes or varieties.

The NIR imaging spectrometer used in this experiment was a MatrixNIR™ Chemical Imaging System (Malvern, Olney, USA) described by Fernandez et al. (2005). Reflectance images on 10 kernels by barley sample were collected in the 900–1,700 nm window. The spectrum of each kernel was the average of the spectra acquired from the total surface of the kernel.

For each sample set (sets 1 and 2), the same methodology was applied to the agronomic and technological data, the classical NIR spectra and the NIR-HIS spectra. PLS-DA models were built in order to assess whether or not the differences observed among varietal classes or among varieties in term of agricultural and technological (agro-technological) data could also be identified with the spectral data. Models were built using samples with known classes and were validated by cross-validation using the leave-one-out method, where each sample was successively left out of the model formulation and independently predicted once. The models were also validated with a test set selected by splitting each sample set into two groups, one for training (model construction) and the other for testing (representing about 15% of the samples). The splitting was done by selecting, at random, four samples by group (6RW, 2RW, 2RS) or two samples by variety, representing the different locations and years. In order to compare the performance of the PLS-DA models with the three sets of data (agro-technological data, classical NIRS spectra and NIR-HIS spectra), the same varieties were selected for the test sets.

In the first step, PLS-DA was used to build models for classifying the three barley classes 6RW, 2RW, 2RS (Sample set 1). Table 1 (left side) shows the sensitivity and specificity of each of the three groups at the calibration, cross-validation and test stages. The classification errors in cross-validation varied between 5 and 9% based on NIR imaging data, as opposed to 0–14% based on classical NIR or 12–17% based on agro-technological data. The performance of the models with the test set was better with NIR-HIS in terms of classification errors (3–14%), as opposed to classical NIR (6–25%) and agro-technological (6–25%) data.

In the second step, PLS-DA was used to build models for classifying the six RW varieties (Sample set 2). Table 3 (right side) shows the sensitivity and specificity for each of the eight varieties at the calibration, cross-validation and test stages. The classification errors in cross-validation varied between 23 and 38% based on NIR imaging data, as opposed to 1–34% based on classical NIR and 5–45% based on agro-technological data. The performance of the models with the test set was poorer, with high classification errors: 7–43% using agro-technological data, 0–50% with classical NIR and 23–54% with NIR-HIS.

In brief, PLS-DA classification models showed that NIR imaging had a classification accuracy of 91% for the three classes (2RS, 2RW, 6RW), as opposed to 87% for classical NIR and 83% for agro-technological data. With

Table 3. Performance of PLS-DA Discrimination Models for Sample Sets 1 and 2 on Agro-Technological Data, Classical NIR Spectra and NIR-HIS Spectra.

	SAMPLE SET 1						SAMPLE SET 2				
	Agro-technological data						Agro-technological data				
Calibration	2RS	2RW	6RW	Carola	Nikel	Seychelles	Sumatra	Palmyra	Jolival	Mandy	Pelican
	84 samples/21 varieties: 20 2RS, 24 2RW, 40 6RW						96 samples, 12 by variety				
Sensitivity	85.0	91.7	87.2	91.7	100	83.3	100	100	83.3	91.7	91.7
Specificity	90.5	89.8	81.8	86.9	86.9	64.3	83.3	78.6	39.3	98.8	91.7
Classification error	12.3	9.3	15.5	10.7	6.6	26.2	8.3	10.7	38.7	4.8	8.3
Cross-validation	84 samples/21 varieties: 20 2RS, 24 2RW, 40 6RW						96 samples, 12 by variety				
Sensitivity	85.0	91.7	84.6	91.7	91.7	83.3	91.7	75.0	75.0	91.7	91.7
Specificity	87.3	84.7	81.8	83.3	82.1	64.3	82.1	76.2	35.7	98.8	90.5
Classification error	13.8	11.8	16.8	12.5	13.1	26.2	13.1	24.4	44.6	4.8	8.9
Test	12 samples/3 varieties: 4 2RS, 4 2RW, 4 6RW						16 samples, 2 by variety				
Sensitivity	100	100	75.0	50.0	50.0	100	50.0	100	100	100	100
Specificity	87.5	62.5	75.0	64.3	85.7	71.4	85.7	71.4	42.9	85.7	85.7
Classification error	6.2	18.7	25.0	42.9	32.1	14.3	32.1	14.3	28.5	7.1	7.1
	Classical NIR spectra						Classical NIR spectra				
Calibration	84 spectra/21 varieties: 20 2RS, 24 2RW, 40 6RW						96 spectra, 12 by variety				
Sensitivity	100	95.8	89.7	91.7	75.0	83.3	83.3	100	75.0	100	100
Specificity	100	96.6	83.7	89.0	65.9	79.3	74.4	95.1	62.2	98.8	94.0
Classification error	0.0	3.8	13.3	9.7	29.6	18.7	21.1	2.4	31.4	0.6	3.0

Cross-validation	84 spectra/21 varieties: 20 2RS, 24 2RW, 40 6RW			96 spectra, 12 by variety							
	2RS	2RW	6RW	Carola	Nikel	Seychelles	Sumatra	Palmyra	Jolival	Mandy	Pelican
Sensitivity	100	91.7	89.7	91.7	66.7	83.3	83.3	91.7	75.0	100	90.9
Specificity	100	91.4	81.4	86.6	64.6	74.4	72.0	90.2	61.0	97.6	94.0
Classification error	0.0	8.5	14.4	10.9	34.3	21.1	22.4	9.0	32.0	1.2	7.5
Test	12 spectra/3 varieties: 4 2RS, 4 2RW, 4 6RW			16 spectra, 2 by variety							
Sensitivity	50.0	100	100	100	100	50.0	50.0	100	100	100	0.0
Specificity	100	87.5	87.5	100	42.9	64.3	71.4	100	57.1	100	100
Classification error	25.0	6.2	6.2	0.0	28.6	42.9	39.3	0.0	21.4	0.0	50.0

NIR-HIS spectra

Calibration	840 spectra/21 varieties: 200 2RS, 240 2RW, 400 6RW			960 spectra, 120 by variety							
	2RS	2RW	6RW	Carola	Nikel	Seychelles	Sumatra	Palmyra	Jolival	Mandy	Pelican
Sensitivity	96.0	91.2	91.2	72.3	76.3	74.8	84.0	78.8	63.0	67.8	77.8
Specificity	95.1	93.0	92.0	57.2	76.5	61.7	75.2	75.3	75.4	72.0	80.6
Classification error	4.4	7.9	8.4	35.2	23.6	31.7	20.4	23.0	30.8	30.1	20.8
Cross-validation	840 spectra/21 varieties: 200 2RS, 240 2RW, 400 6RW			960 spectra, 120 by variety							
Sensitivity	96.0	91.2	90.7	67.2	71.2	68.9	79.8	76.3	60.5	66.1	73.5
Specificity	94.3	91.9	91.3	56.3	75.9	61.2	73.6	74.3	74.6	71.4	79.4
Classification error	4.8	8.4	9.0	38.2	26.5	34.9	23.3	24.7	32.4	31.2	23.5
Test	120 spectra/3 varieties: 40 2RS, 40 2RW, 40 6RW			160 spectra, 20 by variety							
Sensitivity	97.5	89.7	83.8	78.9	70.0	50.0	40.0	60.0	70.0	75.0	15.0
Specificity	97.4	88.3	87.3	49.3	83.5	42.4	56.1	82.0	84.2	72.7	77.7
Classification error	2.6	11.0	14.4	35.9	23.3	53.8	51.9	29.0	22.9	26.2	53.7

regard to classifying varieties in the 6RW class, the results obtained were lower, with a classification accuracy of 78% for agro-technological data, 77% for classical NIR spectra and 63% for NIR-HIS. The classification of some varieties, such as Mandy, however, was better, with rates of 93%, 100% and 74%, respectively. Mandy was clearly different from the other 6RW varieties, being a late variety and having a lower TKW.

A recent study by Zhu et al. (2012) indicated that NIR-HIS could differentiate three types of wheat: strong gluten wheat, medium gluten wheat and weak gluten wheat. The classification accuracy of six wheat cultivars reached 93%. Similarly, Kong et al. (2013) showed the possibility of classifying four rice seed cultivars with a classification accuracy of 100%.

In order to improve classification accuracy, the trend now is to combine several techniques to examine the potential of sensor fusion and data fusion. Zhang et al. (2012) developed classification models to discriminate six maize seed varieties using HIS in the visible and near-infrared (380–1,030 nm) region (VIS-NIR). They showed that by combining textural variables and spectral data, they could achieve a classification accuracy of 98.9%. Yang et al. (2015) achieved a classification accuracy of 98.2% for four varieties of maize seeds by combining morphological, textural and spectral features extracted from VIS-NIR HIS (400–1,000 nm). Teye et al. (2014) showed that the single sensor NIRS and electronic tongue (ET) used to discriminate five cocoa bean varieties had a classification accuracy of 83–93%, whereas data fusion had a classification accuracy of 100%.

3.2 Distillers' dried grains and solubles fraud in relation to botanical origin, geographical origin and production process

The ban on using processed animal protein in the feed led the feed sector to look for other possible protein sources. Among the various possibilities and apart from soybean meal, which is the main source of proteins in feed, distillers' dried grains and solubles (DDGS) could also be an important source. In the USA, 30% of corn is used for ethanol production and most of the DDGS obtained as a residue of the process are exported to Europe. The use of antibiotics or fermentation supplements to improve ethanol production process poses risks to the feed chain. Usually, the product labelling of an affected feed lot shows origin and the paper documentation shows traceability. Incorrect product labeling is common in embargo situations and alternative analytical strategies for ensuring feed authenticity are therefore needed.

Within the framework of the European QSaffe project (2011–2014), a study was conducted on authenticating the origin of DDGS. A total of 191 DDGS samples were collected from reliable sources in Canada, China, Europe and the USA. They were produced from corn (*Zea mays*) and wheat

(*Triticum aestivum* L.) and obtained during the industrial production of bio-ethanol or alcoholic beverages. Various analytical techniques were used in this study: NIRS (Zhou et al. 2014), NIR microscopy (NIRM) (Tena et al. 2015), MIRS (Nietner et al. 2013, Vermeulen et al. 2015a) and Raman spectroscopy (Haughey et al. 2013), as well as MS-based approaches such as proton transfer reaction–mass spectrometry (PTR-MS) (Tres et al. 2014), DART-Orbitrap MS and liquid chromatograph quadrupole time-of-flight MS (LC/Q-TOF/MS) (Novotna et al. 2012). Two proven techniques in food authenticity, isotope ratio mass spectrometry (IRMS) (Nietner et al. 2014) and DNA analysis using polymerase chain reaction (PCR) (Debode et al. 2012), were also among the methods that could identify the DDGS origin. The methods developed were able to determine the botanical origin of the DDGS (corn vs. wheat), and several of them were able to determine the geographical/production origin of the DDGS. These techniques were compared in terms of their complementarities, and an overall strategy for tracing and confirming DDGS origin was described by Vermeulen et al. (2015b).

The results presented below illustrate another approach for studying variability in DDGS samples based on their composition. The 191 DDGS were initially analyzed using a FOSS XDS NIR spectrometer active in the 400–2,500-nm range. Quality parameters such as moisture, protein, fat, fiber and ash were estimated using equations constructed with historical NIRS databases (Fernandez et al. 2010). Samples were described in detail by Vermeulen et al. (2015b). For this study, some samples were removed from the initial dataset because of doubt about their botanical origin based on IRMS analyses. In order to characterize the variability of the retained 181 DDGS samples in terms of production origin, PCA was performed using the five quality parameters, with normalization and autoscale pre-processing applied to the data.

Figure 1 shows the PC1 vs. PC2 scores plot that allowed DDGS samples from corn, wheat, rice and a mixture of wheat and corn to be distinguished. Wheat DDGS from several companies in various European countries were characterized by higher protein content (33.1%) and lower fat content (4.9%) than corn DDGS (28.7% and 8.3%, respectively). Mixtures of wheat and corn DDGS were characterized by medium protein content (30.3%) and low fat content (5.3%). The rice DDGS group was represented by only one sample; it differed considerably from wheat and corn and had a high ash content.

Figure 2 shows the same PC1 vs. PC2 scores plot where the sample marks are coloured according to information on geographical origin and process. Several corn DDGs groups were identified. One group of DDGS samples, residues from a bio-ethanol production company in China (Corn China Origin 1), was characterized by medium protein content (31.1%) and very low fat content (3.2%), which could be explained by fat

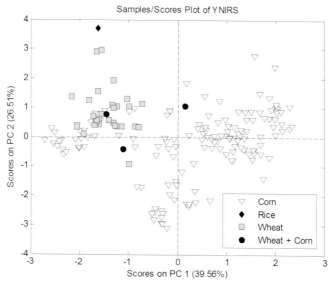

Figure 1. PCA on Predicted Technological Values for Wheat, Corn, Rice and Wheat + Corn Mixtures DDGS Groups: Scores Discriminating the Botanical Groups.

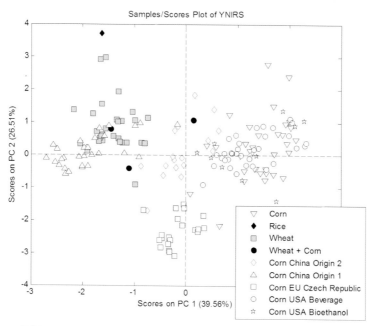

Figure 2. PCA on Predicted Technological Values for Wheat, Corn, Rice and Wheat + Corn Mixtures. DDGS Groups: Scores Discriminating the Industrial Processes.

extraction in the production process. A second group of DDGS samples from another bio-ethanol production company in China (Corn China Origin 2) was characterized by medium fat (8.0%), protein (29.4%), fiber (7.3%) and ash (4.7%) content. A third group of DDGS from a bio-ethanol production company in the Czech Republic (Corn EU Czech Republic) was characterized by medium fat (8.4%) and protein (29.7%) content, but high fiber content (7.9%) and low ash content (3%). DDGS groups from several bio-ethanol or alcoholic beverage production companies in the USA (Corn USA bio-ethanol and Corn USA beverage) were characterized by low protein content (27.3% and 26.7%, respectively) and high fat content (9.8% and 10.1%, respectively).

Figure 3 shows the PC1 and PC2 loadings of PCA. PC1 and PC2 explain 39.6% and 26.5% of the variation, respectively, which are related mainly to protein and fat content (PC1) and ash and fiber content (PC2).

This study showed that PCA gave acceptable results for determining botanical and geographical origin based on compositional profiles. It enabled corn DDGS from three bio-ethanol plants (two in China, origins 1 and 2, and one in the Czech Republic) to be visually distinguished from corn DDGS emanating from bio-ethanol and alcoholic beverage plants in the USA, indicating the potential of each ethanol plant to produce DDGS with consistent compositional characteristics.

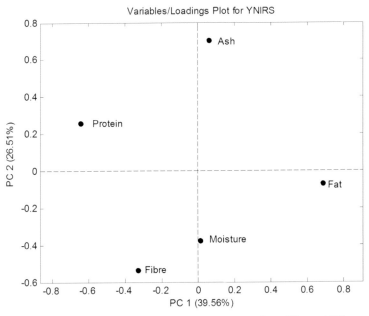

Figure 3. PCA on Predicted Technological Values for Wheat, Corn, Rice and Wheat + Corn Mixtures. DDGS Groups: Loadings Discriminating the Botanical Groups and Industrial Processes.

The study also showed that established analytical approaches in food analysis can be applied to DDGS and could be used for authenticating other materials in the animal feed sector.

3.3 Authentication of fruits and fruit-based products

The authentication and traceability of fruit and fruit-based products pose challenges that need solutions. The authenticity issues relating to these products include assessment of fruit varieties used, assessment of origin of production, assessment of process applied and absence of adulteration with unexpected fruits varieties or exogenous compounds (e.g., sugar, additives). These challenges required fast and reliable methods that can be applied at the level of field production (orchard), transformation unit (artisanal plant or industrial plant) and retailers (local retailer, open market and supermarket). NIRS-based methods could meet these requirements. The authentication and traceability of açai fruits (*Euterpe oleracea*) and cocoa beans (*Theobroma cacao*), both from the Amazonian basin were selected as examples to illustrate the potential of NIRS methods.

3.3.1 Study case 1: açai fruits

More than 200 edible fruits are consumed in the Amazonian basin. Fruits are usually eaten fresh, as juice or puree, and are included in numerous desserts. Fruit production in the region is based on a mixture of extractivism and cultivation. Amongst the Amazonian fruit, açai has a special status. It is produced from a tall, multi-stemmed palm that can reach heights of 3–12 m and is indigenous to the Amazonian basin. The fruit is a dark, spherical berry with a diameter of about 0.7–1.5 cm, and a stone representing 85–90% of the fruit weight (Bichara and Rogez 2011). Fruit production has doubled over the past ten years, being actually over one million tons (IBGE 2016), Most açai production is for export to the USA and Europe, where açai juice is seen as an energy-enhancing drink rich in lipids, fiber and phenolic compounds (Schauss et al. 2006). The palm is native of floodplains ecosystem and has been extensively planted thanks to important irrigation systems in other Amazonian and non-Amazonian lands. In order to ensure traceability and authenticity of açai fruit from this region, it was necessary to develop technological tools for the agricultural industry in post-harvest monitoring and quality improvement. Results from a study conducted by the Federal University of Para (UFPA) in Brazil and CRA-W in Belgium showed that NIRS is an efficient, rapid and non-destructive analytical tool and therefore suitable for the post-harvest monitoring (Amaral 2015a,b). One of the experiments involved assessing the ability of NIRS to determine the geographical origin of açai fruits at a regional level as well as at the level of the entire Amazon basin. In order to conduct this assessment, 106 samples of açai fruit were collected from three municipalities in the north of Pará state in Brazil, all bordering the Amazon river or its tributaries. Twenty fruits of each sample were randomly chosen for the NIR readings, using a handheld device (Phazir, from Polychromix, USA) to collect spectra in the 1,596–2,396 nm range. PCA and LDA were applied to the mean spectra in order to study the potential of NIR to discriminate açai fruits according to their origin. PC-3 and PC-5 managed to partially discriminate Ponta de Pedras fruit samples from those from the other municipalities (Abaetetuba and Muana) (Fig. 4). With LDA permitted to construct discriminant models that could discriminate sample origins with a success rate of 71–90% for locality of origin. A repeatability study showed that NIR has a coefficient of variation < 5%. The first study indicated that NIRS could be used to determine the geographical origin of açai fruits (Amaral et al. 2015a). Another study sought to discriminate açai fruit samples from two agronomically different areas (i.e., floodplains or irrigated lands) based on NIRS combined with chemometrics. The results showed that the methodology was suitable for quality control in the açai industry and could also be used for traceability and authenticity purposes (Amaral et al. 2015b).

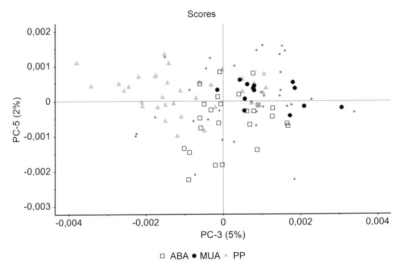

Figure 4. PCA for the Three Municipalities (PP = Ponta de Pedras, ABA = Abaetetuba, MUA = Muana).

3.3.2 *Study case 2: cocoa beans*

Cocoa (*Theobroma cacao*) is a native fruit from Amazonia widely cultivated in many tropical countries. Cocoa fruits, called pods, are collected by the farmers then the beans and the pulp are extracted by breaking the pods and fermented for about six days. The fermented beans are dried before being sold and processed. Cocoa is the basis of all chocolate products and the flavor of the beans depends on variety, cultivation method and conditions (including soil and climate) and post-harvest treatment, which consists mainly of fermentation and drying. In the context of international markets, cocoa-based companies need to manage the supply chain from producers to consumers carefully. Two broad categories of cocoa beans are distinguished by the International Cocoa Organization (ICCO) on the world cocoa market: "fine or flavor" cocoa beans, and "bulk" or "ordinary" cocoa beans. Fine or flavor cocoa beans are produced from Nacional, Criollo and Trinitario cocoa-tree varieties, while bulk cocoa beans come from Forastero trees, normally originating from the lower Amazon region. Currently, only 5% of world cocoa is classified as fine or flavor. Fine or flavor cocoa beans are traded in a niche market that is relatively small but highly specialized and distinct from the bulk cocoa beans market. The unique organoleptic (flavor and aroma) characteristics of fine or flavor cocoa generally attract premium prices for this cocoa. Thus, there is a need for faster, cheaper and more accurate real-time traceability and authentication methods. Traceability to cocoa farms enables marketers to impose liability costs on farms, thereby creating incentives for farms to supply safe and high-quality cocoa beans. Cocoa beans that have been collected safely and have been well fermented and properly dried will attract a higher price. Some studies have described the use of vibrational spectroscopy for determining key quality parameters such as caffeine, theobromine and epicatechin content (Davrieux et al. 2006, 2007a,b, 2009, Hue et al. 2014a,b, Alvarez et al. 2012), as well as for discriminating cocoa beans in terms of variety, genotype or fermentation level.

Recently, the potential of using NIRS to discriminate cocoa genotypes from Ecuador has been assessed (Davrieux et al. 2013). The Ecuadorian cocoa production is 190,000 T per year. Two main clones are grown: Nacional and CCN-51. The Nacional accounts for 80% of the production and is recognized as fine cocoa. This variety is probably indigenous to Ecuador while the CCN-51 cacao variety (hybrid) became widely planted in Ecuador since 1997. Nowadays this variety represents 20% of the production and is classified as of poor quality.

In this study conducted by the Instituto Nacional de Investigaciones Agropecuarias (INIAP, Ecuador) and CIRAD (France), 641 samples were collected over 3 years (2009–2012), representing six different cocoa producing zones in Ecuador and the two genotypes (i.e., Nacional and CCN51). Roughly ground unshelled beans (nibs) were prepared just before analysis using a Seb 810004-Prepline grinder (Seb, Ecully, France). Cocoa samples were analyzed for their NIR diffuse reflectance spectrum using a XDS monochromator spectrometer (Foss NIRSystems, Silver Spring, USA) with rectangular cells and moving RSA system (Rapid Solid Analyser). About 100 g of cocoa were analyzed per sample. A principal component analysis was done on the spectral matrix (n = 641), using centered data and variance/covariance matrix as metric, then the H Mahalanobis distances to the mean average spectra were calculated for each sample on the base of calculated PCs. According to H distance, five samples, presented H > 3, were considerate as spectral outliers and removed.

Then, the remaining sample set was separated in two sub-sets: learning set and validation set. To be representative of the original repartition, 30% of the samples were selected randomly per year for both genotypes. Doing this way, 191 samples (81 CCN-51 and 110 Nacional) were selected as validation samples and the remaining 445 samples (186 CCN-51 and 259 Nacional) were used for calibration. Different classification methods (LDA, Mahalanobis distance discrimination, and SIMCA) were tested using WinISI (Infrasoft, Port Matilda, USA), Xlstat (Addinsoft, Paris, France) and Pirouette (Infometrix Bothel, USA) software.

The best results, expressed as correct classification rates, were observed using SIMCA method. The correct classification rate for the learning set was 94.4% with 10 CCN51 and 15 Nacional misclassified and the correct classification rate was 94,8% for the validation set. The error in validation was about 5%, with only nine samples out of 191 misclassified: six CCN51 (out of 81) and three Nacional (out of 110). One sample (CCN51) was unclassified.

The scatter plot of validation samples distances to each group centroids defined by the model, showed that few samples were close to the separation line (Fig. 5).

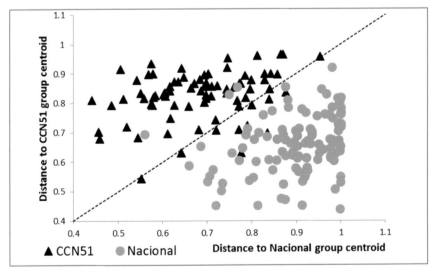

Figure 5. NIRS Discrimination of Cocoa Samples in Terms of Genotype (Nacional Versus CCN51).

The discriminating power calculation highlighted that fat -CH$_2$ absorptions bands (1724 nm and 2308 nm) were prominent in the discrimination.

The potential of NIR-HIS has been also studied for analyzing whole cacao beans from the Amazon basin (Rogez et al. 2015). The objective of the study conducted by UFPA and CRA-W was to discriminate individual beans in terms of their geographical origin and fermentation time. More than 2,000 cocoa beans from 147 samples from Para state were collected. Samples that had been fermented or dried and came from different areas and producers over two harvest years (2012–2013) were tested. Hyperspectral images were collected using an NIR hyperspectral line scan or push-broom imaging system combined with a conveyor belt (Burgermetrics, Latvia). Each image consisted of 320-pixel lines acquired in the 1,100–2,400 nm wavelength range. Figure 6 gives the PCA results (PC1 versus PC4), showing the discrimination of the beans in terms of process applied (fermented vs. non-fermented; sun-dried vs. dark-dried).

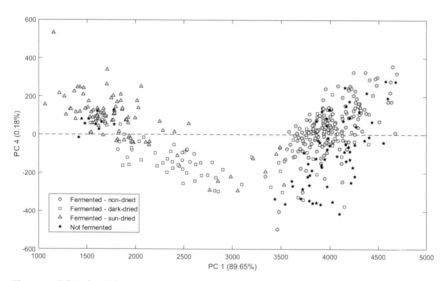

Figure 6. PCA plot (PC1 vs. PC4) of the 147 Cocoa Beans from Tome-Açu Municipality (Para State, BR) Analyzed, Showing the Discrimination of the Beans in Terms of Process Applied (fermented vs. non-fermented; sun-dried vs. dark-dried).

3.4 *Early detection of fraud in food/feed ingredients: the case of unapproved protein enhancement with melamine*

Among the many crises in the food and feed industries in recent years, one of the most serious in terms of health and economic effects was the use of melamine (WHO/FAO 2008, Tyan et al. 2009). In 2007, the US Food and Drug Administration (FDA) found melamine in pet food and in samples of wheat gluten imported from China (FDA 2007). In 2008, almost 300 metric tons of soybean meal destined to organic chickens in France were withdrawn from the market after the authorities discovered melamine levels in them that were 50 times higher than the permitted standard (Adams 2008). Also in 2008, milk and infant formula in China was found to be adulterated with melamine, affecting more than 300,000 children, with six infants dying from kidney stones or other kidney damage (Branigan 2008). Melamine was deliberately added at milk-collecting stations to diluted raw milk, ostensibly to boost its protein content. Subsequently, melamine was detected in many milk and milk-containing products, as well as other food and feed products, exported to many countries worldwide. These crises illustrated the need for a sensitive, reliable and rapid procedure for detecting melamine in both food and feed products (Chan et al. 2008, Dobson et al. 2008, Chen 2009, Gossner et al. 2009). With this aim, in recent years public and private researchers have been focusing on the development of suitable screening methods. Currently, most of the available procedures use LC or gas chromatographic (GC) methods combined with MS. Alternative methods available include the use of antibodies, molecularly imprinted polymers, capillary electrophoresis or gold nanoparticles (Ai et al. 2009, Yan et al. 2009, Sun et al. 2010, Rovina et al. 2015). A complete list of methods was presented by Lin et al. (2009) and Liu et al. (2012). Most of these methods, however, are expensive, matrix dependent, destructive and time-consuming, and require extensive sample preparation. A possible alternative could be the use of vibrational spectroscopy techniques, such as NIRS, which has been used for many years as a quality control tool in the food and feed sectors (Norris et al. 1976, Murray 1986, 1993, Barton and Windham 1988, Shenk and Westerhaus 1995). Only recently, with the development of multivariate calibration procedures, NIRS has been used to detect melamine adulteration in food/feed matrices (Dong et al. 2009, Lu et al. 2009, Mauer et al. 2009, Balabin et al. 2011, Smirnov 2011, Haughey et al. 2012, Abbas et al. 2013, Fernández Pierna et al. 2014, 2015, Baeten et al. 2016). In most of these studies, information from the spectra was obtained using classical and innovative chemometric tools in a targeted way (i.e., knowing in advance the fraudulent substance [melamine] to be detected). More studies are now focusing on the development of non-targeted procedures for characterizing certain products and detecting the presence of possible known or unknown contaminants or fraudulent substances, before the food/feed chain is reached (Baeten et al. 2014, de Jong et al. 2016, Fernández-Pierna et al. 2016). The use of statistical tools to interpret multivariate data obtained from spectra should lead to the establishment of rules for checking compliance against product specifications. In industries already equipped with NIR technology for quality control, it would be easy to adapt it in order to simultaneously check possible contamination at both the start and end of the production chain. Spectra could be combined with chemometric tools to simultaneously check whether or not a product adheres to fixed specifications in terms of composition and quality parameters. Classical chemometric tools could be useful for both tasks, but one tool will usually not be enough for characterizing a product because most of these tools are problem-oriented, meaning that it would be difficult to create thresholds that are useful in tackling future problems. Fernández Pierna et al. (2015) proposed a combination of chemometric techniques with individual characteristics and orientations. This combination includes pattern-recognition techniques that provide adequate differentiation, as well as regression methods evaluated according to their ability to handle the available dataset and predict the status of new samples. The combination therefore facilitates decision-making about product acceptance or rejection. A new technique known as local window PCA (LWPCA), based on a moving window criterion, was proposed and considered as an untargeted method. In this method, a moving window was selected along the wavelength axes in vibrational spectroscopic data and then individual PCA analyses are performed. A calibration set was selected in a localized way from a historical data in order to characterize products that are the most spectroscopically similar to the one to be predicted. Spectral score residuals in this calibration set were extracted and used to build thresholds applied to spectral score residuals of the sample to be predicted. When a residual, at a certain wavenumber, did not meet the defined thresholds, the sample was viewed as abnormal, indicating the possible presence of unusual ingredients and therefore allowing non-targeted analysis. In the case of melamine contamination of milk, this technique was successfully applied (Fernández Pierna et al. 2016). The work was based on the FT-MIR spectra of milk contaminated with melamine. A dataset of 300 samples of UHT liquid milk was used as an historical and clean dataset. Another 12 UHT liquid milk samples were contaminated with melamine at various levels ranging from 0.01% to 1% (100–10,000 ppm). Visual observation of the spectra did not enable clear conclusions to be drawn. GH values allowed abnormalities detection at levels higher than 500 ppm. LWPCA, however, allowed contamination at levels up to 100 ppm to be detected, but at those levels, the detection of melamine in milk became unstable, suggesting that the technique had probably reached its limit of detection. LWPCA technique can also be used for detecting adulterants in soybean meal. Fernández Pierna et al. (2015) devised a complete procedure based on chemometrics and the use of NIRS at the entrance of a feed mill in order to provide early evidence of non-conformity and unusual ingredients. The study focused on the characterization of pure soybean meal with the aim of creating an early control system for detecting and quantifying any unusual ingredients that might be present in the soybean meal, such as melamine, cyanuric acid or whey powder (milk serum). Results showed that the use of NIR, combined with some simple chemometric tools based on distances and residuals from regression equations, was appropriate for authenticating important feed products (soybean meal) and detecting the presence of abnormal samples or impurities in the laboratory and at the feed mill. LWPCA can also be used to address this problem. Table 4 shows the results of the various criteria used to determine the presence of abnormal samples in the data. The first three datasets, collected directly at the feed mill (Fernández Pierna et al. 2015), contained 75, 66 and 57 samples of pure soybean meal, respectively, as well as 59 and 43 mixtures of soybean meal and whey for datasets 1 and 2, respectively, and 48 mixtures of soybean meal and DDGS for dataset 3. A fourth dataset contained five samples of pure soybean meal and 60 mixtures of soybean meal, melamine and cyanuric acid at varying concentrations. In Table 4, the results are presented in terms of classification accuracy, with black for pure soybean meal and red for the different mixtures. The methods used corresponded to the application of PLS regression models on historical data for protein and fat, the calculation of GH values and the application of LWPCA. Most of the methods applied enabled the soybean meal to be characterized. When detecting a possible contaminant, higher percentages of samples correctly detected were obtained with the LWPCA method. Applying a combination of the four techniques to the NIR data at the start of a production chain could lead to

Table 4. Classification Accuracy Percentage for Datasets of Soybean Meal and Soybean Meal Contaminated with Whey, DDGS, Melamine and/or Cyanuric Acid Using the PLS Model (Protein, fat), GH and LWPCA.

		Protein (%)	Fat (%)	GH (%)	LWPCA (%)
Dataset 1	Soybean meal (75 samples)	93.3	100	94.7	96.0
	Soybean meal + whey (59 samples)	91.5	81.4	94.9	96.6
Dataset 2	Soybean meal (66 samples)	98.5	98.5	92.4	93.9
	Soybean meal + whey (43 samples)	95.3	95.3	95.3	95.3
Dataset 3	Soybean meal (57 samples)	93.0	100	100	94.7
	Soybean meal + DDGS (48 samples)	43.7	14.6	14.6	72.9
Dataset 4	Soybean meal (5 samples)	100	100	100	100
	Soybean meal + melamine/cyan acid (60 samples)	63.3	66.7	95.0	95.0
Mean	Soybean meal detection	96.2	99.6	96.8	96.2
	Contaminant detection	73.5	64.5	75.0	90.0

important cost-savings by detecting non-conformity and authenticating important food/feed products (in this case, soybean meal). A possible limitation would be the low sensitivity of NIR to minor constituents, which is probably not a major drawback when dealing with significant contamination crises.

4. Conclusions

The application of vibrational spectroscopy methods to agricultural and food product examples has shown the important potential of these analytical tools in traceability and authentication. NIRS is already widely accepted in the food and feed sectors for determining, in a unique analysis, a large variety of quality control parameters. Strategies using analytical NIR techniques combined with dedicated statistical data analysis tools could be easily implemented in both routine laboratories and in industries to address authentication issues. The ability to use this technique on-line in production plants and the possibility of building a network of spectrometers make NIRS a very attractive screening tool for the food and feed sectors. As shown in the species/varieties discrimination examples, the use of sensors and data fusion to identify varieties at the kernel level opens up new analytical approaches to be investigated. Such approaches could be used to improve the potential of grain sorters, depending on the quality required. The example of DDGS authentication at the international level illustrated the analytical tools available for the feed sector. With the complexity of industrial processes used in plant feed companies and the tendency to promote both regional and organic feed production, more work is needed on feed authentication order to ensure animal feed safety. The examples of açai fruit and cocoa bean authentication and traceability demonstrated the potential of NIRS methods using miniature hand-held instruments as well as NIR-HIS. Using the example of melamine fraud, the description of the development of new chemometric tools such as LWPCA showed the possibility of using simple tools for NIR spectral data treatment in order to authenticate food and feed products and detect abnormal samples at an early stage. New initiatives at the European level, such as the FoodIntegrity and Authent-Net projects (2016–2018), enable analytical experts and funding bodies to provide Europe with an up-to-date and integrated ability to detect fraud and ensure the integrity of the food chain, as well as to coordinate inter-disciplinary research aimed at protecting consumers against fraud.

Keywords: Near infrared, mid infrared, raman spectroscopy, vibrational spectroscopy, authenticity, traceability, food, feed

References

Abbas, O., Lecler, B., Dardenne, P. and Baeten, V. 2013. Detection of melamine and cyanuric acid in feed ingredients by near infrared spectroscopy and chemometrics. Journal of Near Infrared Spectroscopy 21: 183–194.

Abbas, O., Dardenne, P. and Baeten, V. 2012. Near-infrared, mid-infrared and Raman spectroscopy. pp. 59–91. *In:* Pico, Y. (ed.). Chemical Analysis of Food: Techniques and Applications. Elsevier Science, Burlington.

Adams, M. 2008. Melamine Found Contaminating Soy Meal Fed to Organic Chickens. *On-line:* http://www.naturalnews.com/news_000571_melamine_organic_chickens_china. html (12/09/2016).

Ai, K., Liu, Y. and Lu, L. 2009. Hydrogen-bonding recognition-induced color change of gold nanoparticles for visual detection of melamine in raw milk and infant formula. Journal of the American Chemical Society 131(27): 9496–9497.

Álvarez, C., Pérez, E., Cros, E., Lares, M., Assemat, S., Boulanger, R. and Davrieux, F. 2012. The use of near infrared spectroscopy to determine the fat, caffeine, theobromine and (–)-epicatechin contents in unfermented and sun-dried beans of Criollo cocoa. Journal of Near Infrared Spectroscopy 20(2): 307–315.

Amaral, R. Jr., Baeten, V. and Rogez, H. 2015a. Discrimination of açaí fruit (*Euterpe oleracea*) according to their geographical origin using near infrared spectroscopy and chemometrics tools. *Poster at:* 17th ICNIRS: NIR2015, Foz do Iguassu, Brazil, 18–23 October 2015.

Amaral, R., Jr., Silva, J.P., Baeten, V. and Rogez, H. 2015b. Discrimination according to the agronomic condition of açaí (*Euterpe oleracea*) samples using hand-held near infrared spectrometer. *Poster at:* 17th ICNIRS: NIR2015, Foz do Iguassu, Brazil, 18–23 October 2015.

AUTHENT-NET 2016–2018. Food authenticity research network, EU-H2020-696371. *On-line:* http://www.authent-net.eu/.

Baeten, V., Fernández Pierna, J.A., Lecler, B., Abbas, O., Vincke, D., Minet, O., Vermeulen, P. and Dardenne, P. 2016. Near infrared spectroscopy for food and feed: a mature technique. NIR news 27(1): 4–6.

Baeten, V., Vermeulen, P., Fernández Pierna, J.A. and Dardenne, P. 2014. From targeted to untargeted detection of contaminants and foreign bodies in food and feed using NIR spectroscopy. New Food 17(3): 2–9.

Balabin, R. and Smirnov, S.V. 2011. Melamine detection by mid- and near-infrared (MIR/NIR) spectroscopy: a quick and sensitive method for dairy products analysis including liquid milk, infant formula, and milk powder. Talanta 85(1): 562–568.

Baranska, H. 1987. An introduction to Raman scattering. pp. 9–31. *In:* Baranska, H., Labudzinska, A. and Terpinski, J. (eds.). Laser Raman Spectrometry: Analytical Applications. Ellis Horwood, Chichester, UK.

Barton II, F.E. and Windham, W.R. 1988. Determination of acid detergent fibre and crude protein in forages by near infrared reflectance spectroscopy: collaborative study. Journal of the Association of Official Analytical Chemists 71: 1162–1167.

Bellon-Maurel, V., Vigneau, J.L. and Sevila, F. 1994. Infrared and near-infrared technology for the food industry and agricultural uses: on-line applications. Food Control 5(1): 21–27.

Bichara, C.M.G. and Rogez, H. 2011. Açaí (*Euterpe oleracea* Mart.). pp. 1–26. *In:* Yahia, E.M. (ed.). Postharvest Biology and Technology of Tropical and Subtropical Fruits Volume 2: Açai to citrus. Woodhead Publishing, Cambridge.

Branigan, T. 2008. Chinese figures show fivefold rise in babies sick from contaminated milk. The Guardian (London). *On-line:* https://www.theguardian.com/world/2008/dec/02/ china (12/09/2016).

Chan, E.Y.Y., Griffiths, S.M. and Chan, C.W. 2008. Public-health risks of melamine in milk products. The Lancet 372: 1444–1445.

Chen, J. 2009. What can we learn from the 2008 melamine crisis in China? Biomedical and Environmental Sciences 22: 109–111.

Coates, J. 2000. Interpretation of infrared spectra, a practical approach. pp. 10815–10837. *In:* Meyers, R.A. (ed.). Encyclopaedia of Analytical Chemistry. John Wiley and Sons Ltd., Chichester.

Crocombe, R.A. 2013. Handheld Spectrometers: The State of the Art. Proceedings of Next-Generation Spectroscopic Technologies VI, Baltimore, Maryland, USA, 8726.

Davrieux, F., Boulanger, R., Assemat, S., Portillo, E. and Cros, E. 2006. Determining fermentation levels and flavan-3-ol contents in dried cocoa by near infrared spectroscopy. Proceedings of 15th International Cocoa Research Conference, San José, Costa Rica, 1521–1528.

Davrieux, F., Assemat, S., Sukha, D., Bastianelli, D., Boulanger, R. and Cros, E. 2007a. Genotype characterization of cocoa through caffeine and theobromine content predicted by near infrared spectroscopy. Proceedings of the 12th International Conference - Near Infrared Spectoscopy. Auckland, New Zealand, 382–386.

Davrieux, F., Boulanger, R., Assemat, S., Portillo, E., Alvarez, C., Sukha, D.A. and Cros, E. 2007b. Determination of biochemistry composition of cocoa powder using near infrared spectroscopy. pp. 463–466. *In:* SFC (ed.). Proceedings of International Conference Euro Food Chem on Chemistry of Food, Paris, France.

Davrieux, F., Assemat, S., Boulanger, R., Sukha, D.A., Eskes, A., Paulin, D. and Cros, E. 2009. Characterization of cocoa clones from different origins for purine contents predicted by NIRS. Proccedings of 16th International Cocoa Research Conference Bali, Indonesia, 823–829.

Davrieux, F., Jimenez, J.C., Assemat, S., Hue, C., Kapitan, A. and Amores, F. 2013. Ecuador nacional fine cocoa authentication using NIR spectra and multivariate classification methods. Proceedings of ICNIR 2013 - 16th International Conference on Near Infrared Spectroscopy, La Grande Motte, France, 48–52.

Debode, F., Janssen, E., Marien, A. and Berben, G. 2012. DNA detection by conventional and real-time PCR after extraction from vegetable oils. Journal of the American Oil Chemists' Society 89(7): 1249–1257.

De Jong, J., López, P., Mol, H., Baeten, V., Fernández Pierna, J.A., Vermeulen, P., Vincent, U., Boix, A., von Holst, C., Tomaniova, M., Hajslova, J., Yang, Z., Han, L., MacDonald, S., Haughey, S.A. and Elliott, C.T. 2016. Analytical strategies for the early quality and safety assurance in the global feed chain. Trends in Analytical Chemistry 76: 203–215.

Dobson, R.L., Motlagh, S., Quijano, M., Cambron, R.T., Baker, T.R., Pullen, A.M., Regg, B.T., Bigalow-Kern, A.S., Vennard, T., Fix, A., Reimschuessel, R., Overmann, G., Shan, Y. and Daston, G.P. 2008. Identification and characterization of toxicity of contaminants in pet food leading to an outbreak of renal toxicity in cats and dogs. Toxicological Sciences 106(1): 251–262.

Dong, Y.W., Tu, Z.H., Zhu, D.Z., Liu, Y.W., Wang, Y.N., Huang, J.L., Sun, B.L. and Fan, Z.N. 2009. Feasibility of NIR spectroscopy to detect melamine in milk. PubMed 29(11): 2934–2938.

Ebermann, M., Hiller, K., Kurth, S. and Neumann, N. 2009. Design, Operation and Performance of a Fabry–Perot-Based MWIR Microspectrometer. Proceedings of IRS2, SensorþTest Conferences. Nürnberg, Germany, 26–28.

Escarnot, E., Jacquemin, J.M., Agneessens, R. and Paquot, M. 2012. Comparative study of the content and profiles of macronutrients in spelt and wheat, a review. Biotechnology, Agronomy, Society and Environment 16(2): 243–256.

[FDA] U.S. Food and Drug Administration. 2007. Melamine pet food recall – frequently asked questions. *On-line:* http://www.fda.gov/AnimalVeterinary/SafetyHealth/RecallsWithdrawals/ucm129932.htm (12/09/2016).

Fernández Pierna, J.A., Vermeulen, P., Lecler, B., Baeten, V. and Dardenne, P. 2010. Calibration transfer from dispersive instruments to handheld spectrometers. Applied Spectroscopy 64(6): 644–648.

Fernández Pierna, J.A., Vincke, D., Dardenne, P., Yang, Z., Han, L. and Baeten, V. 2014. Line scan hyperspectral imaging spectroscopy for the early detection of melamine and cyanuric acid in feed. Journal of Near Infrared Spectroscopy 22: 103–112.

Fernández Pierna, J.A., Abbas, O., Lecler, B., Hogrel, P., Dardenne, P. and Baeten, V. 2015. NIR fingerprint screening for early control of non-conformity at feed mills. Food Chemistry 189: 2–12.

Fernández Pierna, J.A., Baeten, V., Michotte Renier, A., Cogdill, R.P. and Dardenne, P. 2005. Combination of Support Vector Machines (SVM) and Near Infrared (NIR) imaging spectroscopy for the detection of meat and bone meat (MBM) in compound feeds. Journal of Chemometrics 18(7-8): 341–349.

Fernández Pierna, J.A., Vincke, D., Baeten, V., Grelet, C., Dehareng, F. and Dardenne, P. 2016. Use of a multivariate moving window PCA for the untargeted detection of contaminants in agro-food products, as exemplified by the detection of melamine levels in milk using vibrational spectroscopy. Chemometrics and Intelligent Laboratory System 152: 157–162.

Food Integrity. 2014–2018. Ensuring the integrity of the European food chain, EU-FP7-613688. *On-line:* http://www.foodintegrity.eu (12/09/2016).

Gossner, C.M.E., Schlundt, J., Ben Embarek, P., Hird, S., Lo-Fo-Wong, D., Beltran, J.J.O., Teoh, K.N. and Tritscher, A. 2009. The melamine incident: implications for international food and feed safety. Environmental Health Perspectives 117: 1803–1808.

Haughey, S.A., Graham, S.F., Cancouet, E. and Elliott, C.T. 2012. The application of Near-Infrared Reflectance Spectroscopy (NIRS) to detect melamine adulteration of soya bean meal. Food Chemistry 136(3-4): 1557–1561.

Haughey, S.A., Galvin-King, P., Graham, S.F., Cancouët, E., Bell, S. and Elliott, C.T. 2013. The use of Raman spectroscopy for the detection of contamination and traceability commodities used in the animal feed sector. *Poster at*: Recent Advances in Food Analysis (RAFA 2013), Prague, Czech Republic, 5–8 November 2013.

Hue, C., Brat, P., Gunata, Z., Samaniego, I., Servent, A., Morel, G., Kapitan, A., Boulanger, R. and Davrieux, F. 2014a. Near infrared characterization of changes in flavan-3-ol derivatives in cocoa (*Theobroma cacao* L.) as a function of fermentation temperature. Journal of Agricultural and Food Chemistry 62(41): 10136–10142.

Hue, C., Gunata, Z., Bergounhou, A., Assemat, S., Boulanger, R., Sauvage, F.X. and Davrieux, F. 2014b. Near infrared spectroscopy as a new tool to determine cocoa fermentation levels through ammonia nitrogen quantification. Food Chemistry 148: 240–245.

Huang, H., Yu, H., Xu, H. and Ying, Y. 2008. Near infrared spectroscopy for on/in-line monitoring of quality in foods and beverages: a review. Journal of Food Engineering 87(3): 303–313.

IBGE. 2016. Instituto Brasileiro de Geografia e Estatistica. *On-line:* http://www.ibge.gov.br/english/ (27/09/2016).

Kong, W., Zhang, C., Liu, F., Nie, P. and He, Y. 2013. Rice seed cultivar identification using near-infrared hyperspectral imaging and multivariate data analysis. Sensors 13(7): 8916–8927.

Li-Chan, E. 1996. The application of Raman spectroscopy in food science. Trends in Food Science and Technology 7: 361–370.

Lin, M. 2009. A review of traditional and novel detection techniques for melamine and its analogues in foods and animal feed. Frontiers of Chemical Science and Engineering 3: 427–435.

Liu, Y., Todd, E.E.D., Zhang, Q., Shi, J.R. and Liu, X.J. 2012. Recent developments in the detection of melamine. Journal of Zhejiang University Science B 13(7): 525–532.

Luykx, D.M.A.M. and van Ruth, S.M. 2008. An overview of analytical methods for determining the geographical origin of food products. Food Chemistry 107: 897–911.

Lu, C., Xiang, B., Hao, G., Xu, J., Wang, Z. and Chen, C. 2009. Rapid detection of melamine in milk powder by near infrared spectroscopy. Journal of Near Infrared Spectroscopy 17(2): 59–67.

Massart, D.L., Vandeginste, B.G.M., Deming, S.N., Michotte, Y. and Kaufman, L. 1988. Chemometrics: a textbook, in the series: Data Handling in Science and Technology. Elsevier Science Publishers B.V., Amsterdam, The Netherlands, volume 2.

Mauer, L.J., Chernyshova, A.A., Hiatt, A., Deering, A. and Davis, R. 2009. Melamine detection in infant formula powder using near- and mid-infrared spectroscopy. Journal of Agricultural and Food Chemistry 57: 3974–3980.

McClure, W.F. 2001. Near-infrared instrumentation. pp. 109–127. *In:* Williams, P. and Norris, K. (eds.). Near-infrared Technology in the Agricultural and Food Industries. American Association of Cereal Chemists, St. Paul, MN.

Monfort, B., Herman, J.L., Couvreur, L., Vancutsem, F., Bodson, B., Henriet, F., Weickmans, B., Moreau, J.M., De Proft, M., Steyer, S., Meeùs, P., Frankinet, M. and Falisse, A. 2006. Escourgeon et Orge d'hiver fourragers. pp. 1–13. *In:* Livre blanc "Céréales" – Gembloux, Information avant les semis, September 2006.

Murray, I. 1986. The NIR spectra of homologous series of organic compounds. pp. 13–28. *In:* Hollo, J., Kaftka, K.J. and Gonczy, J.L. (eds.). Proceedings of the International NIR/NIT Conference. Akademiai Kiado, Budapest, Hungary.

Murray, I. 1993. Forage analysis by near infrared reflectance spectroscopy. pp. 285–312. *In:* Davies, A., Baker, R.D., Grant, S.A. and Laidlaw, A.S. (eds.). Sward Management Handbook. British Grassland Society, Reading, UK.

Nietner, T., Haughey, S.A., Ogle, N., Fauhl-Hassek, C. and Elliott, C.T. 2014. Determination of geographical origin of distillers dried grains and solubles using isotope ratio mass spectrometry. Food Research International 60: 146–153.

Nietner, T., Pfister, M., Glomb, M. and Fauhl-Hassek, C. 2013. Authentication of the botanical and geographical origin of DDGS by FT-IR spectroscopy. Journal of Agricultural and Food Chemistry 61(30): 7225–7233.

Norris, K.H., Barnes, R.F., Moore, J.E. and Shenk, J.S. 1976. Predicting forage quality by infrared reflectance spectroscopy. Journal of Animal Science 43: 889–897.

Novotna, H., Cajka, T., Tomaniova, M. and Hajslova, J. 2012. Metabolomic fingerprinting/profiling employing DART coupled with TOF-MS and orbitrap-MS for DDGS authenticity and traceability. *Poster at:* 1st European Workshop on Ambient Mass Spectrometry and Related Mass Spectrometry-Based Techniques in Food/Natural Products Control: Safety, Authenticity, Forensics, Metabolomics, Prague, Czech Republic, 18–20 June 2012.

Osborne, B.G., Fearn, T. and Hindle, P.H. 1993. Practical NIR Spectroscopy with Application in Food and Beverage Analysis. Longman Scientific and Technical, Singapore, pp. 120–141.

Pasquini, C. 2003. Near infrared spectroscopy: fundamentals, practical aspects and analytical applications. Journal of the Brazilian Chemical Society 14(2): 198–219.

Pügner, T., Knobbe, J. and Grüger, H. 2016. Near-infrared grating spectrometer for mobile phone applications. Applied Spectroscopy 70(5): 734–745.

QSAFFE. 2011–2014. Quality and safety of feed and food for Europe, EU-FP7-265702. *On-line:* http://www.qsaffe.eu/ (12/09/2016).

RASFF. 2015. Rapid Alert System for Food and Feed: Preliminary Annual Report 2015. *On-line:* http://ec.europa.eu/food/safety/docs/rasff_annual_report_2015_preliminary.pdf (12/09/2016).

Riedl, J., Esslinger, S. and Fauhl-Hassek, C. 2015. Review of validation and reporting of non-targeted fingerprinting approaches for food authentication. Analytica Chimica Acta 885: 17–32.

Rogez, H., Fernández Pierna, J.A., Souza, J.N.S. and Baeten, V. 2015. Application of near infrared hyperspectral imaging spectroscopy for the analysis of cocoa beans. *Poster at:* 17th ICNIRS: NIR2015, Foz do Iguassu, Brazil, 18–23 October 2015.

Rovina, K. and Siddiquee, S. 2015. A review of recent advances in melamine detection techniques. Journal of Food Composition and Analysis 43: 25–38.

Schauss, A.G., Wu, X., Prior, R.L., Ou, B., Patel, D., Huang, D. and Kababick, J.P. 2006. Phytochemical and nutrient composition of the freeze-dried Amazonian palm berry, *Euterpe oleracea* Mart. (Acai), Journal of Agricultural and Food Chemistry 54: 8598–8603.

Shenk, J.S. and Westerhaus, M.O. 1995. The application of near infrared reflectance spectroscopy (NIRS) to forage analysis. Chapter 10. pp. 406–449. *In*: Fahey, G.C., Collins, M., Mertens, D.R. and Moser, L.E. (eds.). Forage Quality, Evaluation, and Utilization. American Society of Agronomy, Madison, WI, USA.

Smirnov, S.V. 2011. Neural network based method for melamine analysis in liquid milk. Proccedings of International Conference on Intelligent Computation Technology and Automation (ICICTA) 2: 999–1002.

Suman, M. 2016. Verification of the effectiveness on the field of reliable industrial rapid screening technologies and confirmatory methods. Deliverable report in FoodIntegrity project, 10 pp.

Sun, F., Ma, W., Xu, L., Zhu, Y., Liu, L., Peng, C., Wang, L., Kuang, H. and Xu, C. 2010. Analytical methods and recent developments in the detection of melamine. Trends in Analytical Chemistry 29(11): 1239–1249.

Tena, N., Boix Sanfeliú, A. and von Holst, C. 2015. Identification of botanical and geographical origin of distillers dried grains with solubles by near infrared microscopy. Food Control 54: 103–110.

Teye, E., Huang, X., Takrama, J. and Haiyang, G. 2014. Integrating NIR spectroscopy and electronic tongue together with chemometric analysis for accurate classification of cocoa bean varieties. Journal of Food Process Engineering 37(6): 560–566.

Tres, A, Heenan, S.P. and van Ruth, S. 2014. Authentication of dried distilled grain with solubles (DDGS) by fatty acid and volatile profiling. LWT-Food Science and Technology 59(1): 215–221.

Tyan, Y., Yang, M., Jong, S., Wang, C. and Shiea, J. 2009. Melamine contamination. Analytical and Bioanalytical Chemistry 395: 729–735.

Vermeulen, P., Fernández Pierna, J.A., Abbas, O., Dardenne, P. and Baeten, V. 2010. Authentication and traceability of agricultural and food products using vibrational spectroscopy pp. 609–630. *In:* Li-Chan, E.C.Y., Griffiths, P.R. and Chalmers, J.M. (eds.). Applications of Vibrational Spectroscopy in Food Science. John Wiley and Sons Ltd., Chichester, UK.

Vermeulen, P., Fernández Pierna, J.A., Abbas, O., Dardenne, P. and Baeten, V. 2015a. Origin identification of dried distillers' grains with solubles using attenuated total reflection Fourier transform mid-infrared spectroscopy after *in situ* oil extraction. Food Chemistry 189: 19–26.

Vermeulen, P., Herman, J., Lecler, B., Sinnaeve, G., Baeten, V., Dardenne, P. and Fernández Pierna, J.A. 2007. Barley varieties discriminated by the near infrared hyperspectral imaging technique. Proceedings of 13th International Conference on Near Infrared Spectroscopy, UMEA, Sweden, 15–21 June 2007.

Vermeulen, P., Nietner, T., Haughey, S.A., Yang, Z., Tena, N., Chmelarova, H., van Ruth, S., Tomaniova, M., Boix, A., Han, L., Elliott, C.T., Baeten, V. and Fauhl-Hassek, C. 2015b. Origin authentication of distillers' dried grains and solubles (DDGS) – Application and comparison of different analytical strategies. Analytical and Bioanalytical Chemistry 407(21): 6447–6461.

Wang, N., Zhang, X., Yu, Z., Li, G. and Zhou, B. 2014. Quantitative analysis of adulterations in oat flour by FT-NIR spectroscopy, incomplete unbalanced randomized block design, and partial least squares. Journal of Analytical Methods in Chemistry 2014: 1–5.

Williams, P.J., Geladi, P., Fox, G. and Manley, M. 2009a. Maize kernel hardness classification by near infrared (NIR) hyperspectral imaging and multivariate data analysis. Analytica Chimica Acta 653(2): 121–130.

Williams, P.J. 2009b. Near infrared (NIR) hyperspectral imaging for evaluation of whole maize kernels: chemometrics for exploration and classification. Thesis, Stellenbosch University, South Africa.

[WHO/FAO] World Health Organization/Food and Agriculture Organization of the United Nations. 2008. Expert Meeting to Review Toxicological Aspects of Melamine and Cyanuric Acid, 1–10.

Yan, N., Zhou, L., Zhu, Z. and Chen, X. 2009. Determination of melamine in dairy products, fish feed, and fish by capillary zone electrophoresis with diode array detection. Journal of Agricultural and Food Chemistry 57(3): 807–811.

Yang, X., Hong, H., You, Z. and Cheng, F. 2015. Spectral and image integrated analysis of hyperspectral data for waxy corn seed variety classification. Sensors 15(7): 15578–15594.

Zhang, X., Liu, F., He, Y. and Li, X. 2012. Application of hyperspectral imaging and chemometric calibrations for variety discrimination of maize seeds. Sensors 12(12): 17234–17246.

Zhou, X., Yang, Z., Haughey, S.A., Galvin-King, P., Han, L. and Elliott, C.T. 2014. Classification the geographical origin of corn distillers' dried grains with solubles by near infrared reflectance spectroscopy combined with chemometrics: a feasibility study. Food Chemistry 189: 13–18.

Zhu, D., Wang, C., Pang, B., Shan, F., Wu, Q. and Zhao, C. 2012. Identification of wheat cultivars based on the hyperspectral image of single seed. Journal of Nanoelectronics and Optoelectronics 7(2): 167–172.

17

Biocaptors and Barcodes
New Devices for the Traceability and Authenticity

Didier Tousch[1],* and *Benoit Charlot*[2]

1. Introduction

The fabulous development of food markets during the last twenty years, is undoubtedly the origin of companies' mutation particularly in the new management of the production and distribution. Today, mass production and a fierce competition can always lead to excess of production that obliges companies to diversify their products in order to respond quickly to specific markets. The control of the production and distribution flows becomes essential to the companies. Thus they have defined a management tool based both on the computer's power (a great collector of information) and on the smart tag (source of information); the logistic was born. Logistic development coincides with the development of Internet Data Exchanges (IDE), which has reduced the cost of transmitting and processing data to virtually zero. These procedures of identification which associate the information between the products and virtual data become a real economic challenge and give an advantage to penetrate new markets: "welcome to the world of *track and trace* procedures". Today, automatic identification

[1] UMR 95 Qualisud, 15 Avenue Charles Flahault, BP 14491, 34093 Montpellier Cedex 5, France.
[2] IES, Institut d'Electronique et des Systèmes, UMR 5214 CNRS Université de Montpellier, Montpellier, France; E-mail: benoit.charlot@umontpellier.fr
* Corresponding author: didier.tousch@umontpellier.fr

of food products is composed of a set of techniques: barcodes, optical character recognition, and radio tags named RFID tags (Radio-Frequency IDentification). Other techniques such as magnetic cards and pattern recognition are used for different applicative domains. We think that barcode tags possess the fundamental characteristics expected for a traceability tool: encrypted tags enabling real time reading.

During the last ten years, the perception of the food market in the European community has changed, between the necessity to preserve the free-exchange's market and the necessity to preserve the consumer's safety. For this, laws like the «EU regulation 178–2002» constrain the food companies to assure the complete traceability of their products: control of quality must be done not only at the end of the chain but also throughout the supply chain. So, traceability permits to remake history of a product in order to determinate a dysfunction in the supply chain.

Food quality and safety are related in part to the hygienic practices during the industrial processes to obtain the conformity of products for their organoleptic properties, the absence of pathogens, the absence of GMO (Genetically Modified Organism) and other prejudicial substances. Here, the only question is combining analysis (biological, biochemical and/or chemical) with logistic. Automation of the analytical tools permits to work in real time: nanotechnologies will allow these advances. DNA analysis seems to be effective in the evaluation of the presence of a GMO or allergenic species, both in fresh and in processed food (Galimberti et al. 2013). The automation of DNA chip technology combined to the power of the PCR becomes an efficient serial analysis.

Finally, globalized free trade generates a significant increase in counterfeiting. Under the pressure brought by states, companies are actively searching for anti-counterfeiting methods. The cryptology science has been revisited by the modern technologies including the nanotechnologies with the nanometer tags and/or the biotechnologies with the molecular markers. Two approaches have been developed these last year's leading to a new layer of traceability; the first is the "extrinsic" tag apposed on the product for authentication and the second is the "intrinsic tag", own of products for identification where DNA is considered as the most emblematic example (Fig. 1).

The future technology proposed as an innovative tool of traceability will be based on the combination of their properties. The smart tag which should both allow the identification (*track and trace*), the authentication and also a control of another parameter as the cold chain or the provenance of products is a novelty.

Today, many miniaturized systems of coding and labeling are being studied: nano-QRcode, barcode and others, also particles or fibers coded by various molecules, rare earth, fluorophores, quantum-dot, metals, etc. However, none of them meet all the criteria demanded by market needs. So

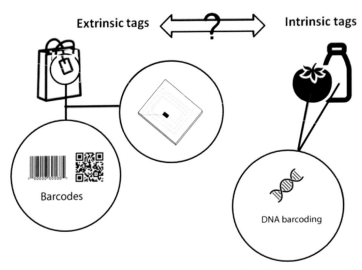

Figure 1. Examples of tags for the agro-food industries: Two identification tags coexist for agro-food products; the extrinsic tags essentially apposed on the bag of products and the intrinsic tags own of products. The printed barcodes essentially used are linear or multidimensional, the QR-code. More recently, Radio Frequency IDentification (RFID) tag with an electronic chip and an antenna allowing a transmission and a reception of information.

we still have to wait for the technologies that will integrate the properties of multiple nanosystems.

In this chapter, we discuss our improving of sensors by the hybrid technologies able to transform a biological signal into measurable physical signal.

2. Usual Tools for Traceability (track and trace)

Traceability means the ability to track and trace any food, feed, food producing animal or substances that will be used for consumption, through all stages of production, processing and distribution in order to manage the risks and crisis. It is why, each product is given a unique identifier which accompanies it and will record at all stages of the process through its supply chain.

2.1 Traceability automatic identification: barcodes

A barcode integrates only Identification Data (ID) represented by a juxtaposition of white and black bars that composes a series of number where each number is predefined as the composition of the juxtaposition of arrested lines number; it is able to have software to produce tags for large serial products. Worth Data (Santa Cruz, California) has been and remains a pioneer in this field, by providing equipment and barcode printing software for computer users. A barcode reader allows capturing a signal emitted by the barcode itself and transmits this information to a terminal which translates the signal into data specific/intrinsic for each product. Today, three portable or fixed barcode readers are used: the laser, the Charge Coupled Devices (CCD) or a light beam emitted by a scanner. The light reflected and received by a photocell detector converts into an electrical signal. In the last step, a computer translates the ID information in a number previously entered in a databank associated on the ID (Palmer 2007, Pearce and Bushnell 2010).

With the matricial-code (or datamatrix), which has very well integrated the notion of pixel, the barcodes are constituted of small black squares (pixels) at high density. The arrangement of these squares defines the information that contains the code: this code is named Quick Response Code (QR Code).

2.2 *Radio frequency iDentification (RFID)*

RFID based on the radio-tag (Zhu et al. 2012) corresponds to a radio antenna associated with an electronic chip incorporated together in a suitable packaging (essentially polymer). The RFID-tags are transmitters - receivers (Fig. 2). The electronic chip contains the IDs and also many data throughout the production chain (as quality control data). Data travel *via* radio waves (radio frequency). Radio frequencies will be chosen depending on the type

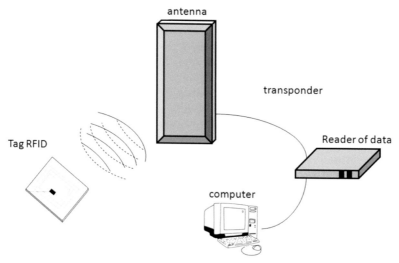

Figure 2. RFID Tag and the Peripheral Devices. The RFID-tag is a transmitter - receiver. A transponder receives the data via an antenna and is in communication with a computer (Zhu et al. 2012).

of application that requires more or less to be worn, a few millimeters to a few hundred meters. There are two major families of RFID tags: (1) Active tags with a power source (a battery) have a great range but at a high cost and a limited life, (2) Passive tags, without battery used the energy propagated at short distance by the transmitter. Today, the food industry uses the RFID technology in a large spectrum of applications for the traceability (Kumari et al. 2015), the cold chain monitoring, livestock management and for the shelf life prediction and quality monitoring.

Kumar et al. (2009) and Kumari et al. (2015) cited that presently some food factories are using the RFID in their supply chain, temperature monitoring. For example, RFID technology has been used by cheese manufacturers to trace cheese along the supply chain. Customers can find the entire history of the cheese. Also, this tracking system could result in rapid product recall in case of a problem. The company eProvenance (Bordeaux, France) has developed an RFID based tracking system to preserve the quality of fine wines and trace their origin.

2.3 *The intrinsic markers of foods (DNA, protein and metabolites)*

The traceability of food according to European law 178/2002 requires, beyond the automatic identifiers, intrinsic identifiers that are components of product themselves. For this, DNA carry identifiers which are specific to species, individual or global populations: molecular analysis establishes identity cards or molecular IDs from heritable characters or not of the organisms. A lot of DNA markers appear with genomics programs and sequencing of eukaryotic genomes: Microsatellites or STR (short tandem repeat) or even the SNP (single nucleotide polymorphism), both are distributed over the genome (Sauer et al. 2000). The PCR technologies (polymerase chain reaction) and the high throughput sequencing coupled with the capillary electrophoresis as well as the mass spectrometry allows for a speedier and more precise analysis. Earlier a DNA barcoding was developed at the University of Guelph (Canada) that integrates the analysis of nucleotide polymorphism within a standard region of the genome (Herbert et al. 2003). DNA barcode regions should have a high interspecific and low intraspecific variability (Kress and Erickson 2008, Herbert et al. 2004). DNA barcoding is used both for origin and quality determination of raw materials notably to detect adulterations in the industrial food chain. Based on a genetic map built using SSR (simple sequence repeat) equivalent to STR, several varieties of vines are currently characterized at the molecular level (Doligez et al. 2006a, Cadot et al. 2006).

The meat industry has launched several systematic genotyping tests of livestock to ensure full traceability from farm to fork. The use of microsatellites in particular allowed the genotyping of several species (Applied Biosystems) including bovine, where 6 microsatellites STR are on the list of 11 STR and are recommended by ISAG (International Society of Animal Genetics) to identify individuals in multiplexing (Brenig and Schütz 2016).

Proteins are also good biological markers for identifying the presence or absence of specific genotypes. The European regulation 2001/18/CE requires tracing Genetically Modified Organisms (GMO) by DNA analysis and also by the transgenic proteins (Martins-Lopes et al. 2013). The usual technology of protein analysis is based on antibodies recognition: *western blot* or Elisa (*enzyme-linked immunosorbent assay*) technologies involving fluorescent dyes. The protein analysis approach in the agri-food industry becomes essential today in order to detect and quantify allergens which affect around 2 to 4% of the world population (Baumert 2014, Kuklinska-Pijanka et al. 2015).

Plants synthesize and accumulate many chemical structures (metabolites), such as phenolic and terpenic metabolites, alkaloids and steroids. Therefore they provide good markers of interspecific or intervarietal genetic diversity both qualitatively and quantitatively. Each genotype has its own profile within the secondary metabolites profile (metabolic fingerprints) obtained by liquid chromatographic techniques as thin layer chromatography (TLC), high performance liquid chromatography (HPLC) and gas chromatography (GC) coupled or not with mass spectrometry (MS) (Thangaraj 2015).

Phenolic compounds synthesized and accumulated in the grape vines are used today both to certify wine or cognac "*Appelation d'origine contrôlée*" (AOC French quality label) and to fight the fraud (adulteration) resulting from mixtures (Delgado et al. 1990). Phenolic markers are found to be quality or/and organoleptic markers for controlled original wines, and now also include the specificity of soils (Jasicka-Misiak et al. 2012, Fernandez-Lorenzo et al. 1999). Quantitative traits of flavor including monoterpenic odor characters in the vineyard are available (Doligez et al. 2006b). For wines and cognac aged in oak barrels, phenolic compounds are used to authenticate the origin of the barrels (Matějíček et al. 2005).

3. New Tools for the Authentication: Micro and Nano-barcodes

Aside from the safety of the food during its manufacture, a new task for traceability is also to detect counterfeit products. Counterfeiting not only leads to a loss of profits for brands, but could also become a risk to the consumer since such products do not comply with the legislation.

3.1 Problematic

Beyond the respect of the trademark rights, counterfeiting represents a threat to consumers likely to purchase a product of poor quality that could endanger their health. No longer confined to luxury products and considered "the ransom of the glory" counterfeiting is rampant today in all industries and food markets are not spared. Thus, counterfeiting becomes a huge public safety issue. Developing countries are the most exposed: in Africa, 70% of antimalarial drugs are counterfeit. In the European Union, it is very easy to purchase cheaper products posing as authentic via the internet without any warranty. As mention above, the tags for fighting counterfeiting exist and are already used as electronic chips, RFID tags, etc. But these techniques are still expensive and are confined to the authentication of specific products. For example, wine professionals on a market currently valued at USD 304 billion (http://www.prweb.com/releases/2013/8/prweb11009246.htm) were trying since few years to find authentication tags. Now, a specific and clever labeling is placing on the bottle (on the bottle label or on the bottle neck) and is marketing under different forms, as a *Bubble tags* (Prooftag company, Montauban, France) (http://www.prooftag.net/) and fluorescent tags incrusted in glass of bottles (Atheor company, Montpellier, France) are being marketed. These discrete Tags, that are sometimes invisible to the eyes, still aren't enough to meet the complete encrypted need of authentication. The market is organized but is still unable to provide an ideal control system. The difficulty in this area is that this type of tag does not generate added value for the company and that the additional costs due to innovative technologies remain a major obstacle for businesses.

The minimal properties of an authentication tag must be the non-reproducibility, the unfalsifiable character and has to be inseparable from the product like a fingerprint. This last point favors an intrinsic tag like DNA but PCR analysis is not applicable in great serials (Fig. 3). So, the best way to authenticate a food can be defined by the following criteria: extrinsic encrypted tag, invisible to the eye with a real-time reading and food-authorized to dump both on the packaging and directly into the food product. The extrinsic new tools for a traceability and authentication are certainly the micro and nano-barcodes (Finkel 2004, Sung-Kyoung and Sang Bok 2009). The micro or nano size of these barcodes allows their insert and spread in a material. Micro particles that contain a code read by microscopy or by optical signal (Fig. 3), non eye-visible, permit the tracking of products and protecting against illegal copies or adulteration. However, it is important that the large-scale particle production can be realized at low-cost. The decrypting of the particle code requires the use of a portable and accurate detection system.

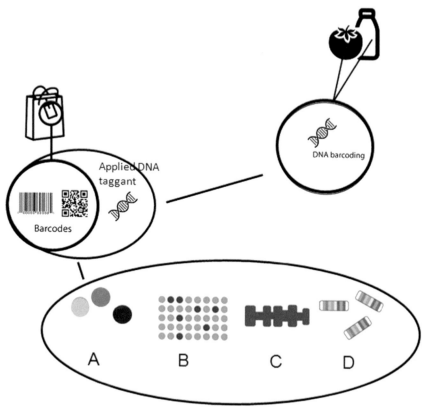

Figure 3. Nano-barcodes for both Identification and Authentication. (A) nano particles with an optical fingerprint, (B) random deposited matrix, (C) shape linear barcode (Sung-Kyoung and Sang Bok 2009), (D) micro-rods with fluorescence pattern.

The next section will present several techniques that are developed for the production of small scales (micron sized) particles that embed a code either from their shape or their optical properties (Fig. 3).

To combine extrinsic and intrinsic markers, it is possible to integrate tags both into food directly and on the packaging. Tags could be DNA, particles or micro barcodes completely healthy and biodegradable, with a food approval. Even if a lot of publications make a projection of traceability nanotechnologies in the food market, it should be noted that the applications of these news nanotechnologies in the food industry are still limited (Chaudhry et al. 2008, Abbas et al. 2009). For a routine use for safety and security, these nanotechnologies need notably showing their economic impact on the market. So, this chapter tries to convince of the great capacity of these nanotechnologies to response of all requirements of the food market even if they not yet ready.

3.2 Holograms

Holography is a technique that consists of recording on a photosensitive material of a light field. It could be compared to photography where an image formed by a lens is recorded upon a material (a light sensitive emulsion) or directly captured with a semiconductor sensor. Holograms are then recordings of interference patterns on a photographic emulsion that are illuminated by a coherent light source: a laser. Holograms can reproduce an impression of 3D surface and illusion of depth. Security holograms are made of a stack of images imprinted on transparent sheets with metallic patterns. One main interest fact about holograms is that the reflected light patterns depend on viewing orientation. This induces a complexity for the scanning of the pattern and its reproduction (Shaked et al. 2009). However, this technique has found many markets, but eventually becomes less secure and is sometimes counterfeited. Moreover, this technical tool is still too expensive for the food industries (Ghaani et al. 2016).

3.3 Applied DNA taggants (Extrinsic DNA tag)

Today the intrinsic DNA tags possess perfectly all the properties rechearch on the market of the fight against food-counterfeiting (Scarano and Rao 2014). This direct identification of biological components of a food remains an effective method to control the origin of a product or a possible adulteration by a lower quality product. However, the fingerprinting-DNA approaches with extrinsic DNA tags have begun to emerge. The principle is easy as it resumes the wonderful coding capacity of DNA by assembling the 4 nucleotides. An immobilized, incorporated or encapsulated DNA on/in a matrix is deposited on the product to be authenticated. Some companies producing this application are emerging today including the company Applied DNA Sciences (USA) (http://www.adnas.com/), which offers many extrinsic labels DNA-based plant DNA with an extensive patent portfolio. However, it is difficult to evaluate objectively the potential of this DNA tag because it is difficult to gain access of the detailed technical papers, which are confidential (Scarano and Rao 2014, http://www.adnas.com/).

3.4 Shape and reflectivity micro barcodes

A key technique for producing micro particles with the ability to encode information is to build particles with a determined shape, being detectable by optical microscopy and image analysis. Encoding information in a particle shape allows using hard, stable and biocompatible materials. Unique or several particles with distinctive shapes are then dispersed in a packaging to act as a micro barcode. The manufacturing of these particles should obviously be done collectively and in a controlled manner, i.e., keeping control of the information stored in the particle form. A technique has recently been developed consisting of producing silicon micro cylinders with the help of dry etching in reactive plasma (Gomez-Martinez 2011, Matthias 2002). This etching technique is very frequently used in the microelectronics industry and microsystems. By controlling the distribution and the pressure of the reactants present in the plasma gas, it is possible to change the isotropic of the chemical physical etching mechanism of silicon. By alternating isotropic etching and anisotropic phases it is possible to create a profile of the silicon rod that encodes information. In the end we obtain micro rods with a key type of profile that are easily detected using microscopy. This technique also allows the production of millions of particles that are then easily incorporated in a polymeric packaging. Figure 4 shows examples of such silicon micro-machined rods that incorporate a code within their profile.

Reflectivity barcoding particles (Nicewarner-Peña 2001) are another type of particles that do not use the geometric shape of particles but the optical properties of reflection of the light of noble metals. Here, the microfabrication techniques are similar to that described in the preceding paragraph in the highly collective character of the process. However the micromachining process does not rely on removing material, but a metal electrochemical growth process through a microporous matrix. These porous membranes are frequently used to produce strips from an electrochemical bath metal growth. The employed membranes are often alumina for their ability to form controlled micro-channels. Growth of metal inside the channels is performed by the reduction of metal ions in solution using a series of baths containing plating to deposit the desired metal sequence. The particles detection is done as in previous particles by using transmission or reflection microscopy with image processing. Here it may be noted that the reflectance of the support may disrupt the analysis of optical signals (Gomez-Martinez 2011, Matthias 2002, Nicewarner-Peña 2001).

(a) (b)

Figure 4. SEM Pictures of Micro-barcodes, (a) Plasma Etched Silicon Micro-rods (Gomez-Martinez 2011) and (b) in Matthias (2002).

3.5 Fluorescent barcode

Fluorescence is light emission caused by the absorption and re-emission of the optical wavelength signal. Absorption of light causes an excitation of molecules of the material conducing to a gain in electrons energy. Recombination of these electrons can happen with several mechanisms as functions of available energy state including radiative recombination that produces a light emission at a wavelength different to that of the excitation light. A fluorescent material is then easily detectable by microscopy if care is taken to filter the excitation wavelength. This technique has revolutionized cell biology allowing to optically tag component of cells. In the field of micro barcodes, fluorescence has great advantages over the techniques described earlier (Finkel et al. 2004). One of these advantages is that particles emit light and which increases their detectability, i.e., the possibility to extract the position and shape of particles against the background and support material. Participle's fluorescence signal is defined as an information carrier and this information can be encoded in many ways such as the fluorescence intensity, shape and position of the particles but also the fluorescence spectrum (Finkel et al. 2004).

3.5.1 Fluorochromes

Several types of materials exhibit fluorescence. Fluorescent organic dyes also called fluorophores or fluorochromes are chemical compounds that reemit light upon excitation, this light emission being at a longer wavelength compared with excitation. These molecules show structure-specific emission profiles; each molecular compound has its own absorption and emission peaks and they also show good quantum yield, e.g., large optical conversion efficiency. When excited by an incoming light, these molecules emit a fluorescent signal readily detectable through the use of a simple microscope or spectrophotometer. Most common fluorescent dyes are fluorescein but also rhodamine and cyanine. Photobleaching, sometimes termed as fading, is the photochemical alteration of a dye or a fluorophore molecule. In this case, fluorescence decreases with illumination and over time. For a dynamic point of view, the fluorescence lifetime of organic dyes ranges about 5 ns, which is too short for temporal discrimination.

Figure 5 shows a typical absorption and emission spectrum of a fluorescent dye. The absorption shows a peak at a given value that correspond at its excitation value and a second peak, at a longer wavelength (smaller energy) that is the emission peak of the fluorochrome. Note that these peaks present a small separation of both absorption and emission peaks and that leads to an eventual cross-talk between different dye molecules. A prominent example of fluorescent encoding was developed by Luminex Corp (Austin, Texas) (https://www.luminexcorp.com/), where

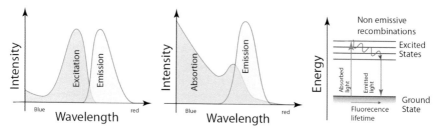

Figure 5. Typical spectrum of Excitation and emission of fluorescent dyes and Quantum dots and sketch of absorption and emissive recombination mechanisms.

two dyes were incorporated at ten different concentrations to obtain 100 unique encoded particles.

3.5.2 *Quantum dots*

Semiconductor nanocrystals or quantum dots are semiconductor nanoparticles (1 to 6 nm) that confine charge carriers (electrons and holes) in the 3 dimensions. The confinement of the charge carriers that can be described as a electron gas induces a change in the bandgap configuration and then in the energy levels inside the materials and hence the recombination mechanisms lead to tunable photoluminescence. This effect is called the quantum size effect. It allows tuning the absorption and emission peaks by the particle size and the width of these peaks by the QD size distribution. This unique tuning property of the photoluminescence is of particular importance in the field of nano-barcoding since these particles can have a unique feature that is closely related to their production process. QDs are typically made from III/V alloys such as InP and InGaP, but also II/VI compounds such as CdSe, CdTe, ZnTe and ZnSe. They are composed of core shell architecture, as for example CdSe core with a ZnS shell. The shell here is of first importance since it rules the surface recombination mechanisms and prevents the core against oxidation, a known issue of quantum dot barcodes. When excited with visible light, quantum dots emit fluorescence in the range of visible to near infrared wavelengths. This emission is generally brighter and relatively less prone to photo bleaching than organic dyes (Ji et al. 2011a). A lot of application for the use of quantum dots have emerged over the last few years as for example the use of quantum dots to detect pesticides in agricultural products (Ji et al. 2011b, Nsibande and Forbes 2016).

3.5.3 *Rare earth*

Fluorochromes and quantum dots, as previously described, emit fluorescence light when they are excited by a higher energy light, mostly in the visible spectrum that imposes the use of filtering for the separation of excitation from the fluorescence light. Rare earth doped nanoparticles can convert invisible infrared light to visible colored light, thus improving the signal to background noise ratio. These particles are called photon upconverting particles since they emit visible light from a lower energy infrared excitation light. Rare earths are composed of seventeen chemical elements that are the fifteen lanthanides plus scandium and yttrium. Rare earth ions are used to dope oxides and silicate glass in order to generate these unique fluorescence properties. This composite material is then shaped into micro particles or rods to build micro barcodes. Compared with fluorochrome, rare earth doped materials are extremely resistant to photobleaching. The silica-based glass matrix of the micro barcodes provides significant advantages compared to polymer-based materials in terms of chemical and harsh environment resistance. Furthermore, the fluorescent lifetime is larger than for fluorochromes and then the dynamic property can also be used to create optical codes in addition with particles shape and emission spectrum (Dejneka 2003).

3.5.4 Fluorescent plant compounds

All the fluorescent particles mentioned in the previous section exhibit interesting emission features to create fluorescence encoded micro-barcodes. But due to their nature or size, these particles are not really biocompatible or inert for the human body, a fact that limits their use for food or packaging. Fluorescence particles containing natural plant compounds are obtained. Under the UV light, these particles can show an original and unique fluorescence spectrum (Fig. 6). With a large collection of natural compounds, it is thus possible to compose a large serial of fluorescence spectrums (Tousch et al. 2011). This metabolic engineering generates edible fluorescent particles, especially for food barcoding that are biocompatible and degradable.

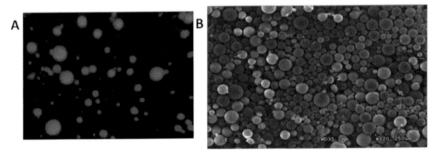

Figure 6. Composite Image (Transmission and fluorescence) of Core Shell Vegetal Particles (GINGKO SFERE). (A) 40X Fluorescence image (400 nm excitation) and (B) 120X scanning electron microscope picture (Tousch et al. 2011).

3.6 Encapsulation and shaping

The new technologies of encapsulation will permit to obtain tracers that are more resistant to environmental stresses, to obtain concentrated tracers and also to obtain food tracers that can be incorporated into the food itself.

3.6.1 Polymer particles

Most of the fluorescent particles or compounds previously described require a shell or an encapsulation matrix to be used as a micro barcode. The role of the encapsulation is mainly to protect the fluorescent material from the environment and to keep it away from chemical aggression, oxidation but also light exposure. In addition to its protective function, this encapsulating material can also participate in the signature of the barcode by its own shape or by a mix of fluorescent particles content. Therefore, many techniques have been developed to incorporate these fluorescent compounds in micro barcodes comprising one or several particles having specific fluorescence properties (Wang et al. 2015).

3.6.2 Microfluidic generated particles

A technique has recently been developed for the production of fluorescent barcode particles using microfluidic circuits. Microfluidics is the science of fluid dynamics in small dimensions, but also a set of microfabrication techniques that are derived from micro sensors production technologies, micro-electro-mechanical systems (MEMS) and microelectronics. Microfluidics aim at using very small amounts of liquid flowing in channels of very small dimensions for biochemical analysis, so called "Lab on a Chip". The main feature of microfluidics is that fluids flow at low Reynolds number depending on density (r), the velocity (v) and the viscosity (μ) of the fluid, according to the formula Re = rvl/μ. In microfluidics, due to the small dimensions of the channels, liquid flows are always in laminar conditions, and that particular flow property has been used for the production of fluorescent barcode particles. One technique, proposed by the P.S. Doyle laboratory (Lee 2014), is based on the use of a photopolymerizable hydrogel to build micro barcodes using a microfluidic circuit. The technique for producing such composite particles uses microfluidic circuits such as depicted in Fig. 7A. Several inputs containing a dispersion of fluorescent particles or solutions of fluorochromes are set together in a main channel. The microfluidic flow properties induce a laminar flow made of a set of parallel sections that all have a different optical signal. The manufacturing system comprises of a multiple input circuit made to line up the fluid hydrogel in a single channel, consisting of parallel ridges that are moving continuously. An ultraviolet light beam passing through an optical slit is

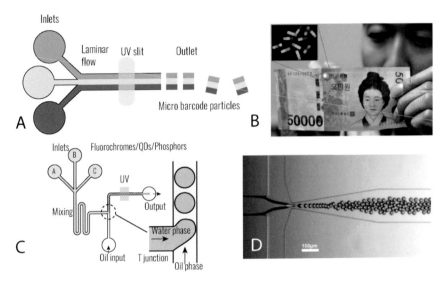

Figure 7. Schematic of microfluidic circuits used to generate color coded particles, using laminar flow properties of small scale fluidics (A) and (B) (Lee et al. 2014) but also diphasic flows to generate droplets that are transformed into particles (C) and (D) (Lee 2014, Zhao 2011a, Ji 2011b, Gerver 2012).

then applied to a section of the central channel. The UV light will induce the photopolymerisation, changing the liquid flow into a series of rods that are collected at the output. These polymer rods carry a code made of a set of different fluorescence colors. These rods can then be integrated to a packaging or paper such as shown in Fig. 7B.

Another interesting feature with microfluidics is the ability to generate a stream of droplets by using non-miscible liquids that interacts in specific junctions. The surface tension between these liquids allows the generation of fluid flow instabilities and the minimization of the surface tending to create a set of droplets rather than cylindrical flows. T junctions in Fig. 7C or cross junctions in Fig. 7D are then frequently used for the generation of droplets of water in oil. This feature has been used to generate spherical polymer droplets that embed a mix of fluorescent particles, this mixture creates a specific optical signature. The schematic of the circuit is shown in Fig. 7C; it starts with several inlets that gathers in a mixing section and then continues toward a T junction that produces a stream of droplets that are photopolymerised to build solid fluorescent barcoding micro particles. These particles can be made of a mix of quantum dots (Zhao 2011, Ji 2011) or lanthanide nanophosphors (Gerver 2012).

3.6.3 Random deposited particles

Encapsulation techniques of fluorescent particles mentioned above can integrate into a micro-barcode a pattern that creates a signature made by a combination of both shape and optical signals, but with a limited number of patterns. These micro-barcodes are produced by batch fabrication techniques that render more complex the possibility to build individual barcodes embedding a unique signature.

There is an alternative to obtain fluorescent particles pattern generation that uses the randomness of a manufacturing process to create micro-barcodes with unique signatures. The principle is to use a physical phenomenon that introduces a random process in the manufacturing so that every micro-barcode embeds a unique signature randomly generated within a very wide range of possibility. This principle is, for example used especially in bubble micro-barcodes of Prooftag company (Montauban, France) (http://www.prooftag.net/). A unique pattern is generated by the random nucleation of microbubbles created during the fabrication process of a polymer tag. The limitation of this method is that each tag must be scanned and recorded in a database before being placed on an item or good and sent on the market. For reading the tag for authentication of the marked item one must refer to this database to display the tag value.

An example of micro-barcode production technique that uses the random process is illustrated in Fig. 8. In this example, vegetable fluorescent particles of a few microns were deposited by a capillary assembly technique onto a micro-structured polymer matrix. This substrate is composed of an array of micro-wells, regularly spaced and slightly than the mean size of fluorescent particles. The dewetting or dip coating of a solution of these particles in suspension onto the polymer matrix will induce the capture of these particles in the wells with a random success rate which will depend

Figure 8. Random deposition pattern of vegetal particles on a micro-structured PDMS substrate. The matrices formed by a dewetting process of a vegetal particle solution produces random matrices with two distinct fluorescence signals. Close up view (fluorescence + transmitted light and 10×10 and 100×100 matrices.

on the position of the particles in the solution at the time of dewetting. The result of this random deposition is visible in Fig. 8 that shows examples of deposition of fluorescent particles on 10×10 and 100×100 matrices. Using successive deposition of particles having different optical signatures it is possible to further increase the uniqueness of the tag signature. The final micro-barcode that has dimensions on the range of 0.1 to 1 mm can be captured with a simple optical system (personal communication).

4. Conclusions

We have presented several classic and new techniques for the production of authentication micro-barcodes. These tiny in size devices are intended to be included in the packaging of objects or goods and to provide information on the nature or the manufacturer of these products. The micro-barcodes have to be inexpensive, stable over time, biocompatible but also easy to detect and decode. Many of these new technologies are based on the production of fluorescent dyes that can be embedded into particles or objects to encode information both in their shape and in their optical properties such as fluorescence spectrum or reflectance. This combination defines their signatures that specifically mark the objects that carry them. However, if these micro-barcode production techniques are attractive because of their precision and difficulty to be falsified, there are still issues that need further developments. The first issue being the harmlessness and biocompatibility. Indeed in order to reach the full traceability of the food, it is neccessary to consider that micro-barcodes will be in contact with food packaging or even will be incorporated directly in the food, so particles and micro-barcodes will have to be ingestible, completely healthy and biodegradable. The authenticated products based on DNA or again on particles viral proteins must first be accepted by the official food instances and then by consumers. Otherwise particles that are based on mineral chemistry and heavy metals are not completely innocuous. Finally, the last point that we consider remains a blockade to detection is that the measuring, reading and decoding functions have to be performed with a low-cost terminal, even directly embedded onto a smart phone. The small size of micro-barcodes often requires enlarger objectives and sources of infrared or ultraviolet light which complicates the reading process. But at the end, one of these micro-barcode technologies may emerge and soon replace the classical barcodes, holograms, RFID tags and QRcodes into packaging and why not in foods directly.

Keywords: Food traceability, track and trace, supply chain, nano-barcodes, particules tags, fluorescence, microfluidic, quantum dots, rare earth

References

Abbas, K.A., Saleh, A.M., Mohamed, A. and MohdAzhan, N. 2009. The recent advances in the nanotechnology and its applications in food processing: a review. Journal of Food Agriculture & Environment 7(3,4): 14–17.

Cadot, Y., Miñana Castelló, M.T. and Chevalier, M. 2006. Flavan-3-ol compositional changes in grape berries (*Vitis vinifera* L. cv Cabernet Franc) before veraison, using two complementary analytical approaches, HPLC reversed phase and histochemistry. Analytica Chimica Acta 563(1-2): 65–75.

Baumert, J.L. 2014. Detecting and Measuring Allergens in Food. Risk Management for Food Allergy. Elsevier Inc. Chapter Thirteen 215–225. http://dx.doi.org/10.1016:B978-0-12-381988-8.00013-0.

Brenig, B. and Schütz, E. 2016. Recent development of allele frequencies and exclusion probabilities of microsatellites used for parentage control in the German Holstein Friesian cattle population. BMC Genetics 17: 18. doi:10.1186/s12863-016-0327-z.

Chaudhry, Q., Scotter, M., Blackburn, J., Ross, B., Boxall, A., Castle, L., Aitken, R. and Watkins, R. 2008. Applications and implications of nanotechnologies for the food sector. Food Additives and Contaminants 25(3): 241–258.

Dejneka, M.J., Streltsov, A., Pal, S., Frutos, A.G., Powell, C.L., Yost, K., Yuen, P.K., Muller, U. and Lahiri, J. 2003. Rare earth-doped glass microbarcodes. Proceedings of the National Academy of Sciences 100-2: 389–393.

Delgado, T., Gomez-Cordoves, C. and Villarroya, B. 1990. Relationships between phenolic compounds of low molecular weight as indicators of the aging conditions and quality of brandies. American Journal Enology and Viticulture 41: 342–345.

Doligez, A., Adam-Blondon, A.F., Cipriani, G., Di Gaspero, G., Laucou, V., Merdinoglu, D., Meredith, C.P., Riaz, S., Roux, C. and This, P. 2006a. An integrated SSR map of grapevine based on five mapping populations. Theoretical and Applied Genetics 113(3): 369–382.

Doligez, A., Audiot, E., Baumes, R. and This, P. 2006b. QTLs for muscat flavor and monoterpenic odorant content in grapevine (*Vitis vinifera* L.). Molecular Breeding 18(2): 109–125.

Fernandez-Lorenzo, J.L., Rigueiro, A. and Ballester, A. 1999. Polyphenols as potential markers to differentiate juvenile and mature chestnut shoot cultures. Tree Physiology 19: 461–466.

Finkel, N.H., Lou, X., Wang, C. and He, L. 2004. Barcoding the microworld. Analytical Chemistry 1, 76, 19: 352A–359A.

Ghaani, M., Cozzolino, C.A., Castelli, G. and Farris, S. 2016. An overview of the intelligent packaging technologies in the food sector. Trends in Food Science & Technology, Volume 51, May 2016, pp. 1–11, ISSN 0924-2244, http://dx.doi.org/10.1016/j.tifs.2016.02.008.

Galimberti, A., De Mattia, F., Losa, A., Bruni, I., Federici, S., Casiraghi, M., Martellos, S. and Labra, M. 2013. DNA barcoding as a new tool for food traceability. Food Research International 50: 55–63.

Gerver, R.E., Gómez-Sjöberg, R., Baxter, B.C., Thorn, K.S., Fordyce, P.M., Diaz-Botia, C.A., Helms, B.A. and DeRisi, J.L. 2012. Programmable microfluidic synthesis of spectrally encoded microsphere. Lab Chip 12: 4716–4723.

Gómez-Martínez, R., Sánchez, A., Duch, M., Esteve, J. and Plaza, J.A. 2011. DRIE based technology for 3D silicon barcodes fabrication. Sensors and Actuators B 154: 181–184.

Hebert, P.D.N., Ratnasingham, S. and De Waard, J.R. 2003. Barcoding animal life: cytochrome *c* oxidase subunit 1 divergences among closely related species. Proceedings of the royal society B. Biological Sciences. 270, 1 doi:10.1098/rsbl.2003.0025.

Hebert, P.D.N., Stoeckle, M.Y., Zemlak, T.S. and Francis, C.M. 2004. Identification of birds through DNA Barcodes. PLOS Biology 2(10): 1657–1663.

Jasicka-Misiak, I., Poliwoda, A., Dereń, M. and Kafarski, P. 2012. Phenolic compounds and abscisic acid as potential markers for the floral origin of two Polish unifloral honeys. Food Chemistry 131, 4, 15: 1149–1156.

Ji, X.H., Cheng, W., Guo, F., Liu, W., Guo, S.S., He, Z.K. and Zhao, X.Z. 2011a. On-demand preparation of quantum dot-encoded microparticles using a droplet microfluidic system. Lab Chip 11(15): 2561–2568.

Ji, X.H., Zhang, N.G., Cheng, W., Guo, F., Liu, W., Guo, S.S., He, Z.K. and Zhao, X.Z. 2011b. Integrated parallel microfluidic device for simultaneous preparation of multiplex optical-encoded microbeads with distinct quantum dot barcodes. Journal of Materials Chemistry 21(35): 13380–13387.

Kress, W.J. and Erickson, D.L. 2008. DNA barcodes: Genes, genomics, and bioinformatics. Proceedings of the National Academy of Sciences USA 105(8): 2761–2762.

Kuklinska-Pijanka, A., Young, E., Chapman, M. and Hindley, J. 2015. Simultaneous quantification of multiple specific food allergen proteins indicates varied allergen content in diagnostic and therapeutic preparations. Clinical and Translational Allergy 5: P29.

Kumar, P., Reinitz, J., Simunovic, J., Sandeep, K.P. and Franzon, P.D. 2009. Overview of RFID technology and its applications in the food industry. Journal of Food Science 74(8): 101–106.

Kumari, L., Narsaiah, K., Grewal, M.K. and Anurag, R.K. 2015. Application of RFID in agro-food sector—a review. Trends in Food Science & Technology 43(2): 144–161.

Lee, J., Bisso, P., Srinivas, R.L., Kim, J.J., Swiston, A.J. and Doyle, P.S. 2014. Universal process-inert encoding architecture for polymer microparticles. Nature Materials 13: 524–529.

Martins-Lopes, P., Gomes, S., Pereira, L. and Guedes-Pinto, H. 2013. Molecular markers for food traceability. Food Technology and Biotechnology 51(2): 198–207.

Matějíček, D., Mikeš, O., Klejdus, B., Štěrbová, D. and Kubáň, V. 2005. Changes in contents of phenolic compounds during maturing of barrique red wines. Food Chemistry 90(4): 791–800.

Matthias, S., Schilling, J., Nielsch, K., Müller, F., Wehrspohn, R.B. and Gösele, U. 2002. Monodisperse diameter-modulated gold microwires. Advanced Materials 14: 1618–1621.

Nicewarner-Peña, S.R., Griffith Freeman, R., Reiss, B.D., He, L., Peña, D.J., Walton, I.D., Cromer, R., Keating, C.D. and Natan, J.M. 2001. Submicrometer metallic barcodes. Science 294(5540): 137–141.

Nsibande, S.A. and Forbes, P.B.C. 2016. Fluorescence detection of pesticides using quantum dots materials—a review. Analytica Chimica Acta 945: 9–22.

Palmer, R.C. 2007. The Bar Code Book, Fifth Edition, Trafford Publishing, ISBN: 978-1-42513-374-0.

Pearce, S. and Bushnell, R.D. 2010. The bar Code Implementation Guide: Using Bar Codes in Distribution. Paperback Edition October 1, ISBN: 0-941668-06-1.

Sauer, S., Lechner, D., Berlin, K., Lehrach, H., Escary, J.L., Fox, N. and Gut, I.G. 2000. A novel procedure for efficient genotyping of single nucleotide polymorphisms. Nucleic Acids Research 1, 28,5: E13.

Scarano, D. and Rao, R. 2014. DNA markers for food products authentication. Diversity 6: 579–596.

Shaked, N.T., Katz, B. and Rosen, J. 2009. Review of three-dimensional holographic imaging by multiple-viewpoint-projection based methods. Applied Optics 48(34): H120–H136.

Sung-Kyoung, K. and Sang Bok, L. 2009. Highly encoded one-dimensional nanostructures for rapid sensing. Journal of Materials Chemistry 19: 1381–1389.

Thangaraj, P. 2015. Pharmacological assays of plant-based. Natural Products Volume 71 of the series Progress in Drug Research, pp. 49–55.

Tousch, D., Taly, F., Olivier, T. and Thierry, A. 2011. Process for Preparing a Chemical Structure having a Phase Partition, Capable of Generating a Specific Fluorescence Spectrum and Uses Thereof. Patent number 13/824,679, with PCT extension.

Wang, M., Duong, B., Fenniria, H. and Su, M. 2015. Nanomaterial-based barcodes. Nanoscale 7-26: 11240–11247. doi:10.1039/c5nr01948f.

Zhao, Y., Shum, H.C., Chen, H., Adams, L.L.A., Gu, Z. and Weitz, D.A. 2011. Microfluidic generation of multifunctional quantum dot barcode particles. Journal of the American Chemical Society 133(23): 8790–8793.

Zhu, X., Mukhopadhyay, S.K. and Kurata, K. 2012. A review of RFID technology and its managerial applications in different industries. Journal of Engineering and Technology Management 29(1): 152–167.

Index